Theory of
AC CIRCUITS

Albert Paul Malvino,
Ph.D., E.E.

MALVINO INC.
Whole-Brain Learning Systems
Mountain View, California
Dublin, Ireland

Textbook Art: ***Joanna M. Malvino***

Cover Photographer: ***David C. Photography***

Cover Designer: ***John Giannotti Design***

Color Separation: ***Creative Color Service Corp.***

Printer: ***McNaughton & Gunn Inc.***

Theory of Ac Circuits

To Cinda,

My beloved daughter

The most courageous person I know

She taught me how to hold on in spite of everything

How to push on in the face of difficulty

How to face defeat again and again without giving up

And how to do whatever is necessary to fulfill my mission.

No man is an island, entire of itself;
every man is a piece of the continent,
a part of the main.
Any man's death diminishes me,
because I am involved in mankind;
and therefore never send to know
for whom the bell tolls;
it tolls for thee.

John Donne

CONTENTS

PREFACE

After reading hundreds of books on ac theory, I decided the world desperately needs one more. Publishers have deluged us with one book after another, each progessively more complicated in its mathematical treatment of ac circuits. All the debris cluttering the subject of ac circuits convinced me a new book was needed; one that would simplify the material as much as possible without losing the crucial concepts.

In fact, I tried to follow Herman Melville's advice about how to discuss science:

> *A man of true science uses but few hard words, and those only when none other will answer his purpose; whereas the smatterer in science thinks that by mouthing hard words, he understands hard things.*

Why do I think this book follows Melville's advice. Because understanding ac circuits is one thing; writing about them is another. Even though I understood an ac concept intuitively, it was often necessary to struggle for weeks looking for the right words and illustrations to nail down the concept. I assure you this book gives the simplest and clearest explanations possible. This doesn't mean the book is easy. It means you never get a complex explanation when a simple one will do. It means the discussion of each topic is no harder than it has to be.

This book is the sequel to *THEORY OF DC CIRCUITS*. The *THEORY OF AC CIRCUITS* continues your introduction to basic electricity. In dc circuits all voltages and currents remain constant. On the other hand, ac circuits introduce a new factor into the discussion. The factor is *time*. In ac circuits, the voltages and currents are continuously changing as time passes. Because of this, the analysis becomes more complicated. But with the additional difficulty comes new opportunities and new devices. I think you will find that ac circuits are far more exciting and fascinating than dc circuits.

As before, this textbook has a companion publication called *QUIK-LAB FOR AC CIRCUITS*. This state-of-art product enhances *THEORY OF AC CIRCUITS* because it shows you computer movies of the basic concepts. Furthermore, *QUIK-LAB* puts ac circuits on a computer screen and allows you to calculate, analyze, and design using nothing more than the keyboard. If at all possible, you should use *QUIK-LAB* with this textbook.

Albert Paul Malvino

OTHER MALVINO INC. BOOKS

Stand-alone Textbooks

Theory of Dc Circuits
Theory of Ac Circuits

IBM PC Text/software Products

Quik-Lab for Dc Circuits (correlated with Theory of Dc Circuits)
Quik-Lab for Ac Circuits (correlated with Theory of Ac Circuits)
Quik-Lab for Basic Electronics (correlated with Electronic Principles, 4th ed.)

Apple II Text/software Products

Dc Circuits
Ac Circuits
Basic Semiconductors

CHAPTER 1
ALTERNATING CURRENT

In dc circuits the source voltage is constant. Because of this, the current is always in the same direction with a constant value. This type of flow is called *direct current* (dc). But there is another kind of flow called *alternating current* (ac). This type of flow reverses its direction at regular intervals. Circuits with alternating current are called *ac circuits.*

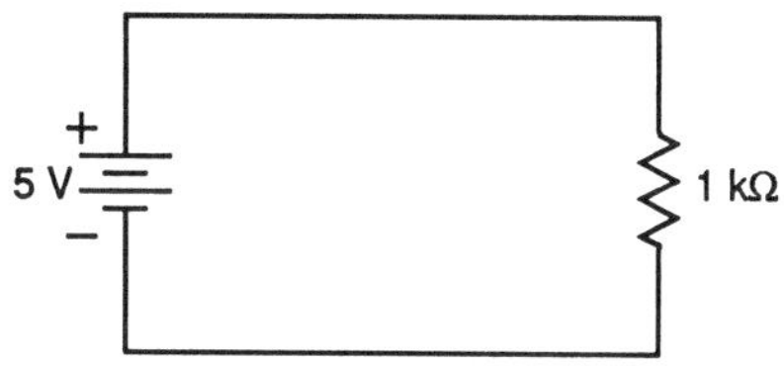

Fig. 1-1.

1-1. AC QUANTITIES

Figure 1-1 is a dc circuit. To calculate the current, we use Ohm's law:

$$I = \frac{V}{R} = \frac{5\text{ V}}{1\text{ k}\Omega} = 5\text{ mA}$$

This current is constant with respect to time. In other words, if we insert and ammeter in the circuit, it will read 5 mA. It will continue to read 5 mA even though time is passing.

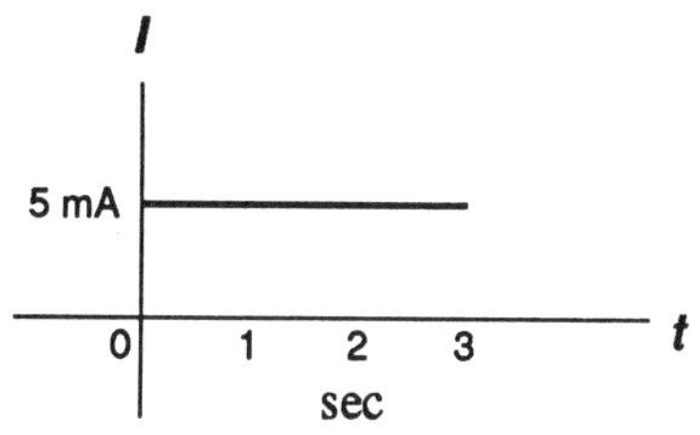

Fig. 1-2.

A direct current, also called a *dc current,* is any current that is constant in time. Figure 1-2 shows the graph of direct current versus time. As you can see, the current is 5 mA at any point in time. The graph begins at $t = 0$, which represents the first instant where current is measured. At the end of the first second ($t = 1$), the current is still 5 mA. Similarly, if we measure the current at $t = 2$, $t = 3$, etc., the current is 5 mA. Because the direct current is constant in value, the graph of direct current is a horizontal line.

Switched Direct Current

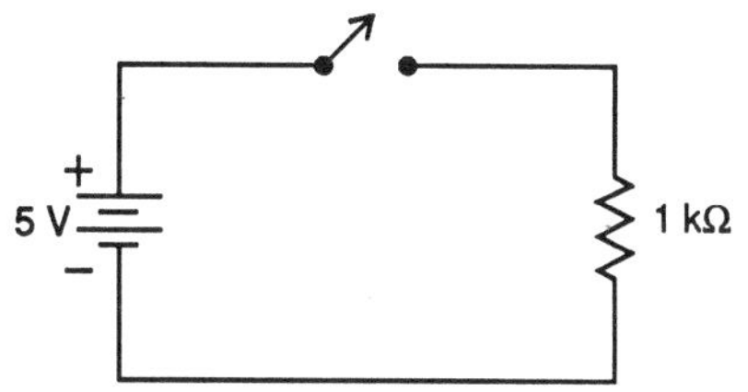

Fig. 1-3.

When a switch is in a dc circuit, the current is a direct current only when the switch is closed. If the switch is opened, the current drops to zero. For instance, Fig. 1-3 shows a battery voltage of 5 V. When the switch is closed, the current is 5 mA. When the switch is open, the current is zero. If the switch is opened and closed at regular intervals, we no longer have a dc current.

For instance, assume the switch is closed at $t = 0$, opened at $t = 1$, closed at $t = 2$, and so on. What does the graph of current versus time look like? Think about this for a few moments before reading on.

The current is 5 mA between $t = 0$ and $t = 1$, as shown in Fig. 1-4. When the switch is opened at $t = 1$, the current immediately drops to zero. Then, it remains at zero between $t = 1$ and $t = 2$. When the switch is closed at $t = 2$, the current immediately increases to 5 mA. Then, it remains at 5 mA between $t = 2$ and $t = 3$.

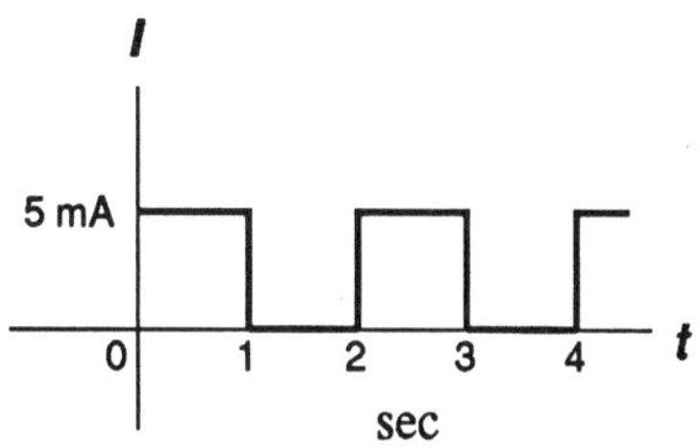

Fig. 1-4.

A graph like Fig. 1-4 is called a *waveform*. This type of graph shows a current or a voltage as a function of time. Waveforms are very useful in the study of ac circuits because they allow us to see how a current or a voltage changes with time. In fact, an instrument called the *oscilloscope* is widely used by technicians and engineers to display the waveforms of ac circuits.

Alternating Current

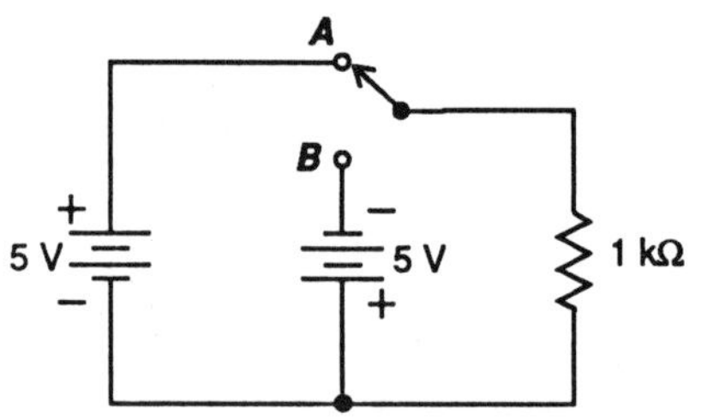

Fig. 1-5.

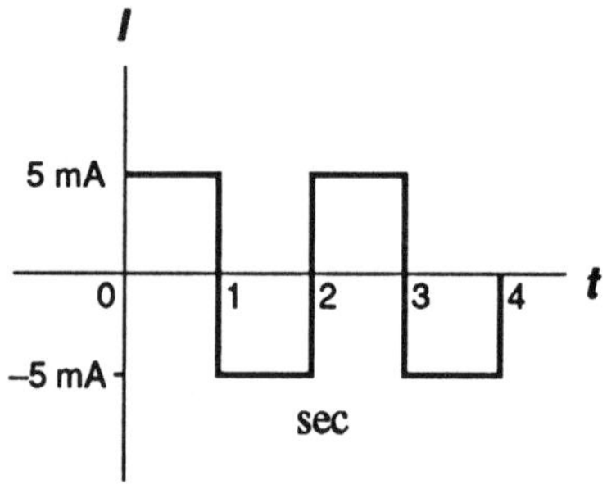

Fig. 1-6.

Figure 1-5 shows a circuit with two dc sources. The switch can be in either of two positions. If the switch is *toggled* (moved to the opposite position) every second in Fig. 1-5, what does the current waveform look like? Think about this for a few minutes before you read on.

With the switch of Fig. 1-5 in position *A*, the current is 5 mA. When the switch is moved to position *B*, the source voltage reverses. In this case, the current has a magnitude of 5 mA, but its direction reverses. To indicate this reversal, we will include a minus sign to get −5 mA.

Either direction can be defined as the positive direction of current. In this example, the positive direction is the direction of current when the switch is in position *A*. The negative direction is the direction of current when the switch is in position *B*.

The reason for using plus and minus signs is because it allows us to graph the current as shown in Fig. 1-6. The current is 5 mA between $t = 0$ and $t = 1$. When the switch is toggled at $t = 1$, the current reverses its direction. Then, its value is −5 mA between $t = 1$ and $t = 2$. When the switch is toggled at $t = 2$, the current again reverses its direction. Then, it remains at 5 mA between $t = 2$ and $t = 3$.

The waveform of Fig. 1-6 is called a *square wave*. It is an example of *alternating current* (ac), defined as any current that reverses its direction at regular intervals. The square wave is only one of the many possible waveforms for alternating current. The next section discusses the most common type of ac waveform.

1-2. AC GENERATOR

Recall the *generator effect*, discussed in your dc circuits book. Whenever a conductor cuts through magnetic lines of force, a voltage is induced between the ends of the conductor. By using a rotating conductor, we can build an *ac generator*, a device that produces alternating voltage.

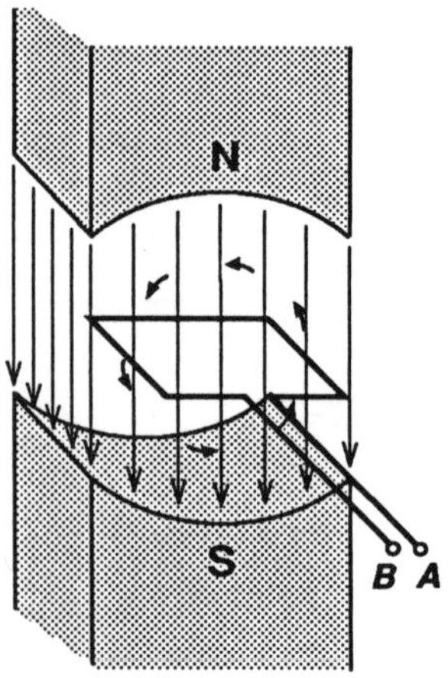

Fig. 1-7.

Conductor Moves in Circular Path

Figure 1-7 shows a conductor loop in a magnetic field. The loop rotates *counterclockwise*. As it rotates, it cuts through lines of force. This produces a voltage between *A* and *B*. If a load resistor were connected between *A* and *B*, a current would exist.

Because of the circular motion, maximum voltage is induced when the conductor is perpendicular to the lines of force, and minimum voltage is induced when the conductor is parallel to the lines of force. For instance, the conductor of Fig. 1-7 is moving parallel to the lines of force. Therefore, it is not cutting the lines of force and no voltage is induced.

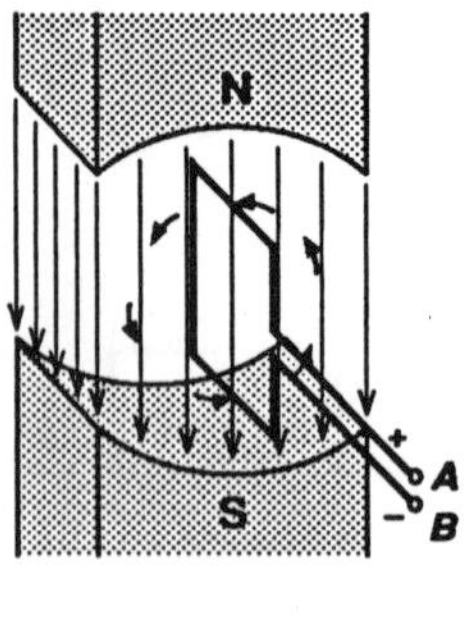

Fig. 1-8.

On the other hand, when the conductor is vertical as shown in Fig. 1-8, it cuts through a maximum number of lines. Because of this, the induced voltage is maximum. As the conductor moves from the horizontal position (Fig. 1-7) to the vertical position (Fig. 1-8), the induced voltage increases from zero to a maximum value.

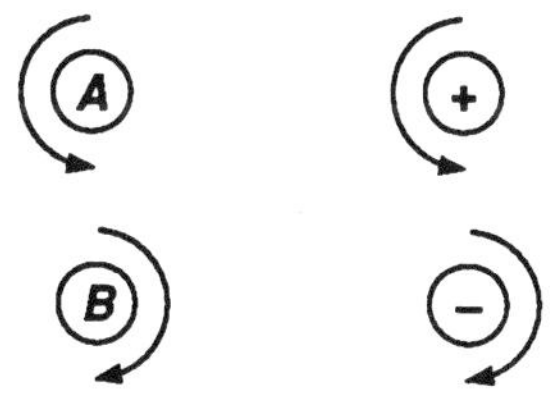

Fig. 1-9.

Voltage is Alternating

Notice the polarity of the induced voltage in Fig. 1-8. The end labeled *A* is positive and the end labeled *B* is negative. You can verify this polarity by applying the left-hand rule for electron flow or the right-hand rule for conventional flow. In either case, the flux curls counterclockwise around the upper conductor and clockwise around the lower conductor as shown in Fig. 1-9a. When you curl your fingers in the direction of the flux, your thumb points in the direction of current. No matter which hand you are using, the upper conductor is positive and the lower one is negative as shown in Fig. 1-9b.

As the conductor continues to rotate, it moves into the horizontal position shown in Fig. 1-10. In this position, it is not cutting any lines of force. Therefore, the induced voltage is zero for this instant. As the conductor continues to rotate, the induced voltage increases until it reaches the vertical position shown in Fig. 1-11. At this point, the voltage is again maximum. But notice the polarity of the voltage. It has reversed. End *B* is now positive, while end *A* has become negative.

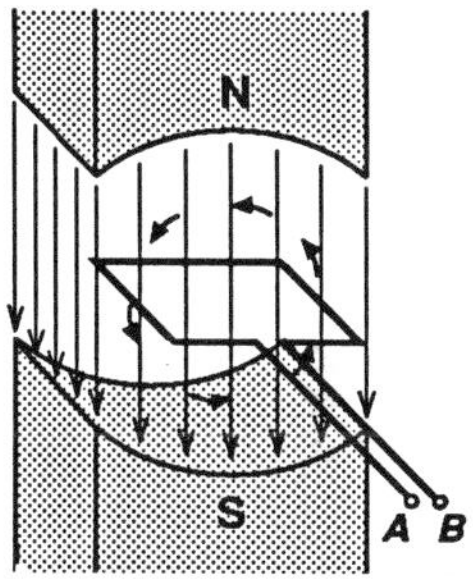

Fig. 1-10.

Again, the proof is to apply the left-hand rule for electron flow or the right-hand rule for conventional flow. In either case, the flux curls counterclockwise around the upper conductor and clockwise around the lower conductor as shown in Fig. 1-12a. No matter which hand you are using, the upper conductor is positive and the lower one is negative as shown in Fig. 1-12b.

Here is a recap of what happens in an ac generator.:

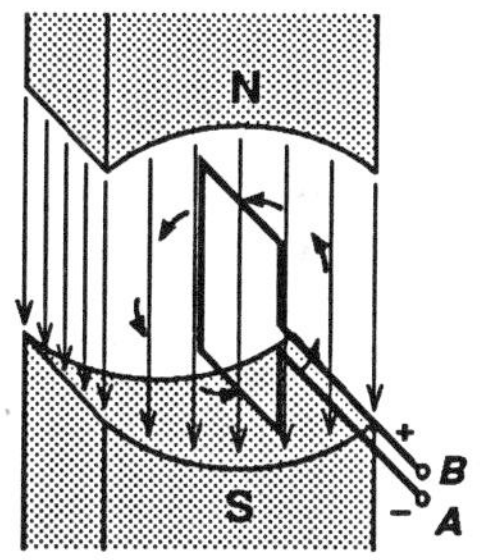

Fig. 1-11.

1. The rotation begins with the conductor in the horizontal position (Fig. 1-7). In this position, the induced voltage is zero because the conductor is moving parallel to the lines of force.

2. The induced voltage increases as the conductor moves toward the vertical position. When the conductor is vertical (Fig. 1-8), the induced voltage is maximum positive on end *A*.

3. The induced voltage decreases from maximum positive as the conductor moves back toward the horizontal position. When the conductor is again horizontal (Fig. 1-10), the induced voltage is zero.

4. The induced voltage again increases as the conductor moves toward the vertical position. When the conductor is vertical (Fig. 1-11), the induced voltage is maximum negative on end *A*.

5. The induced voltage decreases from maximum negative as the conductor moves back toward the horizontal position. When the conductor is again horizontal (Fig. 1-7), the induced voltage is zero. At this point, a new cycle of rotation begins.

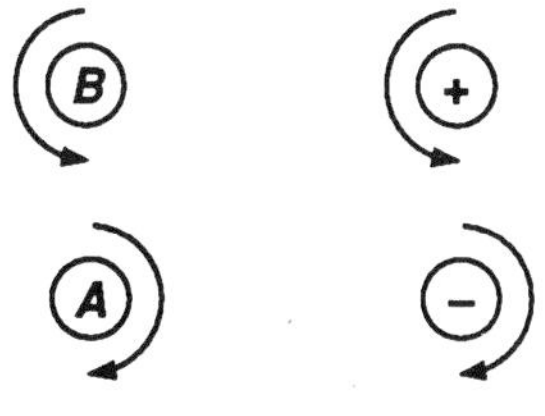

Fig. 1-12.

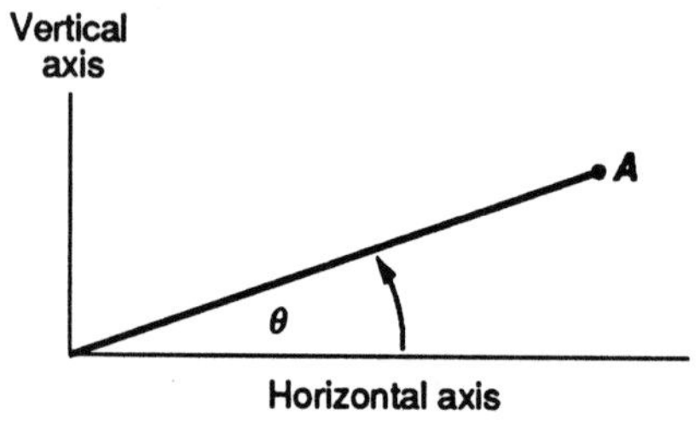

Fig. 1-13.

Angle of Rotation

As the conductor rotates, it forms an angle with the horizontal axis as shown in Fig. 1-13. We will use the Greek letter "theta" to represent this angle of rotation:

θ = angle between conductor and horizontal

This angle determines the amount of flux being cut. When the angle is 0°, no flux is cut because the conductor is moving parallel to the lines of force. When the angle is 90°, maximum flux is cut because the conductor is moving perpendicular to the lines of force.

The induced voltage is positive when the angle is between 0° and 180°. On the other hand, when the angle is between 180° and 360°, the induced voltage is negative. A cycle is one complete rotation of the conductor. In terms of the rotation angle, a cycle represents the motion of the conductor from 0° to 360°. When the conductor passes through 360°, a new cycle begins.

Waveform

Figure 1-14 shows the waveform of the induced voltage. It is a visual summary of what we said earlier. The voltage increases from zero to a maximum value at 90°. As the conductor continues to rotate, the voltage decreases from a maximum to zero at 180°. As the conductor moves from an angle of 180° to 270°, the voltage increases to a maximum value in the negative direction. When the conductor moves from 270° to 360°, the voltage decreases from maximum negative to zero. Then, the next cycle of rotation begins, and the waveform repeats.

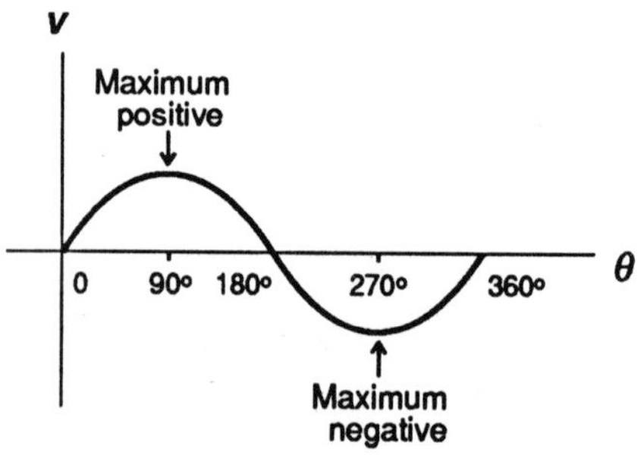

Fig. 1-14.

1-3. THE SINE WAVE

The waveform of Fig. 1-14 is called a *sine wave*. It can be shown that the number of lines being cut at any angle of rotation is proportional to the sine of the angle. Because of this, the formula for induced voltage is

$$v = V_p \sin \theta \qquad (1\text{-}1)$$

where v = voltage at any instant in time
V_p = peak or maximum voltage
θ = angle between conductor and horizontal

In Fig. 1-14, you can see that the voltage is continuously changing throughout the cycle. The *instantaneous voltage* is the voltage at any instant. For example, if the peak value is 100 V, the instantaneous voltage at $\theta = 30°$ is

$$v = (100\ \text{V}) \sin 30° = 50\ \text{V}$$

Equation (1-1) is useful because it allows us to calculate the instantaneous voltage for any angle of rotation.

If you have a scientific calculator, you can quickly and easily calculate the instantaneous values. All you have to do is enter the number of degrees and take the sine. Then, multiply by the peak voltage to get the induced voltage.

A final point. The sine wave is the natural output of any ac generator. The word *sinusoidal* means "varying in proportion to the sine of an angle". You often will see the word "sinusoidal" used to describe the voltage and current from an ac generator. For instance, the induced voltage out of an ac generator is often called a sinusoidal voltage.

EXAMPLE 1-1

Calculate the instantaneous values of a sinusoidal voltage with a peak value of 50 V. Use angles that are 10° apart.

SOLUTION

We will calculate the values for angles of 0°, 10°, 20°, 30° and so on. The general equation to use is

$$v = 50 \sin \theta$$

Here are the calculations for instantaneous voltage between 0° and 90°:

$$v = 50 \sin 0^\circ = 0$$

$$v = 50 \sin 10^\circ = 8.68$$

$$v = 50 \sin 20^\circ = 17.1$$

$$v = 50 \sin 30^\circ = 25$$

$$v = 50 \sin 40^\circ = 32.1$$

$$v = 50 \sin 50^\circ = 38.3$$

$$v = 50 \sin 60^\circ = 43.3$$

$$v = 50 \sin 70^\circ = 47$$

$$v = 50 \sin 80^\circ = 49.2$$

$$v = 50 \sin 90^\circ = 50$$

SINE VALUES

θ	$\sin \theta$
0°	0
10°	0.174
20°	0.342
30°	0.500
40°	0.643
50°	0.766
60°	0.866
70°	0.940
80°	0.985
90°	1

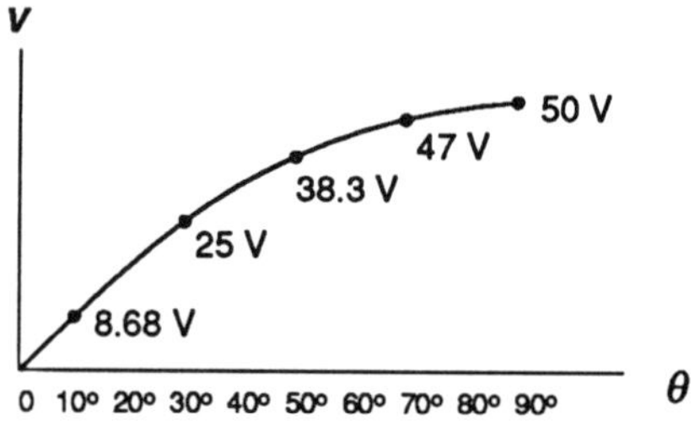

Fig. 1-15.

The foregoing values represent the first quarter cycle of a sine wave. When we graph these values, we get the first quarter cycle shown in Fig. 1-15. For convenience, every other instantaneous value is shown.

The sine function is symmetrical in the sense that the same absolute values appear at larger angles. As discussed in trigonometry, the instantaneous values in the second quarter cycle are given by

$$v = 50 \sin \theta = 50 \sin (180^\circ - \theta)$$

Because of this trigonometric identity, angles between 90° and 180° produce these instantaneous values:

SINE VALUES

θ	$\sin \theta$
90°	1
100°	0.985
110°	0.940
120°	0.866
130°	0.766
140°	0.643
150°	0.500
160°	0.342
170°	0.174
180°	0

$$v = 50 \sin 90^\circ = 50 \sin (180^\circ - 90^\circ) = 50 \sin 90^\circ = 50$$

$$v = 50 \sin 100^\circ = 50 \sin (180^\circ - 100^\circ) = 50 \sin 80^\circ = 49.2$$

$$v = 50 \sin 110^\circ = 50 \sin (180^\circ - 110^\circ) = 50 \sin 70^\circ = 47$$

$$v = 50 \sin 120^\circ = 50 \sin (180^\circ - 120^\circ) = 50 \sin 60^\circ = 43.3$$

$$v = 50 \sin 130^\circ = 50 \sin (180^\circ - 130^\circ) = 50 \sin 50^\circ = 38.3$$

$$v = 50 \sin 140^\circ = 50 \sin (180^\circ - 140^\circ) = 50 \sin 40^\circ = 32.1$$

$$v = 50 \sin 150^\circ = 50 \sin (180^\circ - 150^\circ) = 50 \sin 30^\circ = 25$$

$$v = 50 \sin 160^\circ = 50 \sin (180^\circ - 160^\circ) = 50 \sin 20^\circ = 17.1$$

$$v = 50 \sin 170^\circ = 50 \sin (180^\circ - 170^\circ) = 50 \sin 10^\circ = 8.68$$

$$v = 50 \sin 180^\circ = 50 \sin (180^\circ - 180^\circ) = 50 \sin 0^\circ = 0$$

The instantaneous values in the third quarter cycle are given by this trigonometric identity:

$$v = 50 \sin \theta = -50 \sin (\theta - 180^\circ)$$

Because of this, angles between 180° and 270° produce these instantaneous values:

SINE VALUES

θ	$\sin \theta$
180°	0
190°	−0.174
200°	−0.342
210°	−0.500
220°	−0.643
230°	−0.766
240°	−0.866
250°	−0.940
260°	−0.985
270°	−1

$$v = 50 \sin 180^\circ = -50 \sin (180^\circ - 180^\circ) = -50 \sin (0^\circ) = 0$$

$$v = 50 \sin 190^\circ = -50 \sin (190^\circ - 180^\circ) = -50 \sin (10^\circ) = -8.68$$

$$v = 50 \sin 200^\circ = -50 \sin (200^\circ - 180^\circ) = -50 \sin (20^\circ) = -17.1$$

$$v = 50 \sin 210^\circ = -50 \sin (210^\circ - 180^\circ) = -50 \sin (30^\circ) = -25$$

$$v = 50 \sin 220^\circ = -50 \sin (220^\circ - 180^\circ) = -50 \sin (40^\circ) = -32.1$$

$$v = 50 \sin 230^\circ = -50 \sin (230^\circ - 180^\circ) = -50 \sin (50^\circ) = -38.3$$

$$v = 50 \sin 240^\circ = -50 \sin (240^\circ - 180^\circ) = -50 \sin (60^\circ) = -43.3$$

$v = 50 \sin 250^\circ = -50 \sin (250^\circ - 180^\circ) = -50 \sin (70^\circ) = -47$

$v = 50 \sin 260^\circ = -50 \sin (260^\circ - 180^\circ) = -50 \sin (80^\circ) = -49.2$

$v = 50 \sin 270^\circ = -50 \sin (270^\circ - 180^\circ) = -50 \sin (90^\circ) = -50$

The instantaneous values in the fourth quarter cycle are given by this trigonometric identity:

$v = 50 \sin \theta = -50 \sin (360^\circ - \theta)$

Because of this, angles between 270° and 360° produce these instantaneous values:

$v = 50 \sin 270^\circ = -50 \sin (360^\circ - 270^\circ) = -50 \sin 90^\circ = -50$

$v = 50 \sin 280^\circ = -50 \sin (360^\circ - 280^\circ) = -50 \sin 80^\circ = -49.2$

$v = 50 \sin 290^\circ = -50 \sin (360^\circ - 290^\circ) = -50 \sin 70^\circ = -47$

$v = 50 \sin 300^\circ = -50 \sin (360^\circ - 300^\circ) = -50 \sin 60^\circ = -43.3$

$v = 50 \sin 310^\circ = -50 \sin (360^\circ - 310^\circ) = -50 \sin 50^\circ = -38.3$

$v = 50 \sin 320^\circ = -50 \sin (360^\circ - 320^\circ) = -50 \sin 40^\circ = -32.1$

$v = 50 \sin 330^\circ = -50 \sin (360^\circ - 330^\circ) = -50 \sin 30^\circ = -25$

$v = 50 \sin 340^\circ = -50 \sin (360^\circ - 340^\circ) = -50 \sin 20^\circ = -17.1$

$v = 50 \sin 350^\circ = -50 \sin (360^\circ - 350^\circ) = -50 \sin 10^\circ = -8.68$

$v = 50 \sin 360^\circ = -50 \sin (360^\circ - 360^\circ) = -50 \sin 0^\circ = 0$

SINE VALUES

θ	$\sin \theta$
270°	−1
280°	−0.985
290°	−0.940
300°	−0.866
310°	−0.766
320°	−0.643
330°	−0.500
340°	−0.342
350°	−0.174
360°	0

Figure 1-16 shows a *sketch* (an abbreviated graph) of the calculated values. The graph is a visual reflection of the foregoing calculations. The induced voltage increases from zero to a maximum value of 50 V when the angle increases from 0° to 90°. During the next quarter cycle, the voltage decreases from a maximum of 50 V to zero when the angle increases from 90° to 180°. During the third quarter cycle, the voltage increases from zero to a maximum negative value of −50 V when the angle increases from 180° to 270°. During the final quarter cycle, the voltage decreases from a maximum negative value of −50 V to zero when the angle increases from 270° to 360°.

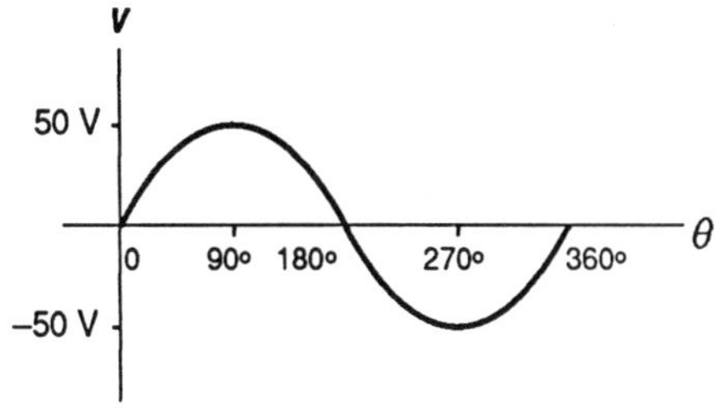

Fig. 1-16.

1-4. FREQUENCY AND PERIOD

The sine wave is very important because it is most common waveform in electricity and electronics. Although all sine waves have the same shape, they may differ in their peak value and other characteristics. This section discusses two properties of a sine wave.

Frequency

The *frequency* of a sine wave is the number of cycles per second. If you know the number of cycles that occur in a given time, you can calculate the frequency with this formula:

$$f = \frac{\text{number of cycles}}{\text{total time}} \qquad (1\text{-}2)$$

For instance, if an ac generator produces 240 cycles in 4 sec, the frequency is

$$f = \frac{240 \text{ cycles}}{4 \text{ sec}} = 60 \text{ cycles/sec}$$

The *hertz* (abbreviated Hz), is the standard unit of measure for frequency. It is defined as

1 hertz = 1 cycle per second

or in abbreviated form,

1 Hz = 1 cycle/sec

Therefore, 60 cycles/sec is usually expressed as 60 Hz.

Period

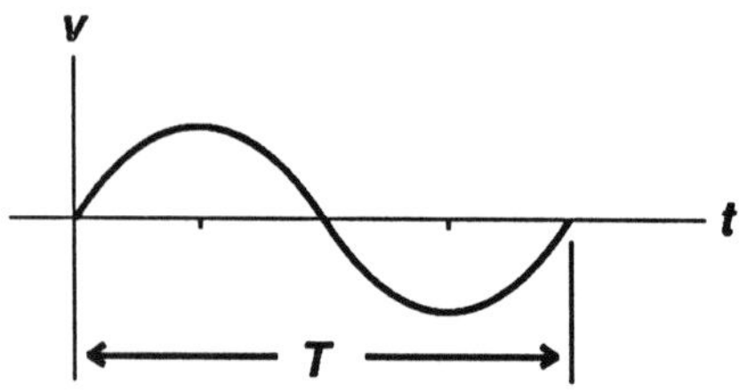

Fig. 1-17.

The *period* of a sine wave is the duration of a complete cycle. In Fig. 1-17, the symbol T represents the period. The variable T stands for time, a reminder that the period is the amount of time it takes to get one complete cycle. For instance, if it takes a conductor 0.2 sec for one complete revolution, the period of the sine wave is

$$T = 0.2 \text{ s}$$

If another generator has a conductor that is rotating 10 times as fast, the period is 10 times smaller:

$$T = 0.02 \text{ s} = 20 \text{ ms}$$

Frequency is the Reciprocal of Period

Recall the meaning of the word *reciprocal*. In mathematics, the reciprocal of a quantity equals 1 divided by the quantity. Here is an important relation between frequency and period:

$$f = \frac{1}{T} \tag{1-3}$$

This says that frequency equals the reciprocal of the period. For instance, if the period is 20 ms, the frequency is

IMPORTANT

$$f = \frac{1}{T}$$

$$f = \frac{1}{20\text{ ms}} = 50\text{ Hz}$$

By rearranging Eq. (1-4), we can derive another useful relation:

$$T = \frac{1}{f} \tag{1-4}$$

This is useful when you know the frequency and want to calculate the period. For instance, if $f = 500$ Hz, the period is

$$T = \frac{1}{500\text{ Hz}} = 2\text{ ms}$$

Remember the reciprocal relation that exists between frequency and period. Whenever you are given one of the quantities, you can automatically find the other quantity by taking the reciprocal of the given quantity. This idea is very important and should be memorized.

EXAMPLE 1-2

What is the frequency in Fig. 1-18?

SOLUTION

Count the number of cycles and divide by the total time. Since 4 cycles occur in 50 ms, the frequency is

$$f = \frac{4\text{ cycles}}{50\text{ ms}} = 80\text{ Hz}$$

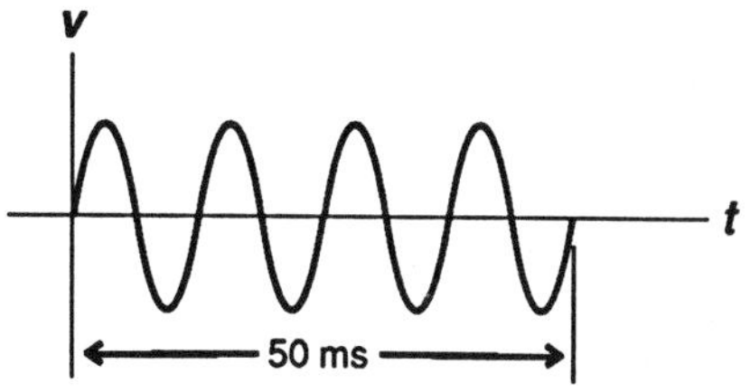

Fig. 1-18.

EXAMPLE 1-3

What is the period in Fig. 1-18?

SOLUTION

There are two ways to find the period. First, you can divide 50 ms by 4 to get the period:

$$T = \frac{50 \text{ ms}}{4} = 12.5 \text{ ms}$$

Second, you can take the reciprocal of frequency to get

$$T = \frac{1}{80 \text{ Hz}} = 12.5 \text{ ms}$$

Either of the foregoing methods is valid.

EXAMPLE 1-3

What is the period in Fig. 1-19? The frequency?

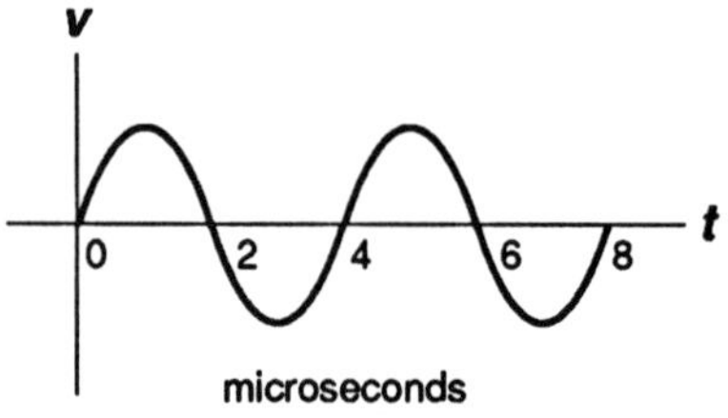

Fig. 1-19.

SOLUTION

One cycle takes 4 μs. Therefore,

$T = 4\ \mu\text{s}$

Take the reciprocal of this to get the frequency:

$$f = \frac{1}{4\ \mu\text{s}} = 250 \text{ kHz}$$

EXAMPLE 1-4

An oscilloscope is an electronic instrument that displays waveforms. It usually includes a grid calibrated in centimeters (cm) as shown in Fig. 1-20. If the vertical scale is 3 V per centimeter and the horizontal scale is 5 ms per centimeter, what is the peak value of the sine wave? The frequency?

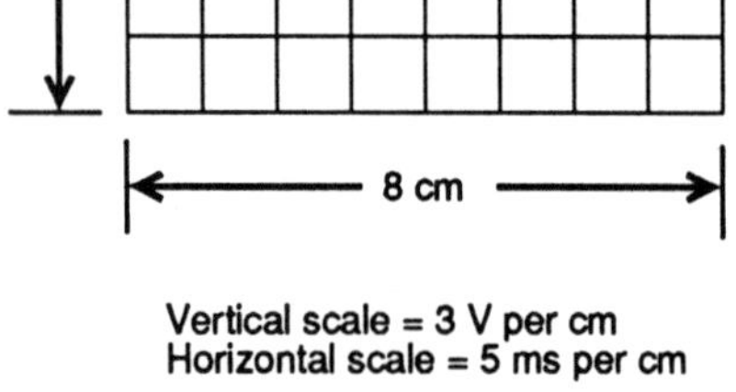

Fig. 1-20.

SOLUTION

The sine wave has a peak value of 2 cm, which translates into

$V_p = (2 \text{ cm})(3 \text{ V/cm}) = 6 \text{ V}$

The period is 4 cm, equivalent to

$T = (4 \text{ cm})(5 \text{ ms/cm})) = 20 \text{ ms}$

Take the reciprocal to get the frequency:

$$f = \frac{1}{20 \text{ ms}} = 50 \text{ Hz}$$

1-5. AVERAGE VALUE

To calculate the *average value* of a quantity, you have to add samples of the quantity, then divide by the number of samples. With this idea in mind, let us find out how to get the average value of a sine wave.

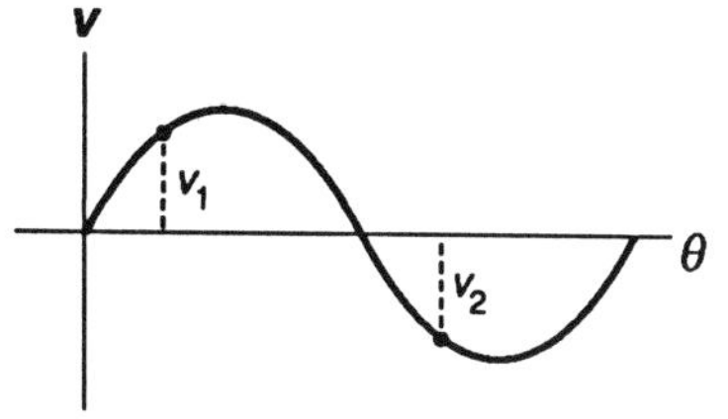

Fig. 1-21.

Average Over a Complete Cycle

The average value of a sine wave over a complete cycle is the sum of all the instantaneous values divided by the number of samples. In Fig 1-21, each positive value v_1 can be matched by an equal and opposite negative value v_2. Therefore, the sum of all instantaneous values throughout the cycle is zero. This means the average value of a sine wave over one cycle is zero. As an equation,

$$V_{av} = 0 \qquad (1\text{-}5)$$

Average Over a Half Cycle

The average value of a sine wave over half a cycle is different. Consider the positive half cycle. When we add all the instantaneous values between 0° and 180°, we are adding only positive numbers. Therefore, we get a sum that is greater than zero. When we divide this sum by the number of samples, we get an average value equal to

$$V_{av} = 0.637V_p \qquad (1\text{-}6)$$

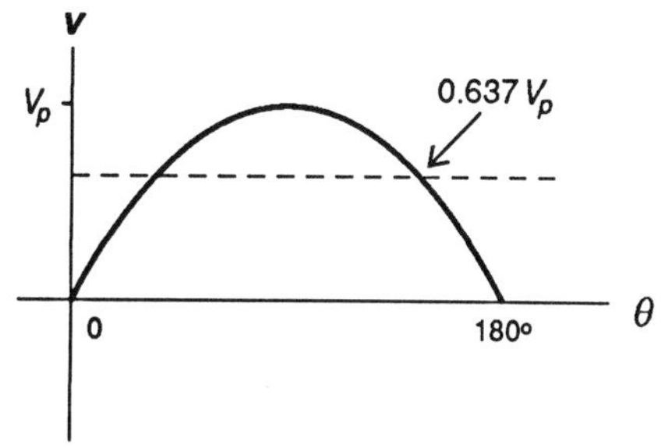

Fig. 1-22.

This says the average value of a sine wave over half a cycle equals 0.637 times the peak value as shown in Fig. 1-22.

Where does the factor 0.637 come from? There are many ways to derive it. For instance, we can add all the instantaneous values of a sine wave taken 1° apart over half a cycle and divide by 180, the number of samples. The result will be 0.637 times the peak value. (Note: the foregoing calculation is easily done with a computer.)

Another way to derive the factor 0.637 is with advanced mathematics. Using a calculus approach, it is possible to derive this equation for the average value of half a cycle:

$$V_{av} = \frac{2V_p}{\pi} \qquad (1\text{-}7)$$

The quantity 2 divided by π equals 0.637.

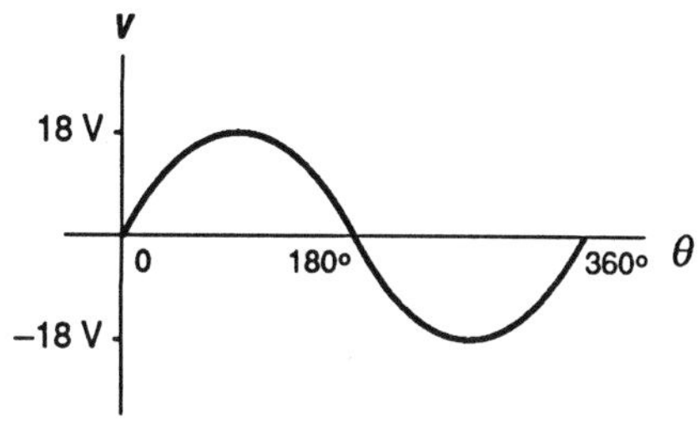

Fig. 1-23.

EXAMPLE 1-5

What is the average value of the sine wave in Fig. 1-23 for the positive half cycle? For the negative half cycle?

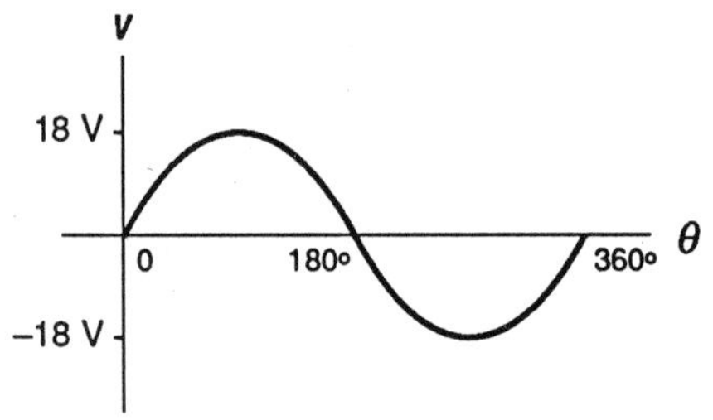

Fig. 1-23.

SOLUTION

The sine wave has a peak value of 18 V. Multiply 0.637 by 18 V to get

$V_{av} = 0.637(18\ V) = 11.5\ V$

In the negative half cycle, all of the instantaneous values are negative. Therefore, the average value of the negative half cycle is equal and opposite the average value of the positive half cycle:

$V_{av} = -11.5\ V$

1-6. RMS VALUES

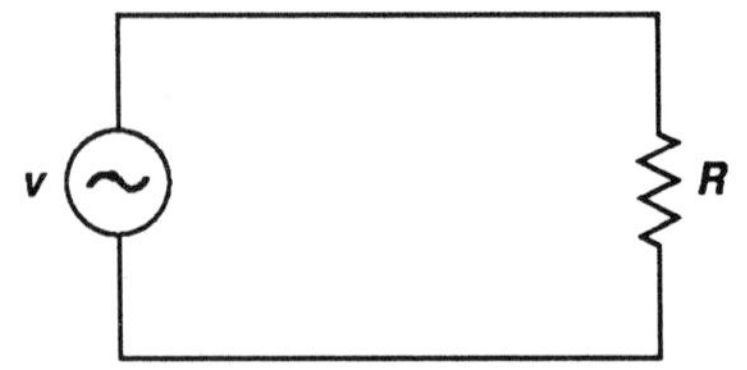

Fig. 1-24.

Figure 1-24 shows the schematic symbol of an ac voltage source. The small sine wave inside the circle is a reminder of the alternating voltage. As previously discussed, the instantaneous voltage varies throughout the cycle and is given by

$v = V_p \sin \theta$

As the angle varies from 0° to 360°, the voltage varies sinusoidally.

What is the power in the resistor? Since the voltage and current change during the cycle, the power also changes. At any instant, the power is given by

$p = vi$

where p, v, and i are instantaneous values. For instance, when the angle is 90°, the instantaneous power is

$p = (V_p \sin 90°)(I_p \sin\ 90°) = V_pI_p$

This is the maximum instantaneous power.

At any other angle that is not an odd multiple of 90°, the instantaneous power is less than this. For example, when the angle is 30°, the instantaneous power is

$p = (V_p \sin 30°)(I_p \sin\ 30°) = 0.25V_pI_p$

When the angle is 60°, the instantaneous power is

$p = (V_p \sin 60°)(I_p \sin\ 60°) = 0.75V_pI_p$

During the negative half cycle, the voltage and the current are both negative, because each reverses its direction. The product of two negative quantities is positive. Therefore, the instantaneous power is a positive quantity during the negative half cycle, as well as the positive half cycle. For instance, when the angle is 270°, the instantaneous power is

$$p = (V_p \sin 270°)(I_p \sin 270°) = (-V_p)(-I_p) = V_p I_p$$

This is the same value of maximum power that we calculated for an angle of 90°.

By calculating the instantaneous power at intervals of 1° over a complete cycle of a sine wave, we can graph the values to get Fig. 1-25. As you can see, the instantaneous power is positive throughout the cycle. Furthermore, the maximum value of instantaneous power occurs at 90° and 270°.

A final point. The waveform of Fig. 1-25 is sinusoidal. It differs from the usual sine wave because it is a sine wave that has been shifted upward until the negative peaks fall on the horizontal axis. One way to prove that the waveform is sinusoidal to use a computer to calculate and plot all instantaneous powers over a complete cycle. Another way is to use trigonometric identities as follows:

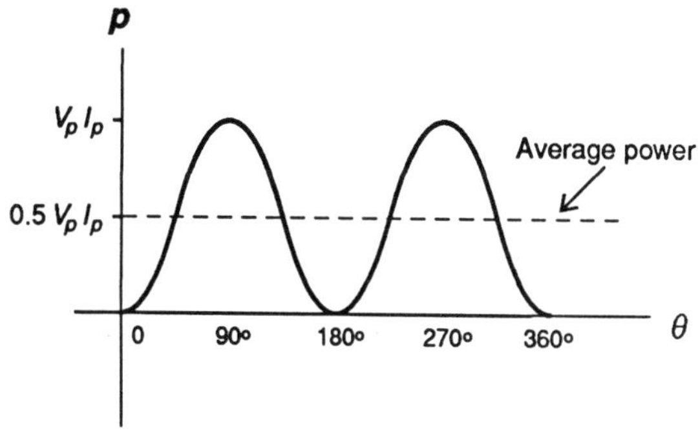

Fig. 1-25.

$$p = vi = (V_p \sin \theta)(I_p \sin \theta) = V_p I_p(\sin \theta)(\sin \theta) = V_p I_p(\sin \theta)^2$$

$$= 0.5 V_p I_p (1 - \cos 2\theta)$$

When this equation is graphed, Fig. 1-25 results.

Average Power Over a Complete Cycle

Because of the symmetry in Fig 1-25, the average power over a complete cycle equals half of the maximum power. In symbols,

$$p_{av} = 0.5 V_p I_p \tag{1-8}$$

For instance, if $V_p = 6$ V and $I_p = 0.4$ mA, the average power over a cycle is

$$p_{av} = 0.5(6\text{ V})(0.4\text{ mA}) = 1.2\text{ mW}$$

Rms Voltage

Figure 1-26a shows an ac source connected to a resistor. This ac source produces an average power of

$$p_{av} = 0.5 V_p I_p$$

Since $I_p = V_p/R$, the equation can be written as

$$p_{av} = \frac{0.5 V_p^2}{R}$$

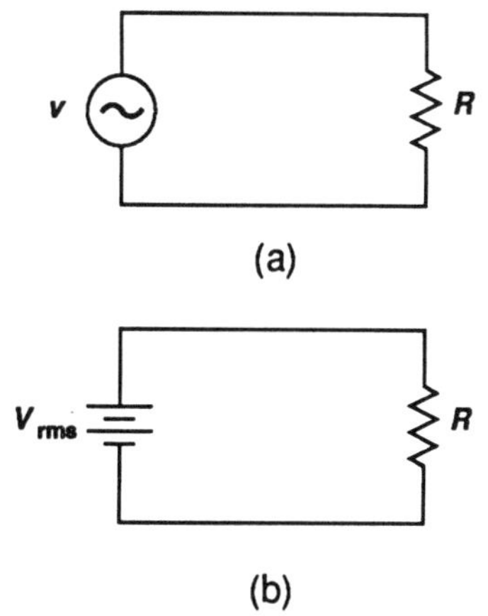

Fig. 1-26.

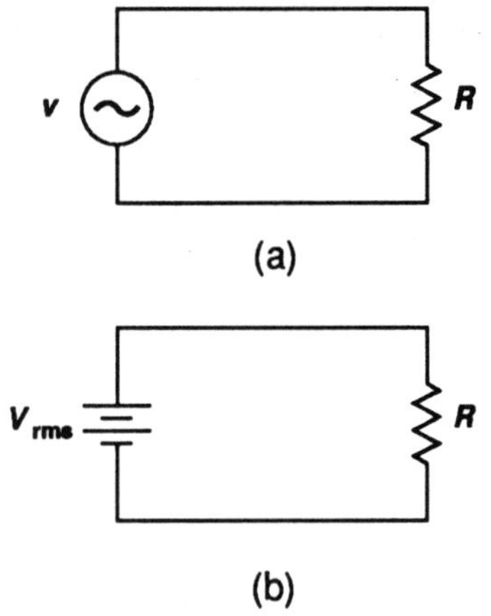

Fig. 1-26.

MEMORIZE

$V_{rms} = 0.707\,V_p$

$I_{rms} = 0.707\,I_p$

Figure 1-26b shows a dc source connected across a resistor of the same value. The *rms voltage* is defined as the dc voltage that produces the same average power as the sinusoidal voltage. Stated another way, the rms voltage is the dc voltage that produces the same amount of heat as the sinusoidal voltage.

We can derive an equation for the rms voltage as follows. The power in Fig. 1-26b is

$$P = \frac{(V_{\text{rms}})^2}{R}$$

Since the two powers are equal, we can write

$$\frac{(V_{\text{rms}})^2}{R} = \frac{0.5V_p^2}{R}$$

or

$$(V_{\text{rms}})^2 = 0.5V_p^2$$

After taking the square root of both sides, we get

$$V_{\text{rms}} = 0.707V_p \qquad (1\text{-}9)$$

This says the rms voltage equals 0.707 times the peak value.

By a similar derivation, the rms current is given by

$$I_{\text{rms}} = 0.707I_p \qquad (1\text{-}10)$$

Equations (1-9) and (1-10) should be memorized because they will be used many times in the remaining chapters. These equations convert a sinusoidal quantity into an equivalent dc quantity. The advantage of this is that we can use dc formulas learned earlier.

Ac Power Formulas

Rms values are used a great deal in the analysis of ac circuits. Why? Because we usually want to know the average power, rather than the instantaneous power. Given a sinusoidal voltage source, we can calculate the average power by multiplying the rms voltage and the rms current. In other words, the three formulas you learned for calculating dc power can now be applied to ac circuits as follows:

$$P = V_{rms}I_{rms} \qquad (1\text{-}11)$$

IMPORTANT

$P = V_{rms}I_{rms}$

$$P = (I_{rms})^2R \qquad (1\text{-}12)$$

$$P = \frac{(V_{rms})^2}{R} \qquad (1\text{-}13)$$

Ac Measurements

Ac voltmeters and ammeters are usually calibrated in rms values. Whenever in doubt, assume that an ac voltmeter indicates rms voltages, because this is true for virtually all ac voltmeters. Once you have measured the rms values of a sine wave, you can calculate the peak values by rearranging Eqs. (1-9) and (1-10) as follows:

$$V_p = 1.414V_{rms} \qquad (1\text{-}14)$$

$$I_p = 1.414I_{rms} \qquad (1\text{-}15)$$

Because these equations are used so often, you should memorize them.

EXAMPLE 1-6

What is the rms voltage in Fig. 1-27? The rms current?

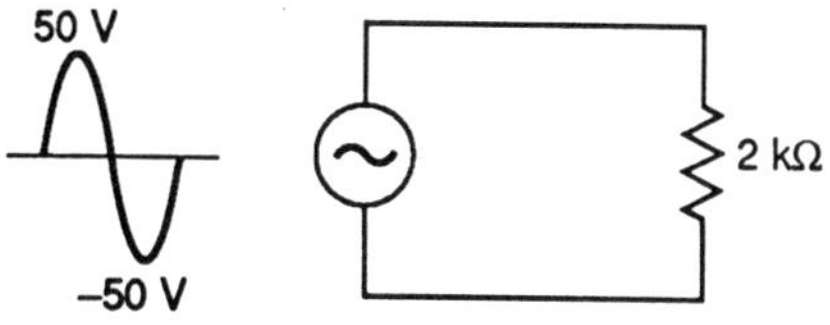

Fig. 1-27.

SOLUTION

Convert the sinusoidal voltage to an equivalent dc voltage. To do this, you multiply 0.707 by the peak voltage to get

$$V_{rms} = 0.707(50\ \text{V}) = 35.4\ \text{V}$$

If a battery of 35.4 V were substituted for the ac source of Fig. 1-27, the resistor would produce exactly the same amount of heat.

Since rms values are equivalent dc values, Ohm's law may be used to calculate the rms current:

$$I_{rms} = \frac{35.4\ \text{V}}{2\ \text{k}\Omega} = 17.7\ \text{mA}$$

EXAMPLE 1-7

What is the average power in Fig. 1-27?

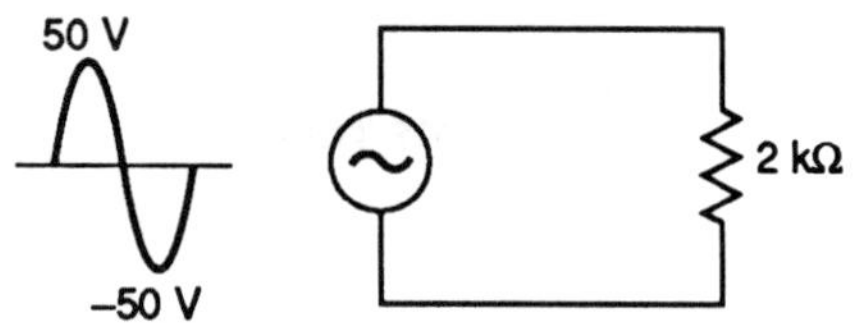

Fig. 1-27.

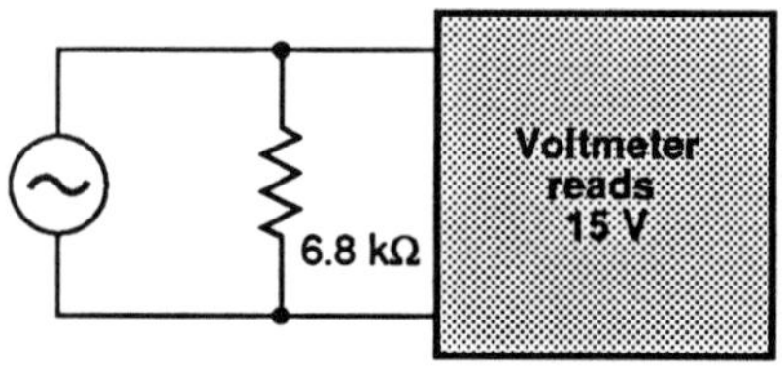

Fig. 1-28.

SOLUTION

You can use any of the three basic power formulas. For instance, rms voltage times rms current gives

$$P = (35.4 \text{ V})(17.7 \text{ mA}) = 0.627 \text{ W}$$

The rms current squared times the resistance also gives the same answer:

$$P = (17.7 \text{ mA})^2(2 \text{ k}\Omega) = 0.627 \text{ W}$$

Similarly, the rms voltage squared divided by the resistance gives

$$P = \frac{(35.4 \text{ V})^2}{2 \text{ k}\Omega} = 0.627 \text{ W}$$

EXAMPLE 1-8

What is the average power in Fig. 1-28?

SOLUTION

You can assume the voltmeter reads rms values because this is true for virtually all ac voltmeters. Then, 15 V rms is across 6.8 kΩ. Square this voltage and divide by the resistance to get

$$P = \frac{(15 \text{ V})^2}{6.8 \text{ k}\Omega} = 33.1 \text{ mW}$$

If you prefer, you can solve the problem like this. First, get the rms current:

$$I_{rms} = \frac{15 \text{ V}}{6.8 \text{ k}\Omega} = 2.21 \text{ mA}$$

Second, multiply rms voltage and current to get

$$P = (15 \text{ V})(2.21 \text{ mA}) = 33.1 \text{ mW}$$

EXAMPLE 1-9

What is the peak voltage in Fig. 1-28? The peak current?

SOLUTION

The rms values are

$$V_{rms} = 15 \text{ V}$$

$$I_{rms} = 2.21 \text{ mA}$$

As derived earlier, the peak values equal 1.414 times the rms values. Therefore,

$V_p = 1.414(15 \text{ V}) = 21.2 \text{ V}$

$I_p = 1.414(2.21 \text{ mA}) = 3.12 \text{ mA}$

EXAMPLE 1-10

Graph the sinusoidal voltage and current for Fig. 1-28.

SOLUTION

All you have to do is draw two sine waves, one for voltage and another for current. Then, label each with the correct peak values as shown in Fig. 1-29.

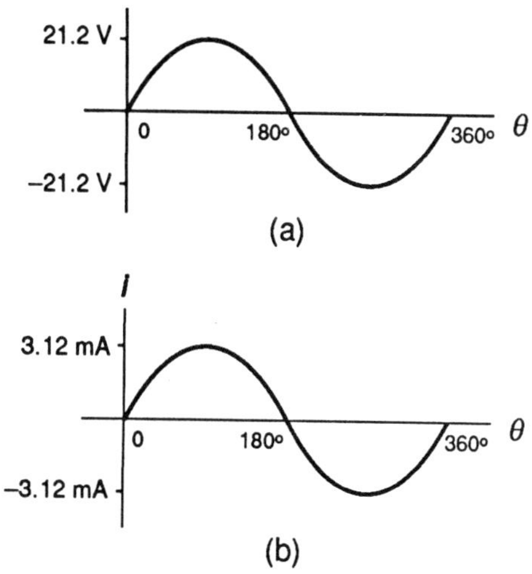

Fig. 1-29.

1-7. AC ANALYSIS

In dc circuits, we can calculate the current with Ohm's law:

$$I = \frac{V}{R}$$

In an ac circuit like Fig. 1-30, we can also use Ohm's law to calculate the instantaneous current:

$$i = \frac{v}{R}$$

To keep dc and ac quantities distinct in our equations, we will use capital letters for dc quantities and lower-case letters for ac quantities.

In Fig. 1-30, a sinusoidal voltage v produces a sinusoidal current i. Since the instantaneous voltage is given by

$$v = V_p \sin \theta$$

the instantaneous current is

$$i = \frac{V_p \sin \theta}{R}$$

When the angle is 90°, the instantaneous current is maximum. Because of this, we can write this equation for the peak current:

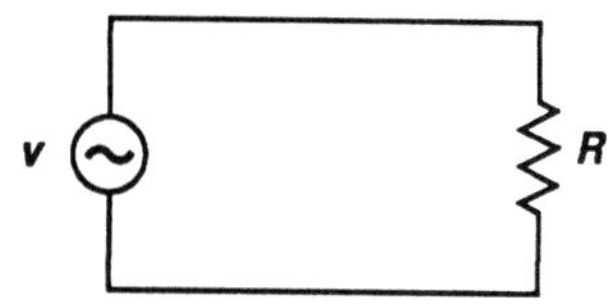

Fig. 1-30.

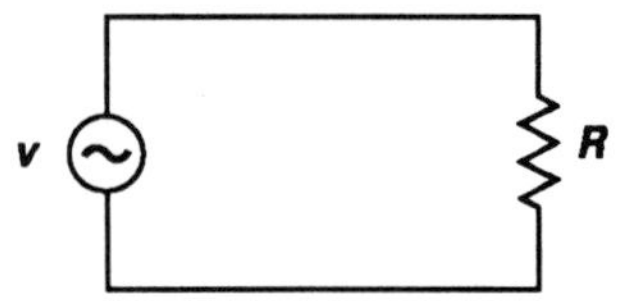

Fig. 1-30.

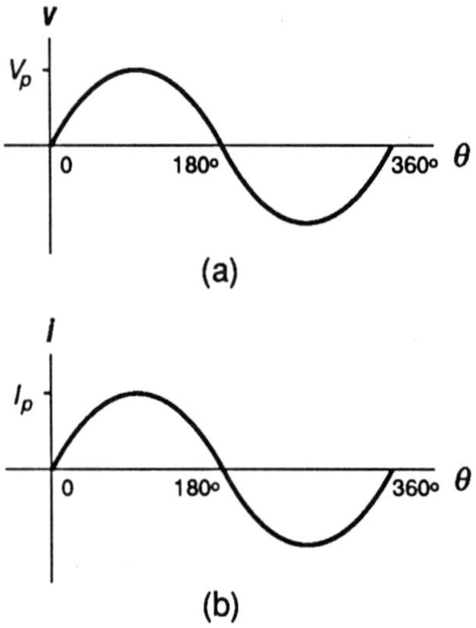

Fig. 1-31.

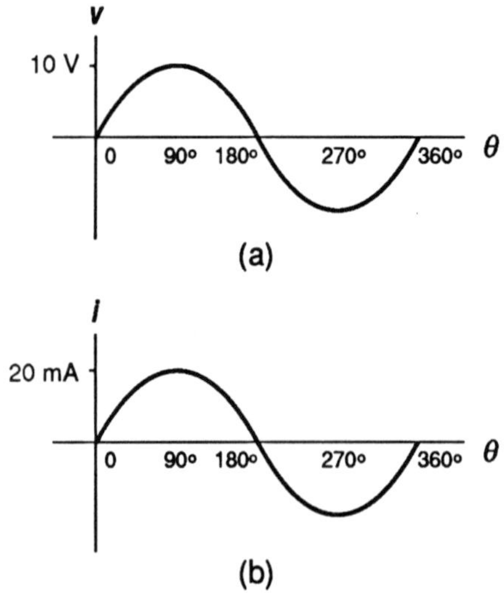

Fig. 1-32.

$$I_p = \frac{V_p}{R} \tag{1-16}$$

This makes sense because it says the peak current equals the peak voltage divided by the resistance.

Sine Waves In Phase

In Fig. 1-30, the instantaneous voltage is given by

$$v = V_p \sin \theta \tag{1-17}$$

Because of Ohm's law, the instantaneous current is given by

$$i = I_p \sin \theta \tag{1-18}$$

where $I_p = V_p / R$.

Figure 1-31 shows the graphs of these two equations. As you see, each sine wave starts at the same instant. Both sine waves reach a positive maximum at the 90°, then decrease to zero, and so on. Waveforms that are aligned like this are described as being *in phase*.

EXAMPLE 1-11

The voltage across a load resistor of 500 Ω is given by

$$v = 10 \sin \theta$$

Sketch the sinusoidal voltage and current.

SOLUTION

The peak voltage is 10 V. Therefore, the peak current is

$$I_p = \frac{10\text{ V}}{500\ \Omega} = 20\text{ mA}$$

The voltage and current waveforms are in phase as shown in Fig. 1-32.

1-8. RADIANS

Distance can be measured in feet, meters, miles, etc. The unit of measure chosen by a scientist depends on what he is trying to do. Up to now, we have measured the angle of rotation in degrees. But this is not the best unit of measure for all applications. This section discusses the *radian*, a unit of angular measure that allows us to include time in our calculations of instantaneous values.

Definition of Radian

A radian is defined as the angle whose arc equals the radius (see Fig. 1-33). The approximate value of this angle is

$$1 \text{ radian} = 57.3^\circ \qquad (1\text{-}19)$$

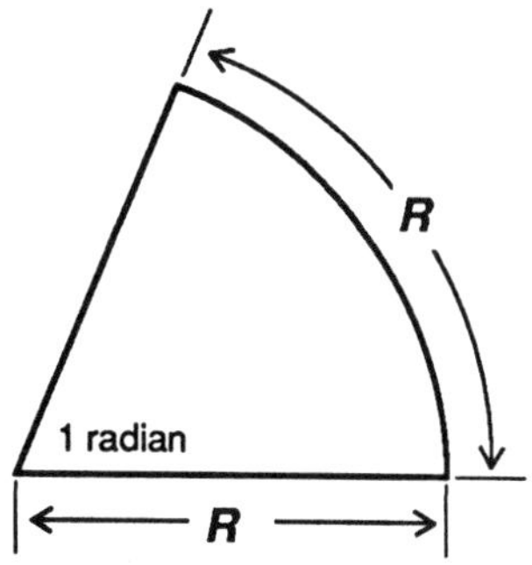

Fig. 1-33.

One way to prove this is to draw Fig. 1-33 with an arc equal to the radius. If you can measure the angle with a protractor, you will get 57.3°.

Here is another way to prove Eq. (1-19). We can set up this proportion with Fig. 1-33:

$$\frac{1 \text{ radian}}{R} = \frac{360^\circ}{2\pi R}$$

This proportion makes sense. It says the ratio of a radian to the arc length equals the ratio of 360° to the circumference. Solving the foregoing equation gives

$$1 \text{ radian} = \frac{360^\circ}{2\pi} = \frac{360^\circ}{6.28} = 57.3^\circ$$

Converting from Degrees to Radians

The angle of rotation may be measured in degrees or in radians. To keep the two types of measurements distinct in our formulas, we will use this notation:

θ_D = angle in degrees

θ_R = angle in radians

Given any angle, we can set up this proportion:

$$\frac{\theta_D}{360^\circ} = \frac{\theta_R}{2\pi}$$

We can rearrange this to get the following conversion formula:

$$\theta_R = \frac{\pi}{180^\circ}\theta_D \qquad (1\text{-}19)$$

This is useful whenever you have the angle in degrees and want to convert it to radians.

Converting from Radians to Degrees

We can rearrange the foregoing equation to get

$$\theta_D = \frac{180^\circ}{\pi}\,\theta_R \qquad (1\text{-}20)$$

This is useful whenever you have the angle in radians and want to convert it to degrees.

EXAMPLE 1-12

Convert each of the following to radians: 0°, 90°, 180°, 270°, and 360°.

SOLUTION

Use the conversion formula to calculate each angle as follows:

$$\theta_R = \frac{\pi}{180^\circ}\,0^\circ = 0$$

$$\theta_R = \frac{\pi}{180^\circ}\,90^\circ = 1.57 \text{ radians}$$

$$\theta_R = \frac{\pi}{180^\circ}\,180^\circ = 3.14 \text{ radians}$$

$$\theta_R = \frac{\pi}{180^\circ}\,270^\circ = 4.71 \text{ radians}$$

$$\theta_R = \frac{\pi}{180^\circ}\,360^\circ = 6.28 \text{ radians}$$

EXAMPLE 1-13

Repeat the foregoing example but leave the angle in terms of π.

SOLUTION

Instead of working out the answers to the final form shown in the preceding example, we can stop short as follows:

$$\theta_R = \frac{\pi}{180^\circ} 0^\circ = 0$$

$$\theta_R = \frac{\pi}{180^\circ} 90^\circ = \frac{\pi}{2} \text{ radians}$$

$$\theta_R = \frac{\pi}{180^\circ} 180^\circ = \pi \text{ radians}$$

$$\theta_R = \frac{\pi}{180^\circ} 270^\circ = \frac{3\pi}{2} \text{ radians}$$

$$\theta_R = \frac{\pi}{180^\circ} 360^\circ = 2\pi \text{ radians}$$

It is easy to remember this pattern because each 90° adds another $\pi/2$ to the number of radians.

MEMORIZE

$\pi/2 = 90^\circ$

$\pi = 180^\circ$

$3\pi/2 = 270^\circ$

$2\pi = 360^\circ$

EXAMPLE 1-14

Graph the following using radians for the angle:

$v = 10 \sin \theta$

SOLUTION

Figure 1-34a shows the graph. This graph is equivalent to Fig. 1-32a. In either case, we have a sinusoidal voltage with a peak value of 10 V. Every 90° interval is equivalent to an additional $\pi/2$.

Figure 1-34b is a memory aid for radians and degrees. This figure shows that 2π radians equals 360°, π radians is 180°, $\pi/2$ radians is 90°, and so on.

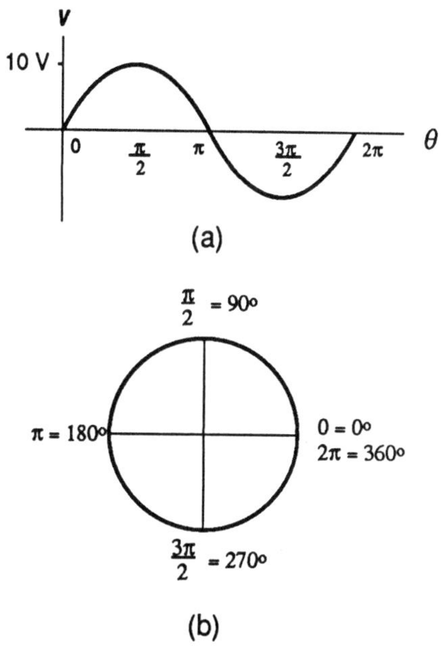

Fig. 1-34.

1-9. TIME

The instantaneous value of a sinusoidal voltage is given by

$v = V_p \sin \theta$

In many applications, we need to express the angle as a function of time. Then, we can calculate instantaneous values at any point in time.

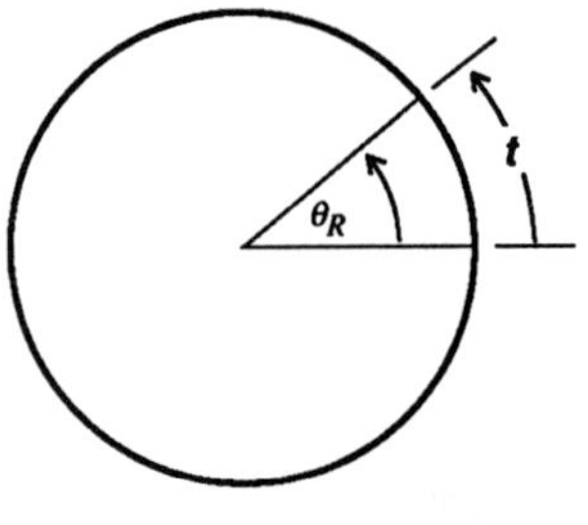

Fig. 1-35.

Here is how derive a relation between the angle of rotation and time. Figure 1-35 shows the conductor at an angle. The conductor has rotated from the horizontal axis to its current position in time *t*. As the conductor continues to rotate, *t* increases linearly. Since the period is the time it takes the conductor to travel through 360°, we can set up this proportion:

$$\frac{\theta_R}{2\pi} = \frac{t}{T}$$

This says the ratio of the angle in radians to 2π radians equals the ratio of elapsed time to the period. Solving for the angle, we get

$$\theta_R = \frac{2\pi t}{T} \quad (1\text{-}21)$$

Since frequency equals the reciprocal of period, we can write the foregoing equation in this equivalent form:

$$\theta_R = 2\pi f t \quad (1\text{-}22)$$

Now, we substitute the foregoing expression for the angle into the equations for sinusoidal voltage to get

$$v = V_p \sin 2\pi f t \quad (1\text{-}23)$$

Given the peak value and frequency of a sinusoidal voltage, we can use this equation to calculate the voltage at any instant during the cycle. A similar expression applies to current:

$$i = I_p \sin 2\pi f t \quad (1\text{-}24)$$

EXAMPLE 1-15

A sinusoidal voltage has a peak value of 10 V and a frequency of 1 kHz. What is the instantaneous voltage when $t = 0.1$ ms?

SOLUTION

Substitute the peak value and frequency into Eq. (1-23) to get

$$v = 10 \sin 2000\pi t$$

When $t = 0.1$ ms, this gives

$$v = 10 \sin (2000\pi)(0.0001) = 10 \sin 0.2\pi = 10 \sin 36^\circ = 5.88$$

For convenience, most people omit the units of measure in the foregoing calculations. In other words, we could have solved the problem by including the units of measure as follows:

$$v = (10 \text{ V}) \sin (2000\pi \text{ Hz})t = (10 \text{ V}) \sin (2000\pi \text{ Hz})(0.1 \text{ ms}) = 5.88 \text{ V}$$

But including the V, Hz, and ms in these calculations clutters the work for no good reason. This is why we usually omit the units with the understanding that the final answer is in volts.

EXAMPLE 1-16

Graph the instantaneous voltage of the preceding example as a function of time.

SOLUTION

Since the frequency is 1 kHz, the period is

$$T = \frac{1}{1 \text{ kHz}} = 1 \text{ ms}$$

Since we know the peak voltage is 10 V and the period is 1 ms, we can graph the voltage as shown in Fig. 1-36.

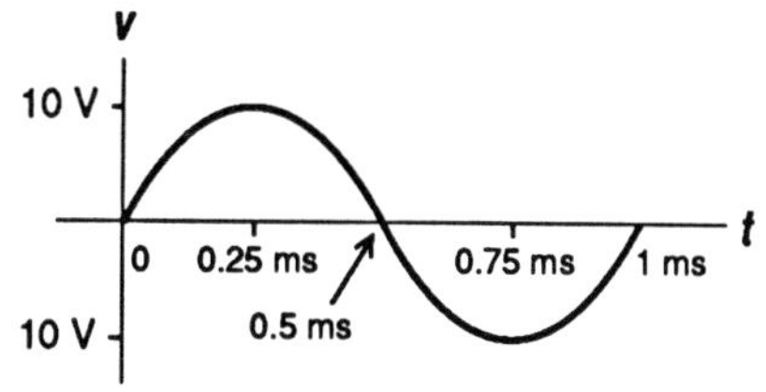

Fig. 1-36.

EXAMPLE 1-17

Here is a comprehensive example. If you can solve problems like the following, you will have mastered all of the basic ideas in this chapter. If necessary, review this example several times until you can perform all of the calculations.

An oscilloscope is connected to a load resistance of 2 kΩ. Figure 1-37 shows the voltage waveform. What is the value of each of these quantities: peak voltage, average voltage, rms voltage, period, frequency, load power, and instantaneous voltage when $t = 0.4$ ms?

SOLUTION

As indicated in Fig. 1-37, the oscilloscope is calibrated to read 5 V per cm on the vertical axis and 1 ms per cm on the horizontal axis. Here are the calculations:

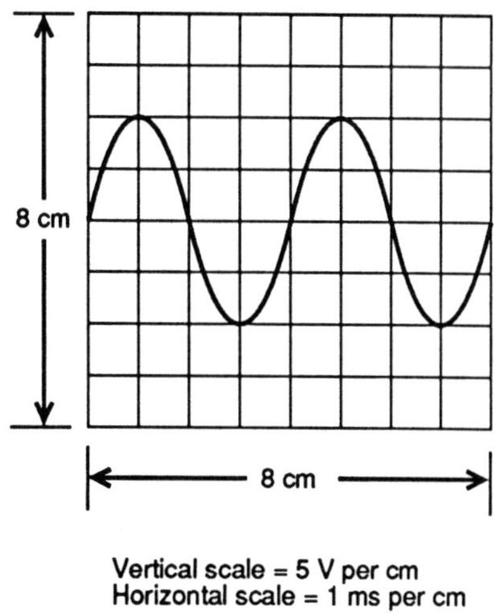

Fig. 1-37.

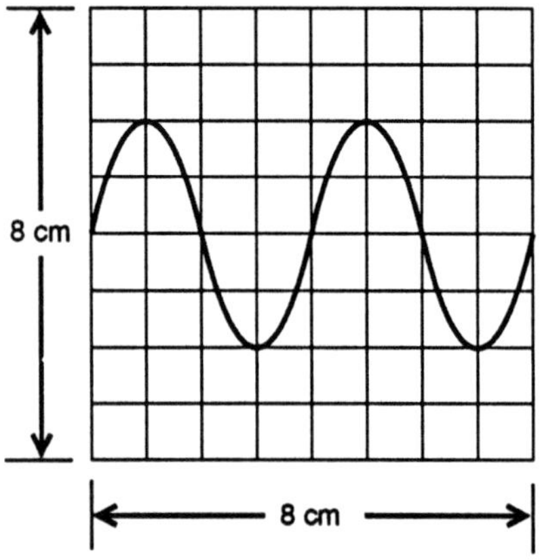

Fig. 1-37.

Peak Voltage:

In Fig. 1-37, you can see that the peak value is 2 cm. When converted to volts, this is

$V_p = (2\text{ cm})(5\text{ V/cm}) = 10\text{ V}$

Average Voltage:

The average voltage depends on how much of the cycle you are talking about. For the complete cycle, the average voltage is zero. For half of the cycle, the average voltage is

$V_{av} = 0.637(10\text{ V}) = 6.37\text{ V}$

Rms Voltage:

The rms value of a sine wave always equals 0.707 times the peak value. In this case,

$V_{rms} = 0.707(10\text{ V}) = 7.07\text{ V}$

Period:

The period is the duration of one complete cycle. In Fig. 1-37, the period is 4 cm, equivalent to

$T = (4\text{ cm})(1\text{ ms/cm}) = 4\text{ ms}$

Frequency:

The easiest way to get the frequency is to take the reciprocal of the period:

$$f = \frac{1}{4\text{ ms}} = 250\text{ Hz}$$

An alternative way to divide the number of cycles by the total time:

$$f = \frac{2\text{ cycles}}{8\text{ ms}} = 250\text{ Hz}$$

Load power:

The waveform of Fig. 1-37 is across a load resistance of 2 kΩ. We must always use rms values when calculating power. One way to calculate the load power is like this. The peak current through the load resistor equals the peak voltage divided by the resistance:

$$I_p = \frac{10\text{ V}}{2\text{ k}\Omega} = 5\text{ mA}$$

The rms current is

$$I_{rms} = 0.707(5\text{ mA}) = 3.54\text{ mA}$$

The load power equals the rms voltage times the rms current:

$$P = (7.07\text{ V})(3.54\text{ mA}) = 25\text{ mW}$$

An alternative and more direct way to calculate the load power is to square the rms voltage and divide by the resistance:

$$P = \frac{(7.07\text{ V})2}{2\text{ k}\Omega} = 25\text{ mW}$$

Instantaneous voltage:

The general formula for instantaneous voltage is

$$v = V_p \sin \theta$$

Since the peak voltage is 10 V and the frequency is 250 Hz, we can write this as

$$v = V_p \sin 2\pi ft = 10 \sin 2\pi(250)t = 10 \sin 500\pi t$$

When $t = 0.4$ ms, the instantaneous voltage is

$$v = 10 \sin 500\pi t = 10 \sin 500\pi(0.0004) = 10 \sin 0.2\pi = 5.88\text{ V}$$

If you have a scientific calculator, the foregoing calculation is easy. The key presses are typically as follows: enter 0.2, multiply by π, press the = key (answer in radians), take the sine, and multiply by 10 to get the answer.

MULTIPLE-CHOICE TEST

The following questions may have more than one right answer. Select the best answer. This is the one that is always true, or covers more situations, or agrees with the material of this chapter, etc.

1. Direct current always
   a. flows in the same direction
   b. is constant with respect to time
   c. has a horizontal graph
   d. all of the above

2. The induced voltage is maximum when the conductor is
   a. parallel to the lines of force
   b. at angle of zero
   c. perpendicular to the lines of force
   d. made of copper

3. When the magnetic lines of force are vertical, the induced voltage is maximum positive at
   a. 0°
   b. 90°
   c. 180°
   d. 270°

4. When the magnetic lines of force are vertical, the induced voltage is maximum negative at
   a. 0°
   b. 90°
   c. 180°
   d. 270°

5. The sine of an angle is maximum positive when the angle is
   a. 0°
   b. 90°
   c. 180°
   d. 270°

6. Frequency is usually measured in
   a. cycles
   b. seconds
   c. cycles per second
   d. hertz

7. If the period is 1 μs, the frequency is
   a. 1 kHz
   b. 10 kHz
   c. 100 kHz
   d. 1 MHz

8. 12 cycles of a sine wave occur in 3 ms. The frequency is
   a. 2.5 kHz
   b. 4 kHz
   c. 2.5 MHz
   d. 4 MHz

9. If a sine wave has a peak value of 5 V, the average value over one cycle is
   a. 0
   b. 3.19 V
   c. 3.54 V
   d. 5 V

10. If a sine wave has a peak value of 25 V, the average value over half a cycle is
   a. 0
   b. 15.9 V
   c. 17.7 V
   d. 25 V

11. If a sine wave has a peak value of 5 V, the rms value is
   a. 0
   b. 3.19 V
   c. 3.54 V
   d. 5 V

12. If a sine wave has a peak value of 25 V, the rms value is
   a. 0
   b. 15.9 V
   c. 17.7 V
   d. 25 V

13. Whenever you calculate average power in an ac circuit, you must use
   a. dc values
   b. average values
   c. rms values
   d. instantaneous values

14. Ohm's law is valid in an ac circuit with ideal resistors if you use
   a. instantaneous values
   b. peak values
   c. rms values
   d. all of the above

15. A radian is closest in value to
   a. 0°
   b. 30°
   c. 60°
   d. 90°

16. The number of radians in a complete cycle is
   a. 0
   b. 1
   c. 3.14
   d. 6.28

17. An angle of $3\pi/2$ radians equals
   a. 0°
   b. 90°
   c. 180°
   d. 270°

18. The angle of rotation is directly proportional to
    a. peak value
    b. frequency
    c. period
    d. π

19. The peak value of a sine wave equals the rms value times
    a. 0.636
    b. 0.707
    c. 1
    d. 1.414

20. Negative current
    a. means protons flow instead of electrons
    b. cannot exist
    c. means the current is in the reverse direction
    d. occurs when a resistance is negative

21. The instantaneous power in an ac circuit with ideal resistors
    a. is negative for half the cycle
    b. may be positive or negative
    c. cannot be negative
    d. equals rms voltage times rms current

22. When the frequency of a sine wave increases, the period
    a. decreases
    b. stays the same
    c. increases
    d. none of the above

23. If the period of a sine wave doubles, the frequency
    a. drops in half
    b. stays the same
    c. doubles
    d. none of the above

24. An angle of 2π radians has the same sine as an angle of
    a. 0°
    b. 90°
    c. 135°
    d. 270°

25. An ac voltmeter usually indicates
    a. dc values
    b. average values
    c. rms values
    d. instantaneous values

PROBLEMS

Sec. 1-3. The Sine Wave

1-1. A sine wave has a peak voltage of 20 V. Calculate the instantaneous voltage for angles of 0°, 15°, 30°, 45°, 60°, 75° and 90°.

1-2. Refer to Fig. 1-16. Draw waveforms like this for each of the following:
a. A sinusoidal voltage with a peak value of 25 V
b. A sinusoidal current with a peak value of 15 mA

1-3. A sinusoidal current has a peak value of 180 mA. What is the instantaneous current when the angle is 45°? (Note; $i = I_p \sin \theta$)

Sec. 1-4. Frequency and Period

1-4. If 9 cycles of a sine wave occur in 2.5 ms, what is the frequency of the sine wave?

1-5. A sine wave has a period of 40 µs. What is its frequency?

1-6. If the frequency of a sine wave is 455 kHz, what is its period?

1-7. What is the peak value and frequency of the sine wave shown in Fig. 1-38?

1-8. An oscilloscope display shows a sine wave with a peak value of 3 cm and a period of 5 cm. If the vertical scale is 50 mV/cm and the horizontal scale is 25 µs/cm, what is peak value and the frequency of the sine wave?

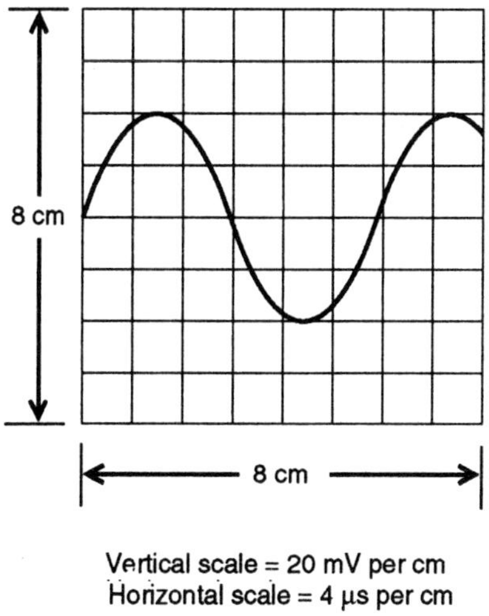

Fig. 1-38.

Sec. 1-5. Average Value

1-9. A sine wave has a peak value of 65 V. What is its average value over half a cycle?

1-10. What is the average value of the sine wave shown in Fig. 1-38 over a complete cycle? Over half of a cycle?

1-11. The average value of a sinusoidal voltage over half a cycle is 12 V. What is the peak value of the sine wave?

Sec. 1-6. Rms Values

1-12. A sine wave has a peak value of 36 V. What is its rms value?

1-13. What is the rms value of the sine wave shown in Fig. 1-38?

1-14. The rms value of a sinusoidal voltage is 4.75 V. What is the peak value of the sine wave?

1-15. If the sinusoidal voltage of Fig. 1-38 is across a load resistance of 300 Ω, what is the average power?

1-16. A resistance of 470 Ω has an rms current of 5 mA. What is the average power?

1-17. What is the average power in Fig. 1-39?

Fig. 1-39.

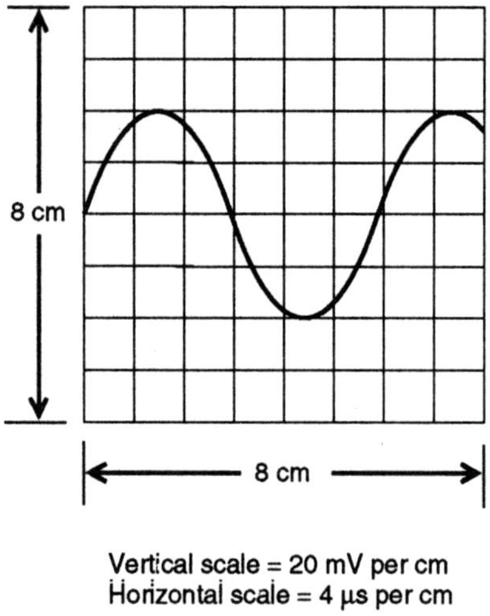

Fig. 1-38.

Fig. 1-39.

Sec. 1-7. Ac Analysis

1-18. What is the peak current in Fig. 1-39?

1-19. A sinusoidal voltage with a peak of 25 V is across a resistance of 470 kΩ. What is the rms current?

1-20. Sketch the voltage and current waveforms for Fig. 1-39.

Sec. 1-8. Radians

1-21. Convert each of the following to radians: 45°, 135°, 225°, and 315°.

1-22. What does 1.25π radians equal in degrees?

1-23. If $v = 25 \sin 0.75\pi$, what does v equal?

1-24. If $i = 0.05 \sin 0.32\pi$, what does i equal?

Sec. 1-9. Time

1-25. A sinusoidal voltage has a peak value of 8 V and a frequency of 2.5 kHz. What is the instantaneous voltage when $t = 125\ \mu s$?

1-26. In Fig. 1-39, what is the instantaneous voltage when $t = 38\ \mu s$?

1-27. Graph a sinusoidal voltage whose peak value is 15 V and whose frequency is 20 kHz. The graph should have a time axis similar to Fig. 1-36.

1-28. In Fig. 1-38, what is the value of each of these quantities: peak voltage, average voltage, rms voltage, period, frequency, and instantaneous voltage when $t = 8\ \mu s$?

1-29. In Fig. 1-39, what is the value of each of these quantities: peak voltage, average voltage, rms voltage, period, frequency, load power, and instantaneous voltage when $t = 65\ \mu s$?

CHAPTER 2 INDUCTANCE

An *inductor* is a coil of wire. When a current flows through the inductor, a magnetic field appears around the inductor. Nothing unusual happens for a dc current. But when an ac current is used, a new phenomenon occurs. This chapter introduces you to inductors.

2-1. TIME-VARYING MAGNETIC FIELDS

When current flows through a wire, a magnetic field appears around the wire as shown in Fig. 2-1. The existence of this magnetic field can be proved by exploring the region around the wire with a compass. The north pole of the compass will always point in the direction of the flux.

If you point your thumb in the direction of current in Fig. 2-1, your fingers curl in the same direction as the magnetic lines of force. Remember to use your left hand if you prefer electron flow, and your right hand if you prefer conventional flow.

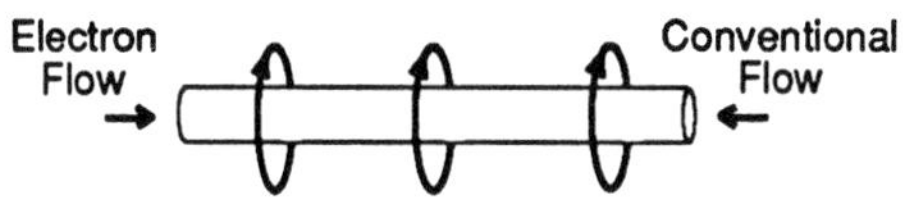

Fig. 2-1.

Sinusoidal Field Around a Wire

Your dc circuits book discussed the direct proportion that exists between flux and current. If you recall,

$$\phi = KI \tag{2-1}$$

where K is a constant of proportionality. Since the magnetic flux is directly proportional to the current, more lines of force appear when the current increases. If the current is sinusoidal in Fig. 2-1,

$$i = I_p \sin \theta$$

Substituting this into Eq. (2-1) gives

$$\phi = KI_p \sin \theta$$

This says the flux varies sinusoidally. It means that a sinusoidal current like Fig. 2-2a will produce a sinusoidal flux like Fig. 2-2b. In other words, the flux increases to a maximum at 90°, decreases to zero at 180°, increases to a maximum in the opposite direction at 270°, and decreases to zero at 360°.

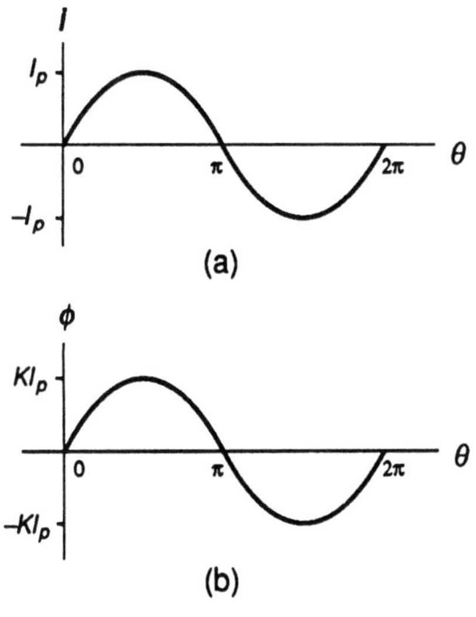

Fig. 2-2.

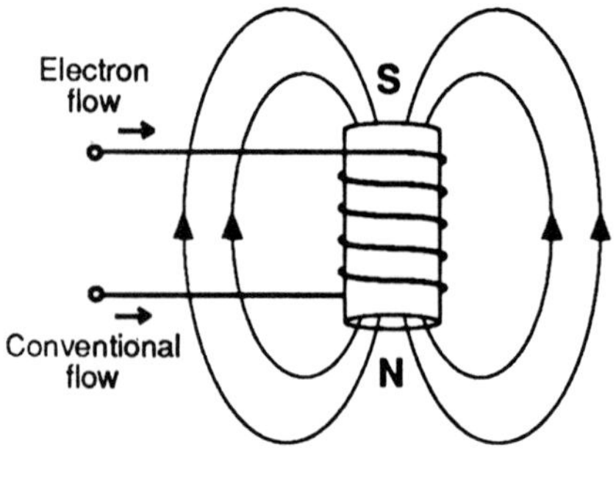

Fig. 2-3.

Varying Field Around an Inductor

To increase the magnetic field, we can use many turns of wire. Figure 2-3 shows an inductor or coil of wire. When charges flow through the inductor, a magnetic field appears around it. The strength of this magnetic field depends on the number of turns, the core material, and other factors. Notice that the lines of force leave the north pole and enter the south pole.

The left-hand or right-hand rule will help you to identify the north and south poles of an inductor. Visualize curling your fingers around the turns of wire in the same direction as the current. Then, your thumb points to the north pole.

The amount of magnetic flux depends on the value of current. When the current increases, more lines of force appear. When the current is sinusoidal, the lines of force increase to a maximum at 90°, decrease to zero at 180°, reach a maximum the opposite way at 270°, and decrease to zero at 360°.

Back EMF in an Inductor

Suppose the current is increasing in Fig. 2-3. Then, the flux is increasing. This changing flux will produce a voltage across the inductor. This voltage, often called the *back emf*, can be quite large. It depends on how many turns of wire there are, how fast the current is changing, etc. (Note: emf is an abbreviation for electro-motive force, the original name for voltage)

Why is a voltage induced across the inductor? Because it is another example of Faraday's law. Your dc circuits book discussed Faraday's law which is mathematically written as

$$v = N\frac{\Delta\phi}{\Delta t} \tag{12-2}$$

This says that the induced voltage equals the number of turns times the change in flux divided by the change in time. In this equation, the change in time should be very small. In fact, the foregoing is an approximation of the this calculus equation:

$$v = N\frac{d\phi}{dt}$$

In this equation, dt represents a Δt that approaches zero.

EXAMPLE 2-1

An inductor has 20 turns. If the change in flux is 0.006 Wb and the change in time is 4 ms, what is the induced voltage?

SOLUTION

Use Eq. (12-2) to calculate the induced voltage:

$$v = 20 \frac{0.006 \text{ Wb}}{4 \text{ ms}} = 30 \text{ V}$$

EXAMPLE 2-2

What happens to the induced voltage in the preceding example for each of these cases:

a. Twice as many turns are used
b. The change in flux is three times smaller
c. The change in time is four times smaller

SOLUTION

a. Equation (2-2) says the induced voltage is directly proportional to the number of turns. When the number of turns doubles, the induced voltage increases to 60 V.
b. Equation (2-2) says the induced voltage is directly proportional to the change in flux. When the change in flux is three times smaller, the induced voltage decreases to 10 V.
c. Equation (2-2) says the induced voltage is inversely proportional to the change in time. When the change in time is four times smaller, the induced voltage increases to 120 V.

2-2. VOLTAGE LEADS CURRENT

The type of current most often used with an inductor is a sinusoidal current, given by

$$i = I_p \sin \theta$$

Since flux is directly proportional to current, the flux is given by

$$\phi = KI_p \sin \theta$$

Because of Faraday's law, the induced voltage is directly proportional to changes in flux, not to the flux itself. Since the flux changes most rapidly at the beginning of the sine wave, the induced voltage is ***maximum at the beginning of the sine wave.***

Out-of-Phase Waveforms

The foregoing idea is very important, so let us discuss it in detail. The change in flux is not the same as the flux. You can see this as follows. A sinusoidal current produces a flux waveform like Fig. 2-4. In a small time interval Δt, the change in flux $\Delta\phi$ is greatest at the beginning of the sine wave. On the other hand, the change in flux becomes very small near the peak of the sine wave. In Fig. 2-4, this means the induced voltage is

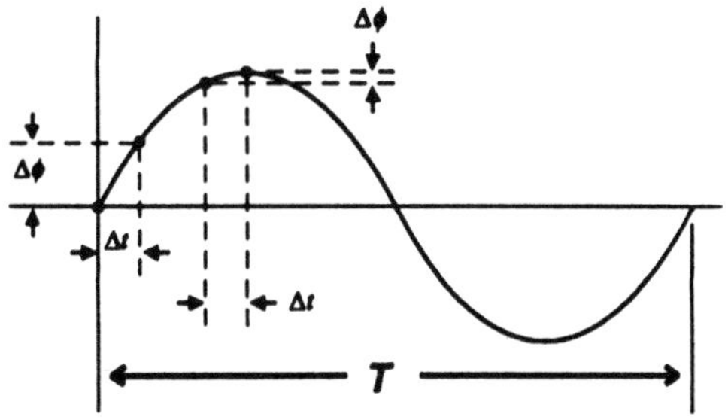

Fig. 2-4.

maximum at the instant the sine wave begins, and zero at the instant the sine wave reaches its peak.

Either with a computer or with a calculus derivation, it is possible to prove that the induced voltage is sinusoidal and shifted in time as shown in Fig. 2-5. This should make sense. Because the flux changes most rapidly at the beginning of the flux waveform (Fig. 2-4), the induced voltage is at its positive maximum (Fig. 2-5). When the flux waveform approaches its peak in Fig. 2-4, the induced voltage passes through zero in Fig. 2-5. When the flux reverses direction, the induced voltage reverses.

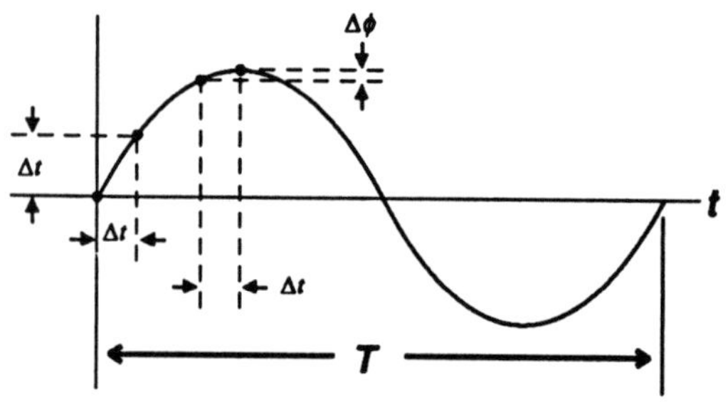

Fig. 2-4.

Trigonometric Identity

The voltage waveform of Fig. 2-5 starts 90° sooner than the flux waveform. Because of this, we can write

$$v = V_p \sin(\theta + 90°) \tag{2-3}$$

As shown in trigonometry, this may be written as

$$v = V_p \cos\theta \tag{2-4}$$

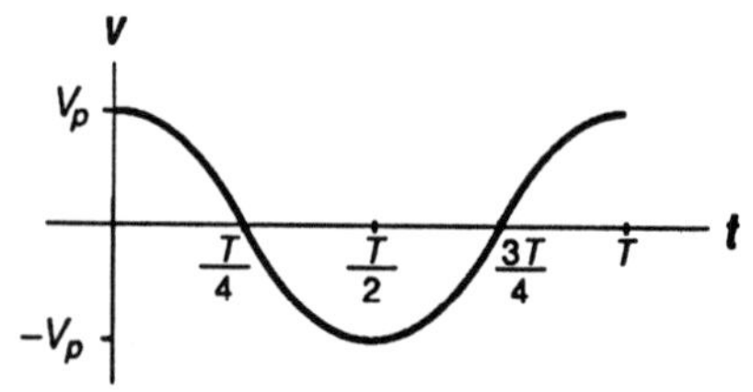

Fig. 2-5.

Whether we use Eq. (2-3) or Eq. (2-4), a major new idea has emerged. The sinusoidal voltage across an inductor leads the sinusoidal current by 90°. This phenomenon is one of the big differences between resistors and inductors. As you know, a sinusoidal current through a resistor produces an in-phase sinusoidal voltage across the resistor. But with an inductor, a sinusoidal current produces an out-of-phase sinusoidal voltage. Figure 2-6 illustrates the concept graphically.

The Exception

If you look hard enough, you usually can find an exception to almost any rule or law. In the case of an inductor, there is an exception that should be mentioned at this time. In the foregoing discussion, we ignored the resistance of the wire. Because of this, the voltage across the inductor was equal to the back emf, which is a magnetically induced voltage. Because of Faraday's law, this magnetically induced voltage is always 90° ahead of the current.

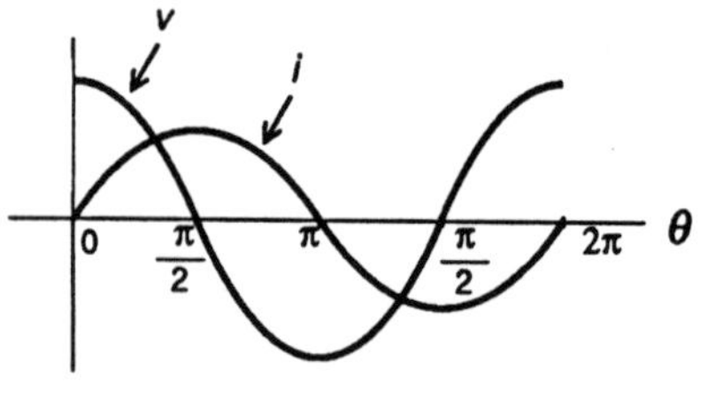

Fig. 2-6.

When the resistance is included, the angle is slightly less than 90°. The difference is usually so small that it's not worth including in our initial discussion of inductors because it will only confuse you. In a later chapter, we will include the effect of resistance. Until then, all inductors are *ideal.* This means that we will approximate the resistance of the inductors as zero.

2-3. INSTANTANEOUS RATE OF CHANGE

If the current is constant, no voltage is induced across an inductor. But when the current is changing in time, a voltage is induced. For a better understanding of this important phenomenon, we will discuss two concepts called the *slope* and *rate of change.*

Slope of a Line

Look at Fig. 2-7. Given any two points on the line, the *rise* is the vertical distance between the points and the *run* is the horizontal distance between the points. The *slope* of the line is defined mathematically as

$$\text{Slope} = \frac{\text{Rise}}{\text{Run}} \qquad (2\text{-}5)$$

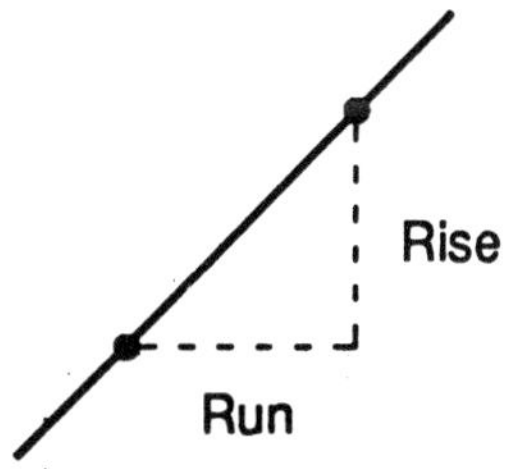

Fig. 2-7.

Slope is useful because it allows us to compare two lines to determine which one is steeper. The line with the larger slope is steeper than the other.

For instance, suppose the rise is 6 inches and the run is 2 inches in Fig. 2-7. Then, the slope is 6 inches divided by 2 inches, which equals 3. This means the line rises 3 inches for every inch of run. As another example, suppose a line has a rise of 8 inches and a run of 2 inches. Then, its slope is 4. In this case, the second line is steeper than the first because it rise 4 inches for every inch of run.

Rate of Change

The *rate of change* between any two points on a graph is defined as the slope of a line drawn through the two points. If the slope is large, the graph is changing quickly with respect to the horizontal coordinate. If the slope is small, the graph is changing slowly. After measuring the rise and run between any two points, you can calculate the rate of change with Eq. (2-5).

When the graph is nonlinear as shown in Fig. 2-8a, we can draw a line through any two points. Then, the slope of line will indicate how fast the graph is changing between the two points. For instance, if points A and B have a rise of 6 inches and a run of 2 inches, the rate of change between A and B is 3. The rate of change between two points on a nonlinear graph is called the *average* rate of change between the two points. This average rate of change will depend on how far apart the two points are.

Something unusual happens when the two points are *very close*. For instance, visualize point A stationary in Fig. 2-8a. Next, visualize point B sliding along the curve toward point A. Eventually, point B will merge with point A. In this case, the line becomes a tangent line as shown in Fig. 2-8b. (*Tangent* means touching without intersecting. The result is a line that touches the curve at only a single point.)

The slope of the tangent line in Fig. 2-8b is called the *instantaneous* rate of change at point A. The instantaneous rate of change is written as

$$\frac{d\phi}{dt} = \text{slope of tangent line}$$

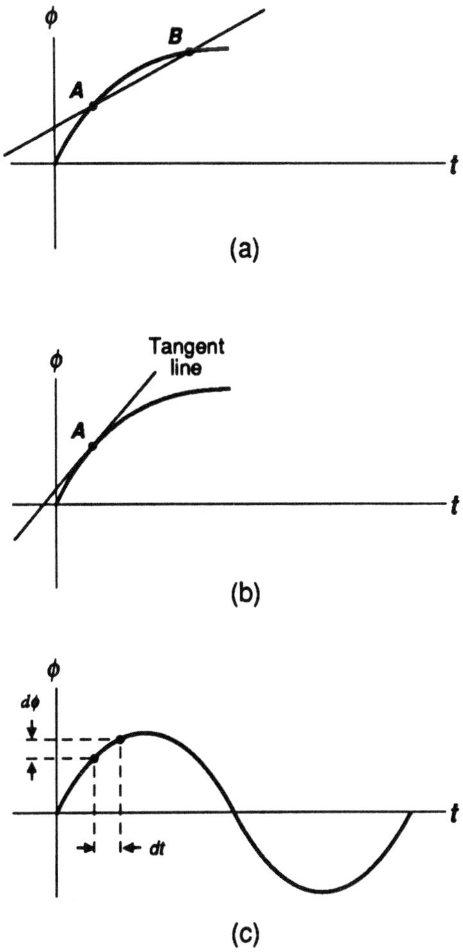

Fig. 2-8.

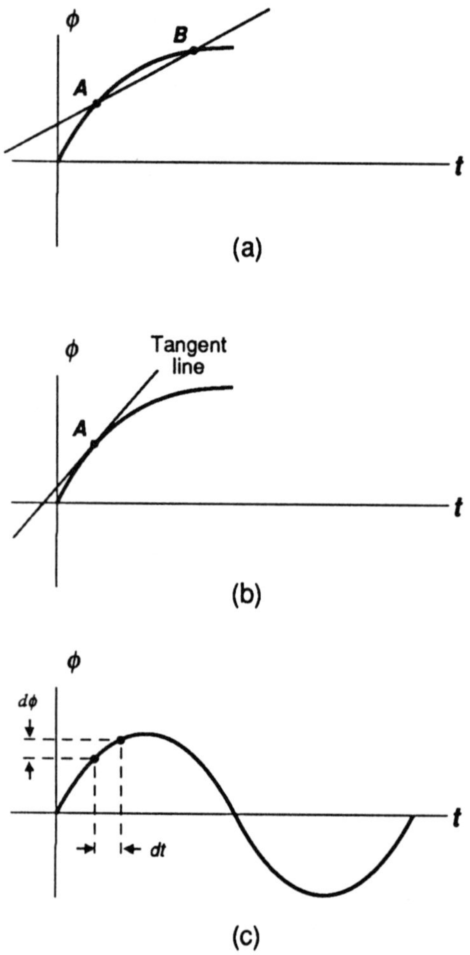

Fig. 2-8.

For instance, if the tangent line has a rise of 10 Wb and a run of 2 s, the instantaneous rate of change is

$$\frac{d\phi}{dt} = \frac{10\text{ Wb}}{2\text{ s}} = 5\text{ Wb/s}$$

In Fig. 2-8b, the tangent line will have a slope of 5 Wb/s at point A. Just to the left of point A, the rate of change is slightly more than 5 Wb/s. Just to the right of point A, the rate of change is slightly less than 5 Wb/s.

Because ac circuits involve time-varying voltages and currents, we need some way to compare how fast the quantities are changing in time. The instantaneous rate of change is the key. By drawing tangent lines to the different points on a waveform, we can see how fast the waveform is changing.

Calculating the Instantaneous Rate of Change

To get the exact value of the instantaneous rate of change, you need to use calculus. But if an approximate value is acceptable, there is a simpler way to find the instantaneous rate of change. This method allows a precalculus student to work out the instantaneous rate of change with an accuracy of better than 1 percent. Here is how the method works.

Figure 2-8c shows two points on a sine wave that are very close to each other. If we take the ratio of $d\phi$ to dt, we get an approximate value for the instantaneous rate of change. For the approximate value to be accurate to within 1 percent, the change in time must be small compared to the period. As a guideline, dt should be at least 100 times smaller than the period. In symbols,

$$dt = 0.01T \tag{2-6}$$

This is the critical condition. If the change in time is less than this, the answer is even more accurate. In later discussions, we will refer to this as the *method of small changes*.

Another Way to Calculate Instantaneous Rate of Change

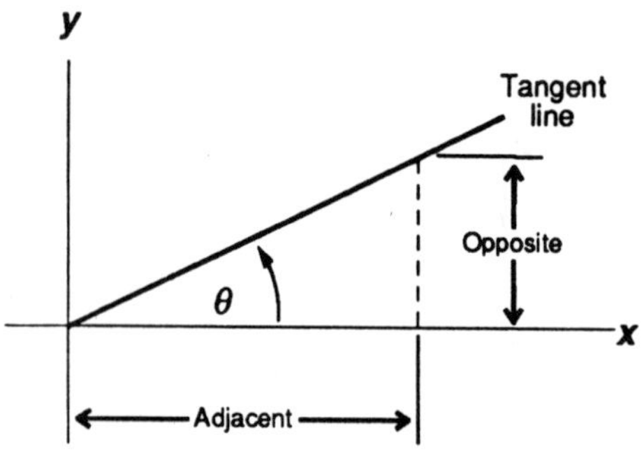

Fig. 2-9.

Dividing the rise by the run is one way to calculate the slope of the tangent line drawn to a point on a curve. There is another way worth mentioning. Figure 2-9 shows a horizontal axis. When you draw a tangent line at a single point on a graph, an angle appears between the horizontal axis and the tangent line. Since the slope of the tangent line equals the opposite side divided by the adjacent side, we can write

$$\text{Slope of tangent line} = \tan\theta \tag{2-7}$$

Two important cases to remember are these. When the angle is 0°, the slope is zero. When the angle is 90°, the slope is infinite. In other words, a horizontal line represents a zero rate of change, whereas a vertical line represents an infinite rate of change.

EXAMPLE 2-3

In Fig. 2-10a, a tangent line (dashed) is drawn through the peak point of a sinusoidal current. What is the instantaneous rate of change at the peak?

SOLUTION

You can see that the tangent line is horizontal right at the peak. Therefore, the angle is 0°, which means the slope is zero. This is equivalent to saying the instantaneous rate of change is zero. In other words, the current is neither increasing or decreasing at the instant the waveform reaches the maximum positive value.

EXAMPLE 2-4

When is the instantaneous rate of change maximum positive on the sine wave of Fig. 2-10a?

SOLUTION

If the instantaneous rate of change is zero at the peaks, then it must be maximum at the zero crossings. In other words, the steepest tangent line that we can draw is at the beginning of the sine wave as shown in Fig. 2-10a. Any other tangent line will have less slope. Our conclusion: The instantaneous rate of change is maximum positive at the beginning of the cycle.

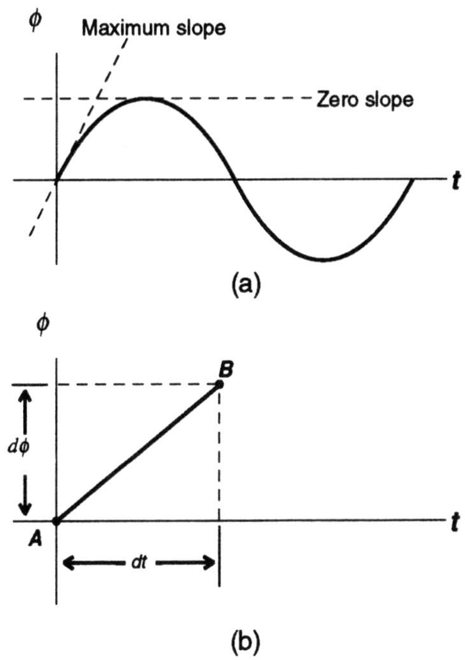

Fig. 2-10.

EXAMPLE 2-5

A sinusoidal flux has a peak value of 5 Wb and a frequency of 60 Hz. What is the instantaneous rate of change at the beginning of the sine wave?

SOLUTION

We can calculate a highly accurate value using the method of small changes. The idea is to take the ratio of very small changes. Figure 2-10b shows the beginning of a sine wave. When the changes are small, the ratio of $d\phi$ to dt equals the instantaneous rate of change.

Here is how to proceed. Since the sinusoidal flux has a peak value of 5 Wb and a frequency of 60 Hz, its instantaneous value is given by

$$\phi = 5 \sin 120\pi t$$

Since the frequency is 60 Hz, the period is

$$T = \frac{1}{60 \text{ Hz}} = 0.0167 \text{ s}$$

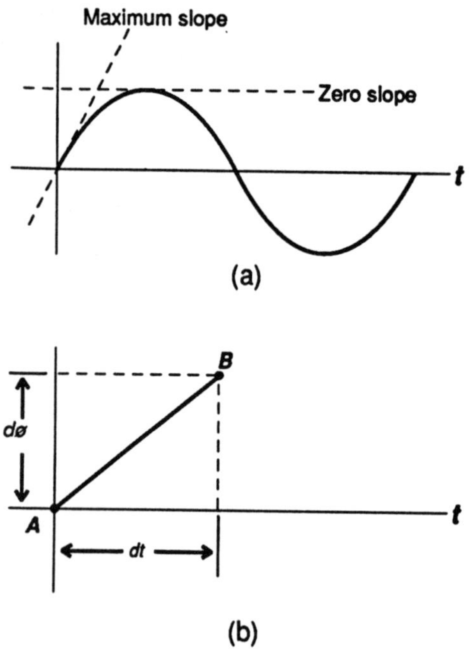

Fig. 2-10.

The change in time has to be at least 100 times smaller than this, which gives

$$dt = 0.01(0.0167\ \text{s}) = 0.000167\ \text{s}$$

In Fig. 2-10b, the instantaneous flux at point *B* is

$$\phi = 5 \sin 120\pi t = 5 \sin 120\pi(0.000167) = 0.315\ \text{Wb}$$

Since the foregoing flux equals the change in flux between point *A* and point *B*, we can write

$$d\phi = 0.315\ \text{Wb}$$

Divide this by 0.000167 s to get the instantaneous rate of change:

$$\frac{d\phi}{dt} = \frac{0.315\ \text{Wb}}{0.000167\ \text{s}} = 1886\ \text{Wb/s}$$

IMPORTANT

$$v = N\frac{d\phi}{dt}$$

2-4. ANOTHER LOOK AT INDUCED VOLTAGE

As mentioned earlier, the exact equation for Faraday's law is

$$v = N\frac{d\phi}{dt} \tag{2-8}$$

Now, you are in a better position to understand this important equation. It says the induced voltage equals the number of turns times the instantaneous rate of change of flux with time. Graphically, it means the induced voltage equals the number of turns times the slope of the tangent line drawn to each point on the flux waveform.

For instance, if the current is sinusoidal, it is given by

$$i = I_p \sin 2\pi ft \tag{2-9}$$

Since flux is directly proportional to current, the flux is given by

$$\phi = KI_p \sin 2\pi ft$$

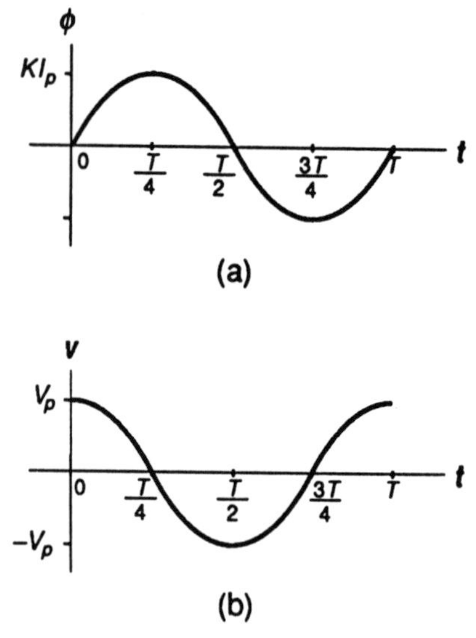

Fig. 2-11.

Figure 2-11a shows the flux waveform. Based on our earlier discussion of instantaneous rate of change, we can summarize the instantaneous rate of change at the critical points as follows:

t	$d\phi/dt$
0	Maximum positive
$T/4$	Zero
$T/2$	Maximum negative
$3T/4$	Zero
T	Maximum positive

In words, the instantaneous rate of change is maximum positive at the beginning of the flux waveform, zero at the positive peak, maximum negative at the half-cycle point, zero at the negative peak, and maximum positive at the beginning of the next cycle.

Therefore, we can draw the sinusoidal voltage as shown in Fig. 2-11b. The peak voltage will depend on the number of turns, the peak flux, and the frequency of the flux waveform. As discussed earlier, the induced voltage is sinusoidal in shape, but shifted a quarter cycle in time, so that voltage leads current. The equation for the induced voltage is

$$v = V_p \sin (2\pi ft + 90°)$$

This is equivalent to

$$v = V_p \cos 2\pi ft \qquad (2\text{-}10)$$

CRITICAL POINTS

t	$d\phi/dt$
0	Max +
$T/4$	Zero
$T/2$	Max –
$3T/4$	Zero
T	Max +

EXAMPLE 2-6

The sinusoidal current through an inductor has a peak value of 250 mA and a frequency of 1 kHz. Write the equation for instantaneous current and voltage. Sketch the waveforms of current and voltage.

SOLUTION

The equation for the instantaneous current is

$$i = I_p \sin 2\pi ft = 0.25 \sin 2000\pi t$$

Given any value of t, we can calculate the instantaneous value of current.

The equation for instantaneous voltage is

$$v = V_p \cos 2\pi ft = V_p \cos 2000\pi t$$

If we had the peak voltage, we could calculate the instantaneous value of voltage. A later section will discuss how to find the peak voltage.

The period of the voltage waveform equals the reciprocal of the frequency:

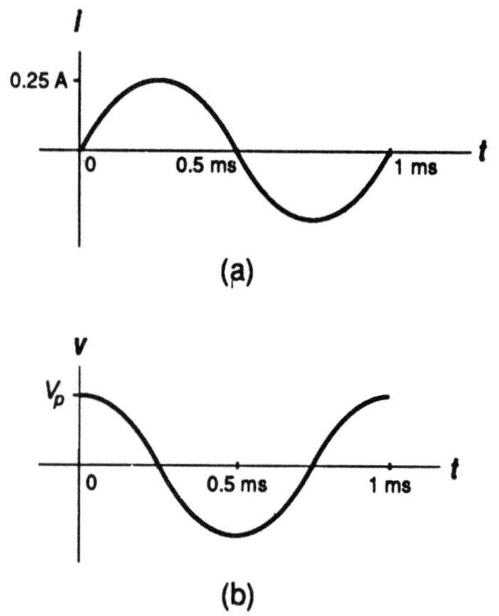

Fig. 2-12.

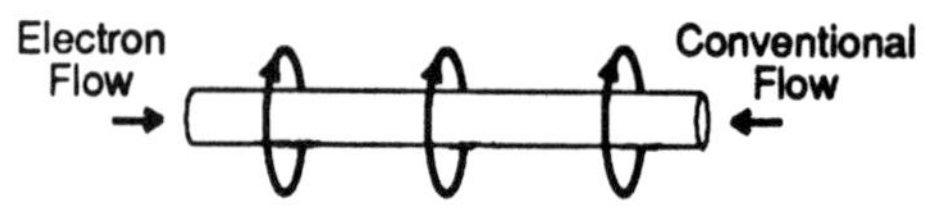

Fig. 2-13.

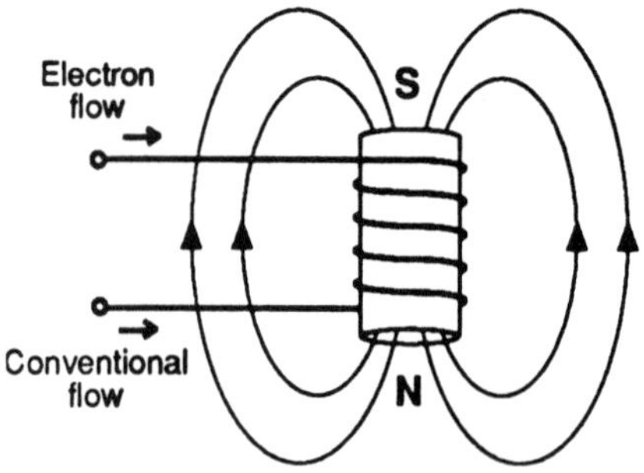

Fig. 2-14.

$$T = \frac{1}{1\ \text{kHz}} = 1\ \text{ms}$$

Figure 2-12a shows the waveform of current with a peak value of 0.25 A and a period of 1 ms. Figure 2-12b shows the induced voltage with a period of 1 ms.

2-5. INDUCTANCE

To calculate the peak voltage across an inductor, we need to define a new quantity called *inductance*. The idea is similar to a resistor. A resistor has resistance, defined as the opposition to current. An inductor has inductance, defined in the following discussion.

Inductance of a Wire

Figure 2-13 shows a current in a wire. As you know, this produces a flux around the wire. The inductance of this wire is defined as

$$L = \frac{\phi}{I} \qquad (2\text{-}11)$$

This says that inductance equals the total flux around the wire divided by the current through the wire. For instance, if the flux is 1 Wb and the current is 2 A, the inductance is

$$L = \frac{1\ \text{Wb}}{2\ \text{A}} = 0.5\ \text{Wb/A}$$

The Henry

The *henry* (abbreviated H) is the metric unit of measure for inductance. It is defined as

$$1\ \text{H} = 1\ \text{Wb/A}$$

Therefore, the preceding answer becomes

$$L = 0.5\ \text{Wb/A} = 0.5\ \text{H}$$

From now on, all inductance values will be given in henries.

Inductance with Many Turns

To increase the magnetic effect, we can use many turns of wire to get an inductor like Fig. 2-14. In this case, the inductance is increased by the number of turns. In other words, the inductance becomes

$$L = \frac{N\phi}{I} \qquad (2\text{-}12)$$

For instance, if $N = 30$, $\phi = 1$ Wb, and $I = 2$ A, the inductance is

$$L = \frac{(30)(1\ \text{Wb})}{2\ \text{A}} = 15\ \text{H}$$

The Toroid

In Fig. 2-14, we are using an air core. To increase the flux, we can use a magnetic core. This type of core provides an excellent path for the lines of force, which increases the inductance.

For instance, Fig. 2-15 shows a toroid, a doughnut-shaped inductor. The core supports the turns of wire. By starting with the defining formula for inductance, we can derive this equation for the inductance:

$$L = \mu \frac{N^2 A}{l} \qquad (2\text{-}13)$$

where μ = a constant that depends on the core material
N = number of turns
A = cross-sectional area of the core
l = length of the core

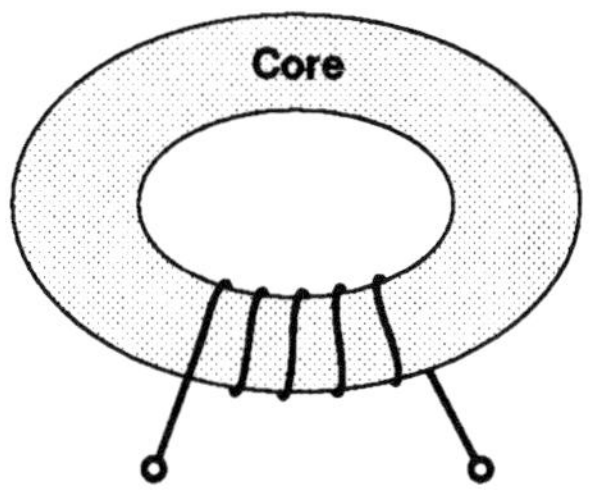

Fig. 2-15.

This equation says the inductance is directly proportional to the area and inversely proportional to the length. If you double the area of the core, you double the inductance. On the other hand, if you double the length of the core, the inductance drops in half.

Permeability

The material used for the core will affect the flux. Some materials result in more flux than others. The quantity μ is called the *permeability*. Its value indicates the effect of the core material. For instance, the permeability of a vacuum has been measured. In the metric system,

$$\mu = 1.26(10^{-6})$$

Most materials including air have a negligible effect on the flux. In other words, they have the same permeability as a vacuum. We refer to all materials with a permeability of $1.26(10^{-6})$ as nonmagnetic.

A few materials are *diamagnetic*. This means they oppose the passage of magnetic flux. Diamagnetic materials have a permeability of less than $1.26(10^{-6})$. Because of this, an inductor with a diamagnetic core has less inductance than one with an air core.

Some materials are called *ferromagnetic*, or simply magnetic. They have permeabilities that are much greater than a vacuum. An inductor with a ferromagnetic core has much more inductance than one with an air core. Examples of ferromagnetic materials are iron, nickel, steel, and cobalt.

EXAMPLE 2-6

A toroidal inductor has a magnetic core with a permeability of $6.3(10^{-3})$. If $N = 100$, $A = 10^{-6}\ \text{m}^2$, and $l = 0.01$ m, what is the inductance?

SOLUTION

Substitute the given values into Eq. (2-13) to get

$$L = 6.3(10^{-3}) \frac{(100^2)(10^{-6})}{0.01} = 6.3\ \text{mH}$$

2-6. RELATIVE PERMEABILITY

Your dc circuits book discussed why some materials are magnetic and others are not. If you recall, the atoms of a magnetic material can align themselves to aid the magnetic field. When these atoms are aligned, the total flux increases, which means the inductance increases.

Relative permeability indicates how easily the atoms of the core can align themselves with the magnetic field. It equals the ratio of the permeability of a material to the permeability of a vacuum. The defining formula for relative permeability is

$$\mu_r = \frac{\mu}{1.26(10^{-6})} \tag{2-14}$$

For instance, if a material has a permeability of $6.3(10^{-3})$, its relative permeability is

$$\mu_r = \frac{6.3(10^{-3})}{1.26(10^{-6})} = 5000$$

Table 2-1 shows the relative permeabilities for some materials. The values are approximate, because exact values depend on the grade of the material, its heat treatment during the manufacturing process, and other factors. As shown, air has a relative permeability that is very close to that of a vacuum. Similarly, aluminum and most other conductors act like nonmagnetic materials.

Table 2-1. Relative Permeability

Material	μ_r
Vacuum	1
Air	1.0000004
Aluminum	1.00000065
Nickel	50
Cobalt	60
Steel	300
Iron	5000
Ferrites	100 to 2500
Permalloy	25,000 to 100,000
Supermalloy	100,000 to 1,000,000

The best known magnetic materials are nickel, iron, steel, and cobalt. These materials have relative permeabilities from approximately 50 to 5000. By mixing these materials, we can get alloys with even higher permeabilities such as ferrites, permalloy, and supermalloy.

By rearranging Eq. (2-14), we can get an equation for the permeability of a core:

$$\mu = 1.26(10^{-6})\, \mu_r \qquad (2\text{-}15)$$

This means you can look up the relative permeability of a material in Table 2-1, then multiply by $1.26(10^{-6})$ to get the permeability of the material. For instance, steel has a relative permeability of 300. Therefore, a steel core has a permeability of

$$\mu = 1.26(10^{-6})300 = 378(10^{-6})$$

EXAMPLE 2-7

An inductor with an air core has an inductance of 10 μH. If the core is changed to a ferrite core with a relative permeability of 400, what is the new value of inductance?

SOLUTION

Since the relative permeability is 400 times larger, the inductance is 400 times larger. Therefore, the inductance is

$$L = 400(10\ \mu\text{H}) = 4\ \text{mH}$$

2-7. VOLTAGE AND INDUCTANCE

Earlier, we discussed Faraday's law, which says the induced voltage equals the number of turns times the instantaneous rate of change of flux with time. In symbols,

$$v = N\frac{d\phi}{dt}$$

Now, we want to derive a new relation for the induced voltage in terms of the inductance.

The Derivation

The inductance of an inductor is given by

$$L = \frac{N\phi}{I}$$

When the current is time-varying, it is customary to use a lower-case letter for the current as follows:

$$L = \frac{N\phi}{i}$$

By cross-multiplying, this becomes

$$Li = N\phi$$

Because of the direct proportion between flux and current, the equation can be written in terms of small changes as follows:

$$L\Delta i = N\Delta\phi$$

We can divide both sides by Δt to get

$$\frac{L\Delta i}{\Delta t} = \frac{N\Delta\phi}{\Delta t}$$

When the changes are infinitesimally small, the average rates of change become instantaneous rates of change:

$$\frac{Ldi}{dt} = \frac{Nd\phi}{dt}$$

The right side of this equation equals the induced voltage v, so the equation may be rearranged to get

$$v = L\frac{di}{dt} \qquad (2\text{-}16)$$

This final equation says the induced voltage across an inductor equals the inductance times the instantaneous rate of change of current with time. Equation (2-16) is of major importance and should be memorized.

IMPORTANT

$$v = L\frac{di}{dt}$$

Even though you may not have studied calculus, you can still calculate the induced voltage using Eq. (2-16). Recall the method of small changes. By working out the small changes between two close points on a waveform, we can get a highly accurate value for the instantaneous rate of change of current with time. Then, multiplying by the inductance gives the induced voltage. (See Example 2-9.)

Voltage Leads Current

Earlier, we learned that a sinusoidal voltage leads a sinusoidal current in an inductor by 90°. Equation (2-16) confirms this result as follows. When the current is sinusoidal, the instantaneous value is given by

$$i = I_p \sin 2\pi ft$$

Figure 2-16a shows the current waveform. The instantaneous rate of change di/dt equals the slope of a tangent line drawn to any point of interest on this waveform. From earlier discussions, we know that the instantaneous rate of change is maximum when the sine wave starts and zero when it reaches its peak.

Equation (2-16) says the induced voltage equals the inductance multiplied by the instantaneous rate of change. By the same logic used in our discussion of changing flux, we can see that the induced voltage is sinusoidal but shifted by a quarter of cycle. Because of this shift, the induced voltage leads the current as shown in Fig. 2-16b. Furthermore, we can write the induced voltage in either of two forms:

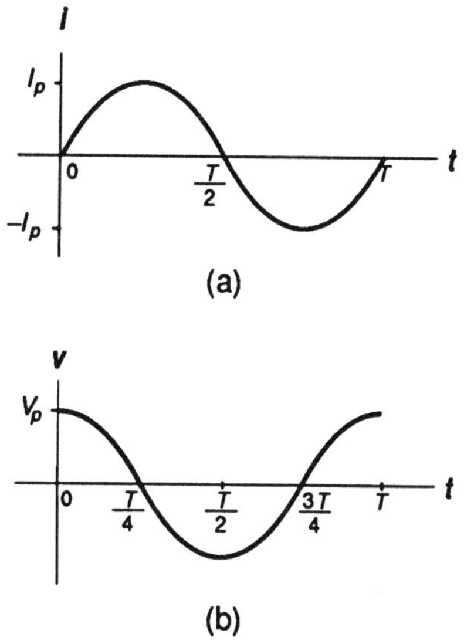

Fig. 2-16.

$$v = V_p \sin (2\pi ft + 90°)$$

or

$$v = V_p \cos 2\pi ft$$

EXAMPLE 2-8

An inductance of 5 mH has a sinusoidal current with a peak value of 25 mA and a frequency of 1 kHz. If the instantaneous rate of change di/dt has a maximum value of 157 A/s, what is the maximum induced voltage? What are the equations for current and voltage? Show the current and voltage waveforms.

SOLUTION

Multiply the inductance by the instantaneous rate of change to get the induced voltage:

$$v = (5 \text{ mH})(157 \text{ A/s}) = 0.785 \text{ V}$$

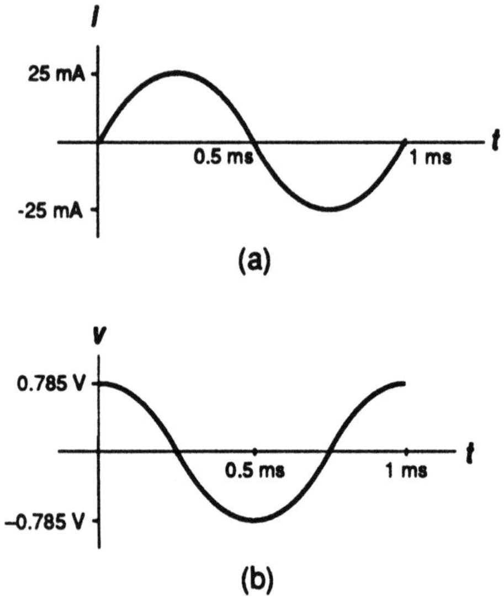

Fig. 2-17.

The current has an equation of

$i = 0.025 \sin 2000\pi t$

The voltage has an equation of

$v = 0.785 \cos 2000\pi t$

Since the frequency is 1 kHz, the period is 1 ms. Figures 2-17a and b shows the waveforms.

EXAMPLE 2-9

An inductance of 2.5 H has a sinusoidal current with a peak value of 200 mA and a frequency of 60 Hz. Use the method of small changes to calculate the maximum induced voltage.

SOLUTION

This is similar to Example 2-5 with a few exceptions. If necessary, review Example 2-5 before proceeding. Since the peak current is 200 mA and the frequency is 60 Hz, the equation for the current is

$i = 0.2 \sin 120\pi t$

The change in time is still the same as in Example 2-5:

$dt = 0.01T = 0.01(16.7 \text{ ms}) = 0.167 \text{ ms}$

The instantaneous current at point B is

$i = 0.2 \sin 120\pi t = 0.2 \sin 120\pi(0.000167) = 12.6 \text{ mA}$

Therefore,

$di = 12.6 \text{ mA}$

Divide this by 0.000167 s to get the intantaneous rate of change:

$$\frac{di}{dt} = \frac{12.6 \text{ mA}}{0.167 \text{ ms}} = 75.4 \text{ A/s}$$

Multiply this by the inductance to get the induced voltage:

$v = (2.5 \text{ H})(75.4 \text{ A/s}) = 189 \text{ V}$

If calculus had been used, the exact answer would be 188.5 V. Our approximate answer of 189 V is well within 1 percent of the exact answer. In the next chapter, you will learn an even easier way to calculate the instantaneous rate of change and the induced voltage at any instant.

2-8. COEFFICIENT OF COUPLING

When the varying magnetic field of an inductor passes through another inductor, the changing flux induces a voltage in the second inductor. For instance, Fig. 2-18 shows a current of I_1 in the left inductor. This creates a flux of ϕ_1 in the left inductor. Some of this flux passes through the right inductor. We symbolize this flux as ϕ_2.

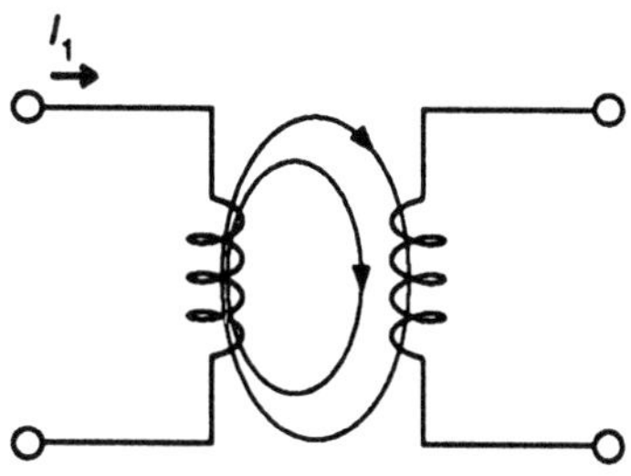

Fig. 2-18.

Definition

The *coefficient of coupling* is the fraction of flux out of the first coil that passes through the second coil. The formula is

$$k = \frac{\phi_2}{\phi_1} \tag{2-17}$$

where k is the coefficient of coupling. It has a value between 0 and 1. For instance, if $\phi_1 = 4$ Wb and $\phi_2 = 0.1$ Wb, the coefficient of coupling is

$$k = \frac{0.1 \text{ Wb}}{4 \text{ Wb}} = 0.025$$

which is equivalent to 2.5%.

Tight Coupling

Tight coupling means two inductors have a value of k that approaches 100 percent. Figure 2-19 shows the usual way to get tight coupling. The two inductors are wound on a magnetic core. Because of this, most of the flux produced by the first inductor passes through the second inductor. In other words, ϕ_2 almost equals ϕ_1. A ferromagnetic core typically has a coefficient of coupling between 0.95 and 1. This means that more than 95 percent of the first flux passes through the second inductor.

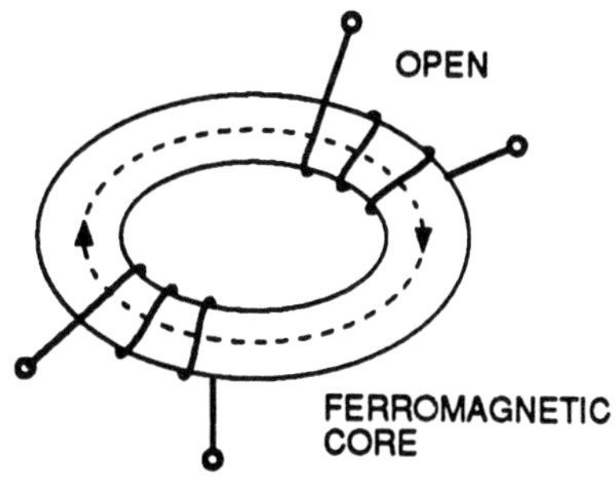

Fig. 2-19.

Loose Coupling

When the core is made of a nonmagnetic material, all it does is provide support for the windings. Because the core has no effect on the magnetic field, the flux is no longer confined to the core. In this case, the coefficient of coupling drops to a low value, typically less than 0.1.

If the two inductors are brought closer together, the coefficient of coupling will increase, but it still remains loose. On the other hand, the further apart the inductors, the smaller the coefficient of coupling. For coils that are far apart, the coefficient of coupling approaches zero.

2-9. MUTUAL INDUCTANCE

When two inductors are in series or in parallel, things get more complicated because the flux of one inductor may pass through the second inductor and induce a voltage. This creates a new kind of inductance called *mutual inductance*.

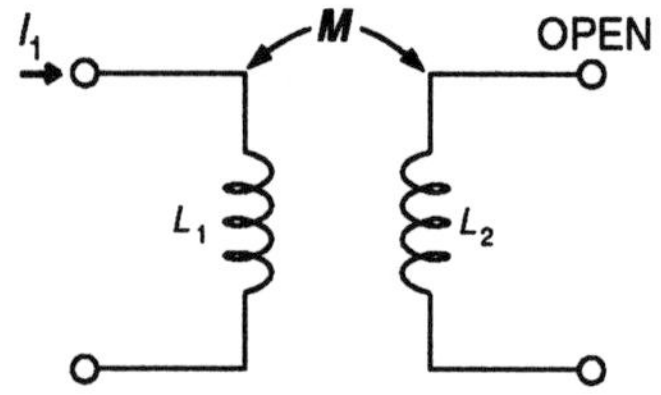

Fig. 2-20.

Definition

Given two inductors like those of Fig. 2-20, the mutual inductance M is defined by this equation:

$$M = \frac{N_2\phi_2}{I_1} \tag{2-18}$$

For instance, if $N_2 = 100$, $\phi_2 = 0.2$ Wb, and $I_1 = 5$ A, the mutual inductance is

$$M = \frac{100(0.2 \text{ Wb})}{5 \text{ A}} = 4 \text{ H}$$

By an advanced derivation, we can derive this valuable relation:

$$M = k\sqrt{L_1L_2}$$

As an example, if two inductors have a coefficient of coupling equal to 0.95, and if the inductances are 100 mH and 400 mH, the mutual inductance is

$$M = 0.95\sqrt{(100 \text{ mH})(400 \text{ mH})} = 190 \text{ mH}$$

Series Inductors

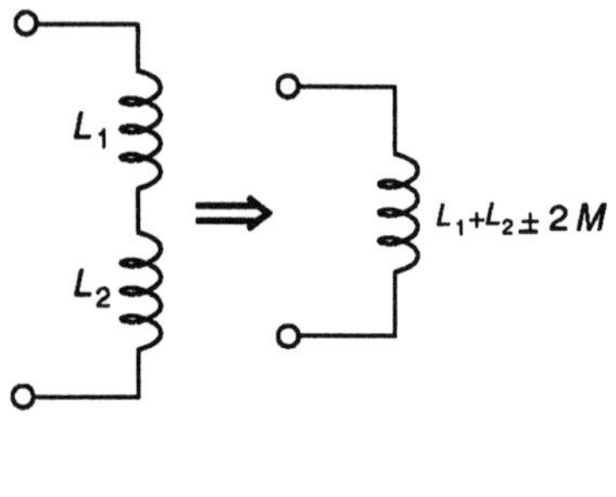

Fig. 2-21.

When two inductors are in series (Fig. 2-21), some of the flux produced by each inductor passes through the other inductor. Because of this, the total inductance depends on each separate inductance and the mutual inductance. The derivation of total inductance is too complicated to show here, so we will state the following formula without proof:

$$L = L_1 + L_2 \pm 2M \tag{2-19}$$

The final term in this equation may be plus or minus, depending on how the poles of the inductors are aligned. Stated another way, it depends on whether the two fluxes aid or oppose each other.

Figure 2-22 shows the case of aiding fluxes. When the poles are lined up as shown here, the two fluxes add together and produce more total inductance. In this case, the total inductance is given by

$$L = L_1 + L_2 + 2M$$

As an example, if L_1 = 100 mH, L_2 = 400 mH, and M = 190 mH, the total inductance is

$$L = 100 \text{ mH} + 400 \text{ mH} + 2(190 \text{ mH}) = 880 \text{ mH}$$

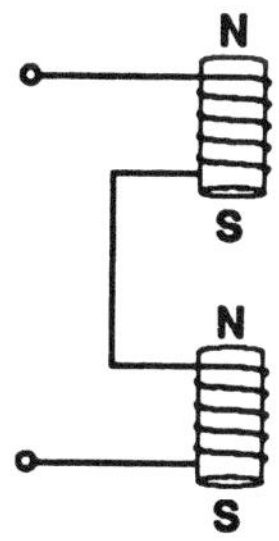

Fig. 2-22.

Figure 2-23 shows the case of opposing fluxes. When the poles are lined up as shown here, the two fluxes oppose each other and produce less total inductance. In this case, the total inductance is given by

$$L = L_1 + L_2 - 2M$$

With the values used in the preceding calculation, the total inductance is

$$L = 100 \text{ mH} + 400 \text{ mH} - 2(190 \text{ mH}) = 120 \text{ mH}$$

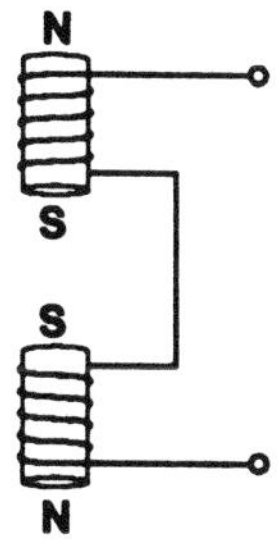

Fig. 2-23.

When Inductors are Far Apart

You don't have to worry about mutual inductance when the inductors are far apart. In this case, the coefficient of coupling is approximately zero, which means the mutual inductance is zero. Then, Eq. (2-19) simplifies to

$$L = L_1 + L_2 \qquad (2\text{-}20)$$

SERIES: FAR APART

$$L = L_1 + L_2$$

This says you can add series inductances when there's negligible coupling between them. As a guideline, you can ignore the mutual inductance of small inductors that are more than 1 inch apart separated by air. With the values used in the preceding calculations, the total inductance is

$$L = 100 \text{ mH} + 400 \text{ mH} = 500 \text{ mH}$$

Parallel Inductors

It can be shown that inductors in parallel have a total inductance of

$$L = \frac{L_1 L_2 - M^2}{L_1 + L_2 \pm 2M} \qquad (2\text{-}21)$$

Again, the $2M$ term must be added or subtracted, depending on whether the fluxes aid or oppose each other.

When the inductors are far enough apart, the coefficient of coupling drops to zero and the equation simplifies to

PARALLEL: FAR APART

$$L = \frac{L_1 L_2}{L_1 + L_2}$$

$$L = \frac{L_1 L_2}{L_1 + L_2} \tag{2-22}$$

With the values used in earlier calculations, the total inductance of two parallel inductors with zero coupling is

$$L = \frac{(100 \text{ mH})(400 \text{ mH})}{100 \text{ mH} + 400 \text{ mH}} = 80 \text{ mH}$$

2-10. THE IDEAL TRANSFORMER

A *transformer* is two or more inductors whose coefficient of coupling is greater than zero. To simplify the discussion of a transformer, we will use the *ideal* transformer. This is a transformer with the following properties:

1. Zero resistance in the inductors
2. Zero power losses
3. Unity coefficient of coupling
4. Relative permeability of infinity

Manufacturers cannot build the ideal transformer, but they can come close. They choose wire sizes to get a very low resistance, very low power losses, very high coefficient of coupling, and very high relative permeability. Because a real transformer approaches the ideal transformer, most people approximate a real transformer as ideal. This greatly simplifies a difficult analysis.

Ideal Voltages

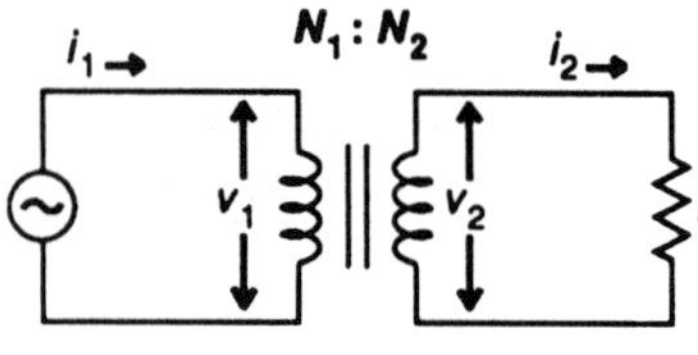

Fig. 2-24.

Figure 2-24 shows a transformer between an ac source and a load resistance of R_2. The inductor on the left is called the *primary winding*, and the one on the right is called the *secondary winding*. The vertical lines between the primary and secondary windings are used on a schematic diagram to indicate a ferromagnetic core. The primary winding has N_1 turns and the secondary winding has N_2 turns.

With an ideal transformer, the only voltages across the windings are the induced voltages given by Faraday's law. In other words, the ac source produces a primary voltage of

$$v_1 = N_1 \frac{d\phi_1}{dt}$$

and a secondary voltage of

$$v_2 = N_2 \frac{d\phi_2}{dt}$$

Since $k = 1$ in an ideal transformer, the two fluxes are equal and the foregoing equations simplify to

$$v_1 = N_1 \frac{d\phi}{dt}$$

and

$$v_2 = N_2 \frac{d\phi}{dt}$$

Take the ratio of the two foregoing equations and you get

$$\frac{v_1}{v_2} = \frac{N_1}{N_2}$$

The right side of this equation is called the *turns ratio* of the transformer. It is the ratio of the primary turns to secondary turns. The equation says the ratio of primary to secondary voltage equals the turns ratio.

By inverting both sides, we get

$$\frac{v_2}{v_1} = \frac{N_2}{N_1}$$

which can be rearranged as

$$v_2 = \frac{v_1}{N_1/N_2} \qquad \text{(2-23)}$$

IMPORTANT

$$v_2 = \frac{v_1}{N_1/N_2}$$

This says the secondary voltage equals the primary voltage divided by the turns ratio.

Step-up and Step-down Transformers

Equation (2-23) has commercial value. It tells us we can transform a source voltage to higher or lower values. For instance, if the primary winding has 100 turns and the secondary winding has 300 turns, the turns ratio is 1/3 and Eq. (2-23) becomes

$$v_2 = \frac{v_1}{1/3}$$

or

$$v_2 = 3v_1$$

This says the secondary voltage is three times as large as the primary voltage. This is an example of a *step-up* transformer, one whose secondary voltage has been stepped up to a higher value than the primary voltage.

On the other hand, if the primary winding has 300 turns and the secondary winding has 100 turns, the turns ratio is 3 and Eq. (2-23) becomes

$$v_2 = \frac{v_1}{3}$$

This says the secondary voltage is one-third of the primary voltage. This is an example of a *step-down* transformer, one whose secondary voltage has been stepped down to a lower value than the primary voltage.

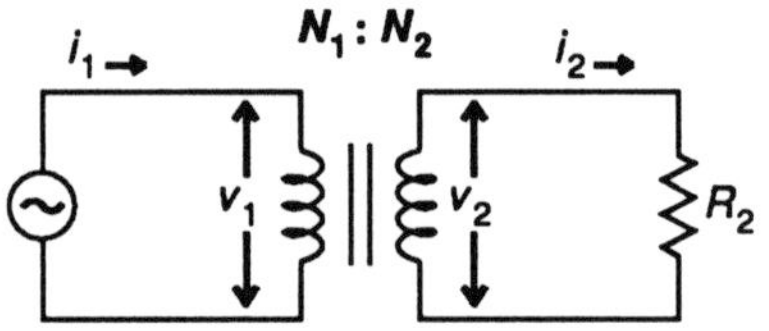

Fig. 2-24.

Ideal Currents

In Fig. 2-24, the load power at any instant is

$$p_2 = v_2 i_2$$

The power into the primary winding is

$$p_1 = v_1 i_1$$

Since an ideal transformer has zero power losses, all of the power into the primary winding appears at the load. This means

$$p_2 = p_1$$

or

$$v_2 i_2 = v_1 i_1$$

We can rearrange this into

$$\frac{i_2}{i_1} = \frac{v_1}{v_2}$$

Since the right side of this equation equals the turns ratio,

$$\frac{i_2}{i_1} = \frac{N_1}{N_2}$$

or

$$i_2 = \frac{N_1}{N_2} i_1 \qquad (2\text{-}24)$$

This says the secondary current equals the turns ratio times the primary current.

Currents in Step-up and Step-down Transformers

Equation (2-24) tells us something important. It says that currents are stepped in the opposite direction of voltage. For instance, if the primary winding has 300 turns and the secondary winding has 100 turns, the turns ratio is 3 and the transformer is the step-down variety where the voltage is decreased by a factor of 3. But watch what happens to current. Since the turns ratio is 3, Eq. (2-24) gives

$$i_2 = 3i_1$$

This says the secondary current is three times the primary current. The current is stepped up, whereas the voltage is stepped down.

The opposite is also true. Given a step-up transformer, the voltage is stepped up but the current is stepped down.

Resistances

Transformers have another unusual ability. They can transform the load resistance into a lower or higher value when measured from the primary side. Here is the idea. Earlier, we derived these two relations:

$$\frac{v_1}{v_2} = \frac{N_1}{N_2}$$

and

$$\frac{i_2}{i_1} = \frac{N_1}{N_2}$$

Let $TR = N_1/N_2$, where TR stands for the turns ratio. Then, the product of the two foregoing equations is

$$\frac{v_1 i_2}{v_2 i_1} = (TR)^2$$

which can be rearranged as

$$\frac{v_1/i_1}{v_2/i_2} = (TR)^2$$

But the ratio of v_1 to i_1 is the primary resistance and the ratio of v_2 to i_2 is the secondary or load resistance. Therefore, the foregoing equation may be written as

$$\frac{R_1}{R_2} = (TR)^2$$

or

$$R_1 = (TR)^2 R_2 \qquad (2\text{-}25)$$

This says the primary resistance equals the square of the turns ratio times the secondary resistance.

What it Means

Equation (2-25) is a classic equation in the study of ac circuits. It tells us something new happens to the load resistance when seen from the primary side of the transformer. In this equation, the load resistance R_2 is multiplied by the square of the turns ratio to get R_1. And what is R_1? It is the load resistance as it appears to the ac source.

In other words, when we connect a transformer between an ac source and a load resistance, the load resistance is transformed to a different value. For instance, suppose we have a step-down transformer with a turns ratio of 3. The square of this is 9. If the load resistance is 1 kΩ, it would be transformed into a new resistance of

$$R_1 = 9R_2 = 9(1\ \text{k}\Omega) = 9\ \text{k}\Omega$$

In effect, the ac source sees a resistance of 9 kΩ instead of 1 kΩ.

Dot Convention

Figure 2-25a shows a transformer with dots on the primary and secondary windings. The ac source voltage is plus-minus to indicate the positive half cycle. During the negative half cycle, these signs reverse. The dots on the transformer are sometimes used on schematic diagrams to indicate the same polarity. In other words, during the half cycle when the dotted end of the primary winding is positive, the dotted end of the secondary winding is also positive. During the negative half cycle, both voltages reverse polarity.

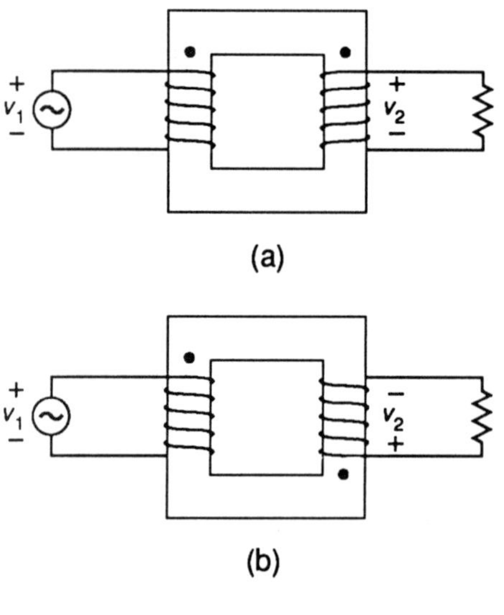

Fig. 2-25.

Figure 2-25b shows a transformer whose secondary turns are wound in the opposite direction. In this case, the dot is placed on the lower end because the induced voltage reverses its polarity. Again, the dotted ends of a transformer are positive during the positive half cycle of input voltage and negative during the negative half cycle of input voltage.

An Easy Way to Find Primary Current

Often, you need to calculate the primary current. The easiest way to do this is to take advantage of the fact that the primary and secondary powers are equal in an ideal transformer. In symbols,

$$p_1 = p_2$$

or

$$v_1 i_1 = p_2$$

Solving for the primary current, we get

$$i_1 = \frac{p_2}{v_1} \qquad (2\text{-}25)$$

IMPORTANT

$$i_1 = \frac{p_2}{v_1}$$

This says the primary current equals the secondary power divided by the primary voltage. In most problems, you can calculate the secondary or load power. Then, you can divide this power by the primary voltage to find the primary current. This is usually the easiest way to get the primary current.

EXAMPLE 2-10

The voltage from an ac outlet in a typical home is sinusoidal with an rms value of 115 V and a frequency of 60 Hz. Suppose this ac voltage is the input to the transformer shown in Fig. 2-26. What is the load voltage? The primary current? The resistance seen by the ac source?

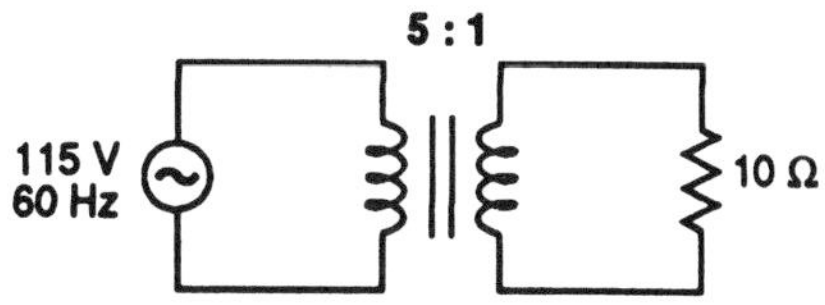

Fig. 2-26.

SOLUTION

Since the turns ratio is 5, we have a step-down transformer. The load voltage is

$$v_2 = \frac{115\text{ V}}{5} = 23\text{ V}$$

The load power equals the square of the load voltage divided by the load resistance:

$$p_2 = \frac{(23\text{ V})^2}{10\ \Omega} = 52.9\text{ W}$$

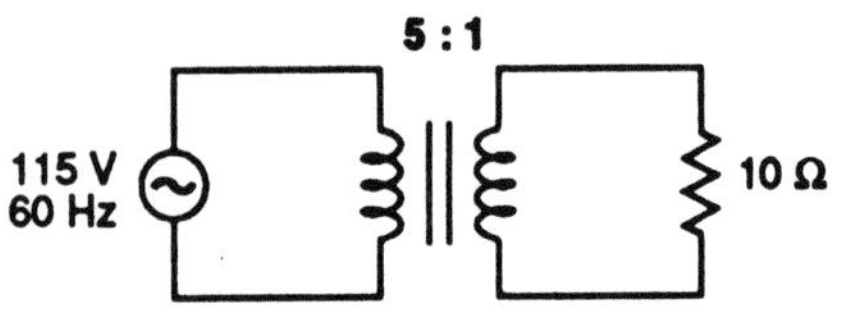

Fig. 2-26.

The primary current equals the load power divided by the primary voltage:

$$i_1 = \frac{52.9\ \text{W}}{115\ \text{V}} = 0.46\ \text{A}$$

The primary resistance equals the square of the turns ratio times the secondary resistance. Since the turns ratio is 5, the square is 25, and

$$R_1 = 25(10\ \Omega) = 250\ \Omega$$

Here is a way to check the work. The primary power is

$$p_1 = (115\ \text{V})(0.46\ \text{A}) = 52.9\ \text{W}$$

This is the same as the load power, which is correct because we are assuming an ideal transformer. Next, calculate the primary current in a different way by dividing the primary voltage by the primary resistance:

$$i_1 = \frac{115\ \text{V}}{250\ \Omega} = 0.46\ \text{A}$$

As you see, this agrees with the value calculated earlier.

EXAMPLE 2-11

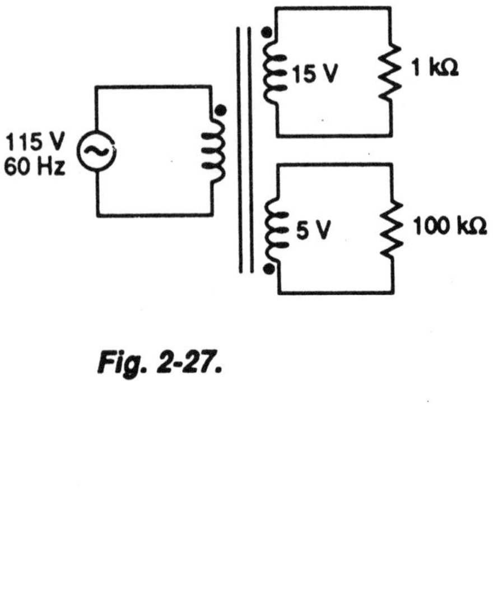

Fig. 2-27.

Figure 2-27 shows a step-down transformer with three windings: one primary and two secondaries. What is the primary current?

SOLUTION

The easiest way to do this is by dividing the total load power by the primary current. The load power in the upper resistor is

$$P_2 = \frac{(15\ \text{V})^2}{1\ \text{k}\Omega} = 225\ \text{mW}$$

and the load power in the lower resistor is

$$P_2 = \frac{(5\ \text{V})^2}{100\ \Omega} = 250\ \text{mW}$$

Assuming an ideal transformer, the input power to the primary equals the sum of the two output powers:

$$P_1 = P_2 + P_3 = 225\ \text{mW} + 250\ \text{mW} = 475\ \text{mW}$$

Since the primary voltage is known, we can calculate the primary current as follows:

$$I_1 = \frac{475\text{ mW}}{115\text{ V}} = 4.13\text{ mA}$$

MULTIPLE-CHOICE TEST

The following questions may have more than one right answer. Select the best answer. This is the one that is always true, or covers more situations, or agrees with the material of this chapter, etc.

1. The flux around an inductor is directly proportional to the
   a. resistance of the inductor
   b. voltage across the inductor
   c. current through the inductor
   d. all of the above

2. If the current is sinusoidal, the flux is
   a. constant
   b. time-varying
   c. rectangular
   d. sinusoidal

3. The voltage induced across an inductor is directly proportional to the
   a. number of turns
   b. instantaneous flux
   c. change in time
   d. instantaneous current

4. The induced voltage across an inductor is
   a. in phase with the current
   b. 90° ahead of the current
   c. 180° behind the current
   d. 90° behind the current

5. The slope of a vertical line is
   a. smaller than the slope of a horizontal line
   b. equal to the slope of a horizontal line
   c. greater than the slope of a horizontal line
   d. none of the above

6. The instantaneous rate of change of a sine wave is maximum positive at
   a. the point where it crosses the horizontal axis
   b. positive peak
   c. negative peak
   d. none of the above

7. To get an approximate value for the instantaneous rate of change of flux with time, you can divide $\Delta\phi$ by Δt if the
   a. flux is small
   b. current is small
   c. change in time is small
   d. all of the above

8. Inductance equals flux divided by
   a. time
   b. current
   c. induced voltage
   d. number of turns

9. The relative permeability of air is
   a. $1.26(10^{-6})$
   b. $2.24(10^{-6})$
   c. $15.3(10^{-6})$
   d. 1

10. Materials with a strong magnetic effect are called
    a. diamagnetic
    b. paramagnetic
    c. hypomagnetic
    d. ferromagnetic

11. The material with the largest permeability is
    a. vacuum
    b. air
    c. aluminum
    d. iron

12. The induced voltage across an inductor is directly proportional to the
    a. inductance
    b. current
    c. voltage
    d. resistance

13. The voltage across an inductor leads the current by
    a. 0°
    b. 90°
    c. 180°
    d. 270°

14. The coefficient of coupling can be
    a. less than zero
    b. no greater than 1
    c. greater than 1
    d. negative

15. Tight coupling occurs with an
    a. air core
    b. plastic core
    c. aluminum core
    d. iron core

16. The secondary voltage from a step-down transformer is
    a. less than the primary voltage
    b. equal to the primary voltage
    c. greater than the primary voltage
    d. none of the above

17. If two series inductors have zero coupling, the total inductance is
    a. less than either inductance
    b. the sum of the inductances
    c. the product over the sum of the inductances
    d. none of the above

18. If the instantaneous rate of change of current with time doubles, the voltage induced in an inductor will
    a. drop in half
    b. stay the same
    c. double
    d. quadruple

19. The back emf across an inductor increases when the
    a. resistance increases
    b. current decreases
    c. number of turns increases
    d. time decreases

20. The instantaneous rate of change refers to the average rate of change between two points that are
    a. infinitesmally close together
    b. far apart
    c. near the beginning of a sine wave
    d. near the peak of a sine wave

PROBLEMS

Sec. 2-1: Time-varying Magnetic Fields

2-1. An inductor has 10 turns. If the change in flux is 10 μWb and the change in time is 8 μs, what is the induced voltage?
2-2. If the flux changes 3 μWb during 8 μs, what is the voltage induced across an inductor of 25 turns?
2-3. What is the induced voltage in Prob. 2-1 if the number of turns is doubled?
2-4. What is the induced voltage in Prob. 2-2 if the change in flux doubles but the time remains the same?
2-5. What is the induced voltage in Prob. 2-1 if the change in flux remains the same but the time doubles?

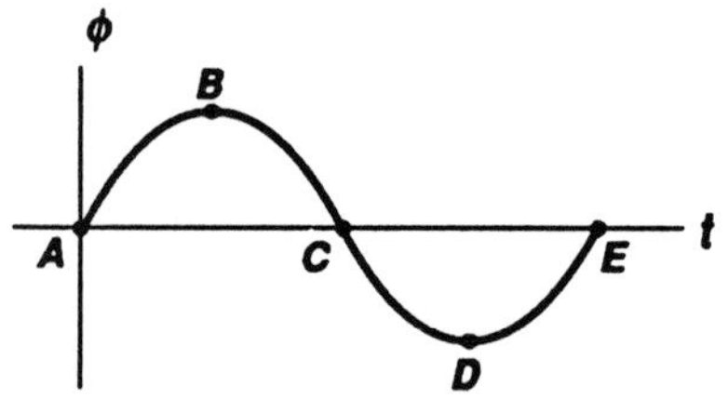

Fig. 2-28.

Sec. 2-3: Instantaneous Rate of Change

2-6. At what points in Fig. 2-28 does the flux have an instantaneous rate of change equal to zero.
2-7. In Fig. 2-28, is the instantaneous rate of change positive or negative in each of the following:

a. Between points A and B
b. Between points B and C
c. Between points C and D
d. Between points D and E

2-8. A sinusoidal flux has a peak value of 25 Wb and a frequency of 3 kHz. What is the instantaneous rate of change at the beginning of the sine wave?

Sec. 2-4: Another Look at Induced Voltage

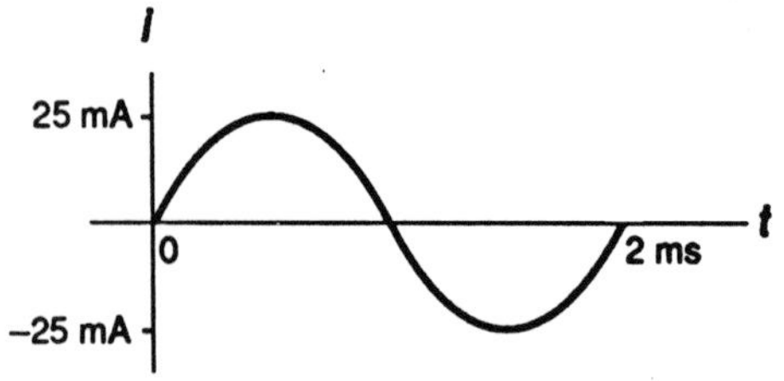

Fig. 2-29.

2-9. The sinusoidal current through an inductor has a peak value of 120 mA and a frequency of 5 kHz. Write the equation for instantaneous current and voltage. Sketch the waveforms of current and voltage.
2-10. The sinusoidal current through an inductor has a peak value of 40 mA and a frequency of 2 kHz. If the peak voltage is 5 V, what is the instantaneous voltage when $t = 0$? When $t = 62.5$ μs?
2-11. Figure 2-29 shows the current through an inductor. What is the equation for instantaneous voltage? If the peak voltage is 12 V, what is the instantaneous voltage when $t = 0$? When $t = 0.5$ μs?

Sec. 2-5: Inductance

2-12. A toroidal inductor has a core with a permeability of $2.1(10^{-3})$. If $N = 250$, $A = 2(10^{-6})$ m^2 and $l = 0.02$ m, what is the inductance?
2-13. If the number of turns doubles in Prob. 2-12, what is the value of inductance?

Sec. 2-6: Relative Permeability

2-14. An inductor with an air core has an inductance of 2.5 μH. If the core is changed to a steel core, what is new value of inductance?
2-15. If the core of Prob. 2-14 is changed to iron, what is the inductance?

Sec. 2-7. Voltage and Inductance

2-16. An inductance of 2 mH has a sinusoidal current with a peak value of 60 mA and a frequency of 300 Hz. If the instantaneous rate of change *di/dt* has a maximum value of 113 A/s, what is the maximum induced voltage? What are the equations for current and voltage. Show the current and voltage waveforms.
2-17. An inductance of 5 H has a sinusoidal current with a peak value of 50 mA and a frequency of 60 Hz. Use the method of small changes to calculate the maximum induced voltage.
2-18. The inductor of Fig. 2-30a has the current of Fig. 2-30b. Use the method of small changes to calculate the maximum induced voltage.

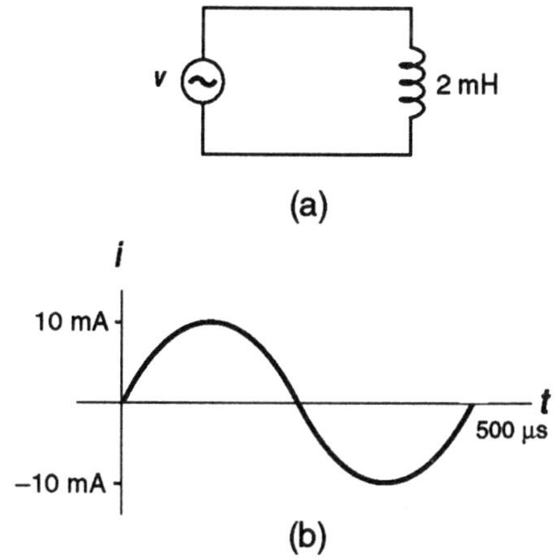

Fig. 2-30.

Sec. 2-8. Coefficient of Coupling

2-19. The flux through an inductor is 12 Wb and the flux through another is 1 Wb. What is the coefficient of coupling between the inductors?
2-20. Two inductors have a coefficient of coupling equal to 0.5. What is the coefficient of coupling if the flux in one of the inductors doubles?

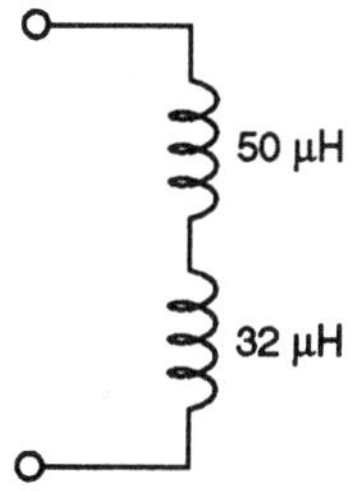

Fig. 2-31.

Sec. 2-9. Mutual Inductance

2-21. Two inductors are in series. The inductances are 25 mH and 40 mH. If the coefficient of coupling is zero, what is the total inductance?
2-22. Two series inductors of 10 mH and 15 mH have a coefficient of coupling equal to 0.5. What is the total inductance if their fluxes aid? If the fluxes oppose?
2-23. What is the total inductance in Fig. 2-31 for each of these conditions:
a. coefficient of coupling is zero
b. *k* equals 0.5
c. *k* is 100%
2-24. Two inductors are in parallel. The inductances are 25 mH and 40 mH. If the coefficient of coupling is zero, what is the total inductance?
2-25. Two parallel inductors of 10 mH and 15 mH have a coefficient of coupling equal to 0.5. What is the total inductance if their fluxes aid? If the fluxes oppose?

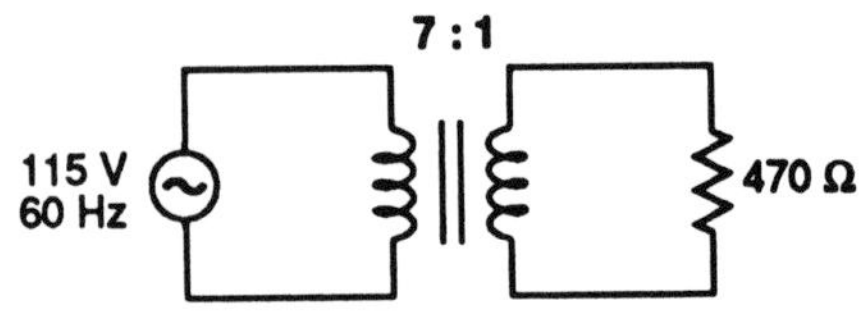

Fig. 2-32.

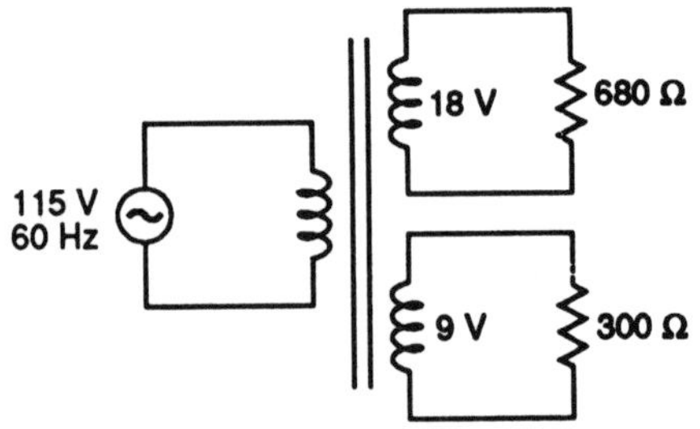

Fig. 2-33.

Sec. 2-10. The Ideal Transformer

2-26. A transformer has a turns ratio N_1/N_2 of 5. If the rms primary voltage is 115 V, what is the rms secondary voltage?

2-27. The ac voltage across a primary winding is 115 V ± 10%. If $N_1/N_2 = 8$, what is the minimum rms secondary voltage? The maximum rms secondary voltage?

2-28. What is the load voltage in Fig. 2-32? The current in each winding? The resistance seen by the ac source? The load power and the primary power?

2-29. Figure 2-33 shows a step-down transformer with three windings: one primary and two secondaries. What is the primary current?

CHAPTER 3
INDUCTIVE REACTANCE

This chapter is about *inductive reactance*, the opposition that a coil has to alternating current. Inductive reactance is different from resistance because it is produced by the back emf of the coil rather than atomic collisions within the material. Since inductive reactance depends on back emf, it changes when the frequency changes. Because of this, a designer can create new effects that are impossible with resistors.

3-1. BASIC IDEAS

In Fig. 3-1, the sinusoidal current through the inductor produces a back emf across the inductor. Because of the back emf, the inductor opposes the current. We call this opposition the inductive reactance, symbolized by X_L.

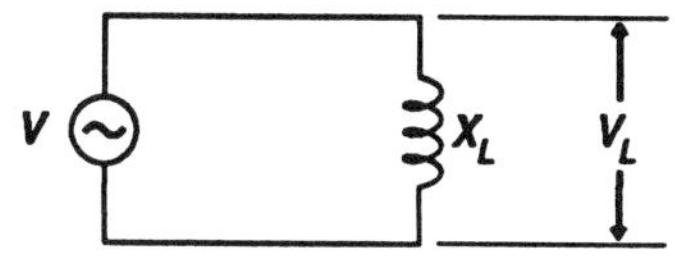

Fig. 3-1.

Definition

In Fig. 3-1, the rms value of the sinusoidal voltage across the inductor is symbolized by V_L, and the rms value of the sinusoidal current is symbolized by I_L. The defining equation for inductive reactance is

$$X_L = \frac{V_L}{I_L} \tag{3-1}$$

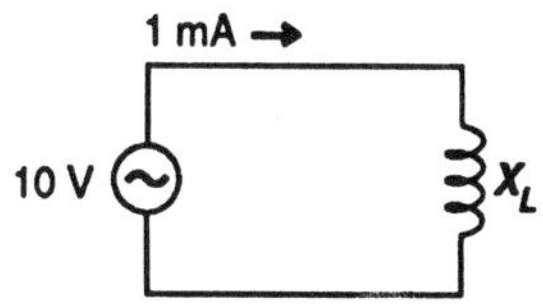

Fig. 3-2.

Since peak values are directly proportional to rms values, Eq. (3-1) may be used with peak values as well as rms values.

For instance, in Fig. 3-2 the rms voltage across the inductor is 10 V. Assume the current is 1 mA. Then, you can divide the voltage by the current to get the inductive reactance:

$$X_L = \frac{10\ \text{V}}{1\ \text{mA}} = 10\ \text{k}\Omega$$

In Fig. 3-2, the peak values are 14.1 V and 1.41 mA. Therefore, another way to calculate the inductive reactance is with peak values:

$$X_L = \frac{14.1\ \text{V}}{1.41\ \text{mA}} = 10\ \text{k}\Omega$$

Ohm's Law

IMPORTANT

$$I_L = \frac{V_L}{X_L}$$

Equation (3-1) is a form of Ohm's law applied to an inductor. We can rearrange the equation to get

$$I_L = \frac{V_L}{X_L} \qquad (3\text{-}2)$$

This form of Ohm's law for an inductor is useful because sometimes we know the values of voltage and inductive reactance. In this case, we can calculate the rms current through the inductor.

By cross-multiplying Eq. (3-2), we get this alternative form of Ohm's law:

$$V_L = I_L X_L \qquad (3\text{-}3)$$

This equation is useful when we have the rms current and the inductive reactance. Then, we can calculate the rms voltage across the inductor.

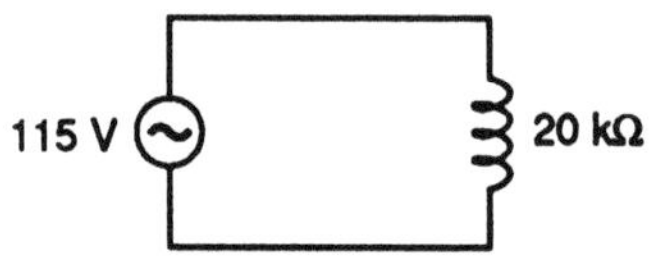

Fig. 3-3.

EXAMPLE 3-1

What is the rms current in Fig. 3-3?

SOLUTION

Divide the rms voltage by the inductive reactance to get

$$I = \frac{115\ \text{V}}{20\ \text{k}\Omega} = 5.75\ \text{mA}$$

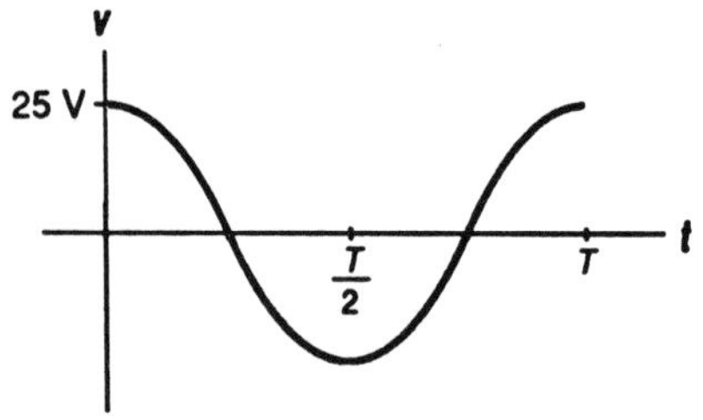

Fig. 3-4.

EXAMPLE 3-2

Suppose the sinusoidal voltage of Fig. 3-4 is across an inductive reactance of 8 kΩ. What is the peak current? The rms current?

SOLUTION

This is the type of waveform you may see on an oscilloscope. In this case, it is easier to work with peak values. Since the peak voltage is 25 V, the peak current is

$$I_p = \frac{25\ \text{V}}{8\ \text{k}\Omega} = 3.13\ \text{mA}$$

The rms current equals 0.707 times the peak current:

$$I_{\text{rms}} = 0.707(3.13\ \text{mA}) = 2.21\ \text{mA}$$

EXAMPLE 3-3

If the current is 2 mA in Fig. 3-5, what is the rms voltage ?
The peak voltage?

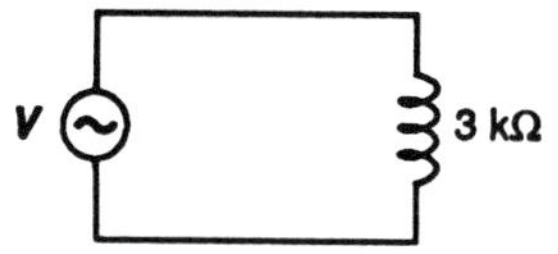

Fig. 3-5.

SOLUTION

Multiply the rms current by the inductive reactance to get the rms voltage across the inductor:

$V_{\text{rms}} = (2\ \text{mA})(3\ \text{k}\Omega) = 6\ \text{V}$

The peak voltage is

$V_p = 1.414(6\ \text{V}) = 8.48\ \text{V}$

EXAMPLE 3-4

An inductive reactance of 10 kΩ has a sinusoidal current through it. If the peak current is 2.25 mA, what is the rms voltage across the inductor?

SOLUTION

Do this in two steps. First, find the peak voltage:

$V_p = (2.25\ \text{mA})(10\ \text{k}\Omega)) = 22.5\ \text{V}$

Second, find the rms voltage:

$V_{\text{rms}} = 0.707(22.5\ \text{V}) = 15.9\ \text{V}$

3-2. THE EFFECT OF *F* AND *L*

Two factors that affect the inductive reactance are the *frequency* and the *inductance*. First, increasing the frequency means the flux changes faster. Since the instantaneous rate of change of flux with time changes the back emf, a change in frequency will affect the inductive reactance. Second, increasing the inductance means a given current will produce more magnetic effect. But a greater magnetic effect changes the back emf, which changes the inductive reactance.

Formula for Inductive Reactance

The following formula can be derived for the inductive reactance:

$$X_L = 2\pi fL \qquad (3\text{-}4)$$

MEMORIZE

$X_L = 2\pi fL$

Memorize this formula because it is used constantly in the analysis of ac circuits. This formula is useful when you know the values of frequency and inductance. For instance, if $f = 60$ Hz and $L = 5$ H, the inductive reactance is

$$X_L = 2\pi(60\text{ Hz})(5\text{ H}) = 1.88\text{ k}\Omega$$

Calculating *L* and *f*

Equation (3-4) contains three variables: X_L, f, and L. Given any two of these quantities, you can solve for the third. For instance, if you know the values of f and L, you can substitute into the equation and calculate X_L.

On the other hand, if you have the values of X_L and f, you can rearrange the equation to get

$$L = \frac{X_L}{2\pi f} \tag{3-5}$$

After substituting the values of X_L and f, you can calculate the value of L.

A third possibility is knowing the values of X_L and L. In this case, the equation can be rearranged as

$$f = \frac{X_L}{2\pi L} \tag{3-6}$$

After substituting for X_L and L, you can solve for f.

Derivation

Where does Eq. (3-4) come from? The standard proof uses calculus. But we can prove the equation using only trigonometry and the method of small changes. If you are not interested in seeing the proof, skip to Example 3-5.

To begin with, one of identities discussed in trigonometry is this relation:

$$\sin x = x - \frac{x^3}{3!} + \frac{x^5}{5!} - \frac{x^7}{7!} + \ldots$$

where x is an angle in radians. When x is small, the higher-order terms drop out, leaving

$$\sin x = x$$

In terms of the rotation angle, this may be written as

$$\sin 2\pi ft = 2\pi ft$$

The equation for any sinusoidal current is

$$i = I_p \sin 2\pi ft$$

When the angle of rotation is small, the foregoing simplifies to

$$i = I_p(2\pi ft) = 2\pi f I_p t$$

Recall the discussion of small changes. Figure 3-6 shows the earliest part of a sine wave. When dt is less than $0.01T$, the earliest part of a sine wave becomes linear. Because of this, variables i and t may be replaced by the changes di and dt. In other words, the foregoing equation may be written as

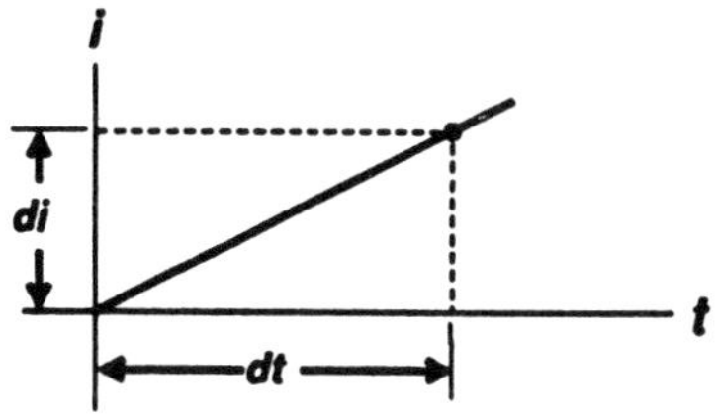

Fig. 3-6.

$$di = 2\pi f I_p dt$$

or

$$\frac{di}{dt} = 2\pi f I_p \tag{3-7}$$

Recall the physical meaning of this. It is the slope of the line in Fig. 3-6. It tells you how the fast the sine wave is changing at the beginning of the cycle.

The relation between induced voltage and inductance is given by

$$v = L\frac{di}{dt}$$

When we substitute Eq. (3-7) into this, we get

$$V_p = L(2\pi f I_p)$$

which rearranges into

$$\frac{V_p}{I_p} = 2\pi fL$$

or

$$X_L = 2\pi fL$$

This final result gives us a way to calculate inductive reactance using frequency and inductance.

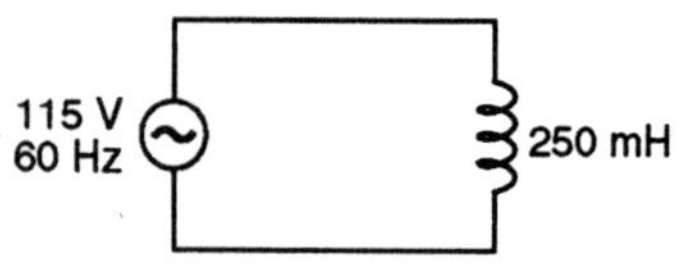

Fig. 3-7.

EXAMPLE 3-5

What is the rms current in Fig. 3-7?

SOLUTION

First, get the inductive reactance:

$$X_L = 2\pi(60\text{ Hz})(250\text{ mH}) = 94.2\ \Omega$$

Second, use Ohm's law to get the rms current:

$$I = \frac{115\text{ V}}{94.2\ \Omega} = 1.22\text{ A}$$

EXAMPLE 3-6

What is the rms current in Fig. 3-7 if the frequency is changed to 600 Hz?

SOLUTION

You can use the same approach as in the preceding example. Or you can use a shortcut. When the frequency increases by a factor of 10, the inductive reactance increases by a factor of 10 and the current decreases by a factor of 10. This means we can divide the current of the preceding example by 10 to get

$$I = \frac{1.22\text{ A}}{10} = 0.122\text{ A}$$

EXAMPLE 3-7

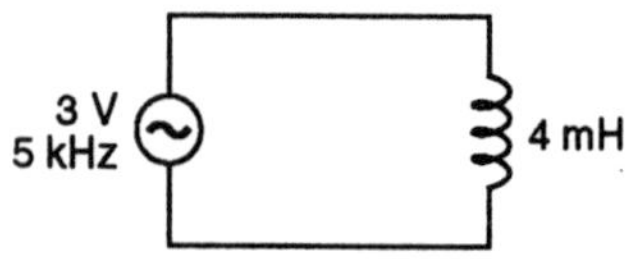

Fig. 3-8.

What is the rms current in Fig. 3-8?

SOLUTION

The solution is similar to Example 3-5, except for the numbers. First, get the inductive reactance:

$$X_L = 2\pi(5\text{ kHz})(4\text{ mH}) = 126\ \Omega$$

Second, use Ohm's law to get the rms current:

$$I = \frac{3\text{ V}}{126\ \Omega} = 23.8\text{ mA}$$

3-3. SERIES AND PARALLEL INDUCTORS

Since inductive reactance is measured in ohms, it makes sense to say that inductive reactances in series can be added to get the total inductive reactance. Likewise, inductive reactances in parallel may be combined by the product-over-sum rule to get the total inductive reactance.

X_L for Series Inductors

Here is the formula for the total inductive reactance of two series inductors:

$$X_L = X_{L1} + X_{L2} \tag{3-8}$$

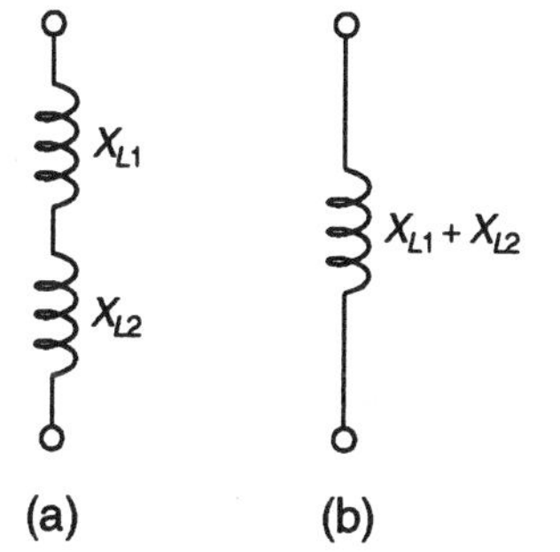

Fig. 3-9.

Figure 3-9a shows two inductors in series. The total inductive reactance is the sum of the two inductive reactances as shown in Fig. 3-9b. For instance, if $X_{L1} = 50\ \Omega$ and $X_{L2} = 75\ \Omega$, the total inductive reactance is

$$X_L = 50\ \Omega + 75\ \Omega = 125\ \Omega$$

This assumes the inductors are far enough apart that $k = 0$. In practice, inductors mounted on a circuit board usually are far enough apart that insignificant flux linkage occurs between the two coils. In future discussions and problems, you may assume that the coefficient of coupling is zero, unless otherwise indicated.

When two inductors are in series, they act the same as a single inductor with a value of

$$L = L_1 + L_2 \tag{3-9}$$

This equation is easy to prove. Equation (3-8) may be written as

$$2\pi fL = 2\pi fL_1 + 2\pi fL_2$$

Divide both sides of the equation by $2\pi f$ and you get Eq. (3-9).

X_L for Parallel Inductors

Inductive reactances in parallel combine by the product-over-the-sum rule:

$$X_L = \frac{X_{L1}X_{L2}}{X_{L1} + X_{L2}} \tag{3-10}$$

For instance, given two inductive reactances as shown in Fig. 3-10a, the total inductive reactance is given by the product over the sum as shown in Fig. 3-10b. This assumes the inductors are far enough apart that $k = 0$.

When two inductors are in parallel, they act the same as a single inductor with a value of

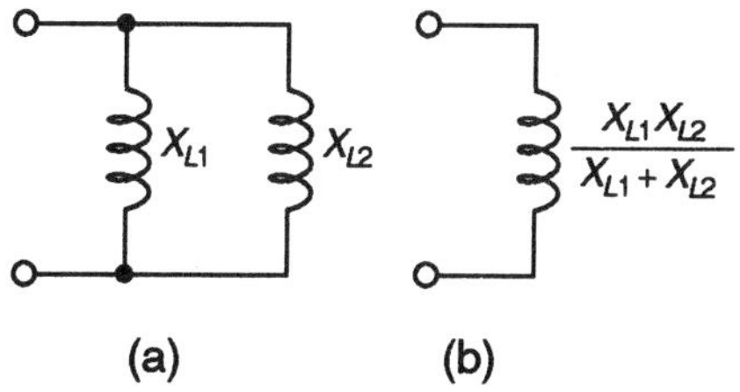

Fig. 3-10.

$$L = \frac{L_1 L_2}{L_1 + L_2} \quad (3\text{-}11)$$

The proof of this is similar to that given for series inductors. Express Eq. (3-10) in terms of $2\pi f$. Then, divide both sides of the equation by $2\pi f$. If you do this correctly, you will get Eq. (3-11).

EXAMPLE 3-8

What is the rms current in Fig. 3-11?

Fig. 3-11.

SOLUTION

The inductors are in series. The total inductance is

$$L = 2\text{ mH} + 4\text{ mH} = 6\text{ mH}$$

The inductive reactance is

$$X_L = 2\pi(2\text{ kHz})(6\text{ mH}) = 75.4\ \Omega$$

Next, use Ohm's law to get the rms current:

$$I = \frac{5\text{ V}}{75.4\ \Omega} = 66.3\text{ mA}$$

EXAMPLE 3-9

What is the rms current in Fig. 3-12?

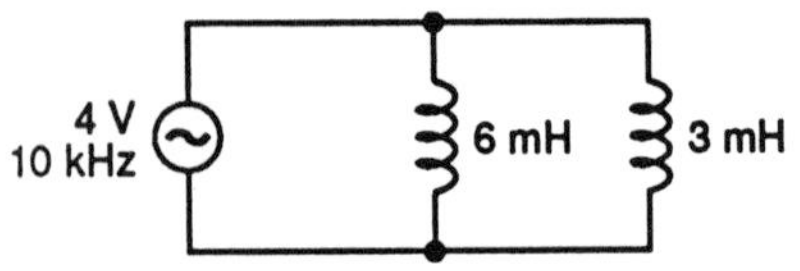

Fig. 3-12.

SOLUTION

The inductors are in parallel. The total inductance is

$$L = \frac{(6\text{ mH})(3\text{ mH})}{6\text{mH} + 3\text{ mH}} = 2\text{ mH}$$

The inductive reactance is

$$X_L = 2\pi(10\text{ kHz})(2\text{ mH}) = 126\ \Omega$$

Now, use Ohm's law to get the rms current:

$$I = \frac{4\text{ V}}{126\ \Omega} = 31.7\text{ mA}$$

3-4. INSTANTANEOUS ANALYSIS

Typical ac voltmeters will indicate the voltages as rms values. But when you use an oscilloscope, you will see sinusoidal voltages on the screen. For this reason, you have to know more about the instantaneous operation of ac circuits.

Additive Property of Sine Waves

Because the sine wave is the most common waveform in ac circuits, it will help our analysis if we know more about it. For instance, here is an important property of the sine wave:

The sum of two or more in-phase sine waves is a sine wave.

For instance, suppose we have two sinusoidal voltages given by

$$v_1 = V_{p1} \sin \theta$$

$$v_2 = V_{p2} \sin \theta$$

If we add the two sinusoidal voltages, we get

$$v = v_1 + v_2 = V_{p1} \sin \theta + V_{p2} \sin \theta$$

Factoring out the sine term gives

$$v = (V_{p1} + V_{p2}) \sin \theta$$

This says the sum of the two sine waves is a new sine wave with a peak value equal to the sum of the two peak values.

Figure 3-13 illustrates the idea graphically. If we add the instantaneous values of v_1 and v_2, we get the new sine wave labeled v. At an angle of 90°, the peak value of the new sine wave is

$$V_p = V_{p1} + V_{p2}$$

For instance, if V_{p1} = 3 V and V_{p2} = 4 V, the new sine wave has a peak value of

$$V_p = 3 \text{ V} + 4 \text{ V} = 7 \text{ V}$$

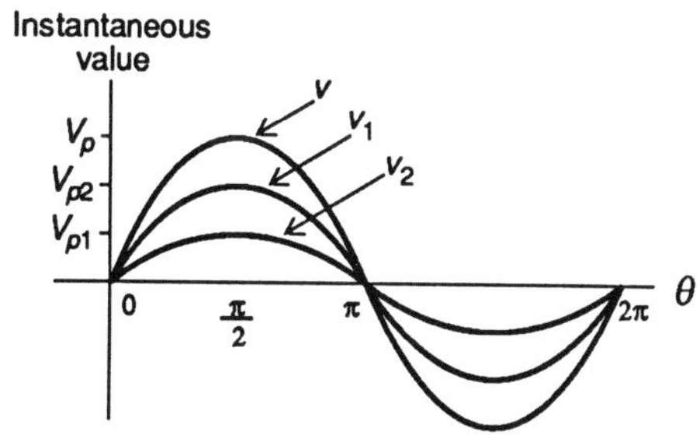

Fig. 3-13.

Inductors in Series

There are two more properties worth discussing at this time. As you know, the current is the same in all parts of a series dc circuit. In ac circuits, the equivalent statement is this:

The instantaneous current is the same in all parts of a series ac circuit.

You also learned that the total voltage in a series dc circuit equals the sum of the individual voltages. In ac circuits, the equivalent statement is

The total instantaneous voltage in a series ac circuit is the sum of the individual instantaneous voltages.

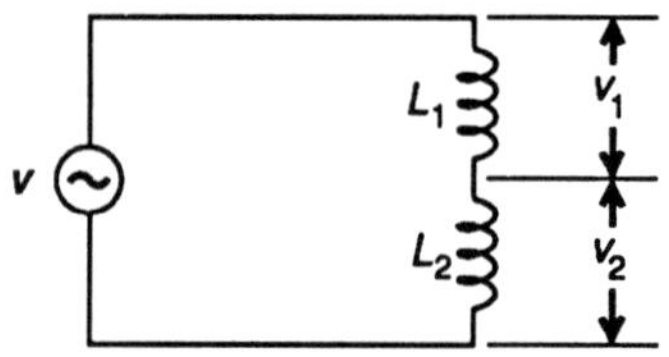

Fig. 3-14.

Here is an example of using the foregoing properties of ac circuits. Figure 3-14 shows an ac voltage source across two inductors. Because of the series connection, the instantaneous current is the same in all parts of the circuit. The values of this current can be calculated at any instant in time with this equation:

$$i = I_p \sin 2\pi ft$$

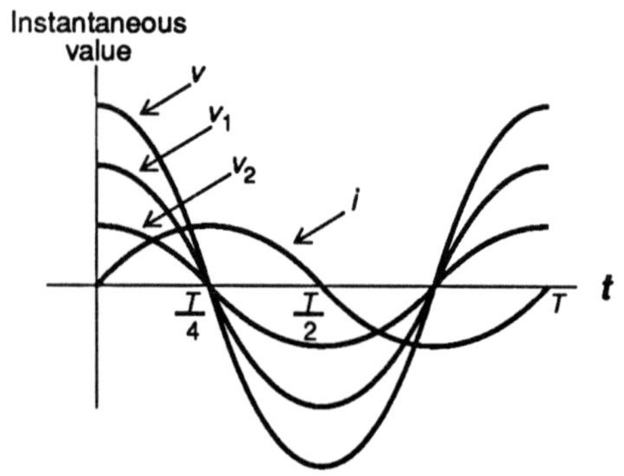

Fig. 3-15.

Figure 3-15 shows a sketch of this sinusoidal current.

The instantaneous current is the same in both inductors. For instance, when the instantaneous current is zero, both inductors have zero current. When the instantaneous current reaches its peak value, both inductors have a maximum current.

Because the instantaneous current is the same in both inductors, each inductor voltage leads the current by 90°. Therefore, the inductor voltages are given by

$$v_1 = V_{p1} \cos 2\pi ft$$
$$v_2 = V_{p2} \cos 2\pi ft$$

We can add these two voltages to get the total voltage:

$$v = v_1 + v_2 = V_{p1} \cos 2\pi ft + V_{p2} \cos 2\pi ft$$

or

$$v = (V_{p1} + V_{p2}) \cos 2\pi ft$$

This says the total voltage is a sine wave shifted by a quarter cycle. The peak value is the sum of the two peak values.

Figure 3-15 shows the inductor voltages and the total voltage. As you can see, v_1 and v_2 lead i by a quarter cycle. Notice that the total voltage v is the sum of the individual voltages v_1 and v_2. Because of this, the total voltage v also leads the current i by a quarter cycle. Make sure you understand the waveforms in this figure. The key ideas are

1. *The same current is in all series components.*
2. *The individual voltages lead this current by a quarter cycle.*
3. *The total voltage is the sum of the individual voltages.*

Inductors in Parallel

There are two more properties worth discussing at this time. As you know, the voltage is the same across all parts of a parallel dc circuit. In ac circuits, the equivalent statement is this:

The instantaneous voltage is the same across all parts of a parallel ac circuit.

You also learned that the total current in a parallel dc circuit equals the sum of the individual currents. In ac circuits, the equivalent statement is

The total instantaneous current in a parallel ac circuit is the sum of the individual instantaneous currents.

Here is an example of using the foregoing properties of ac circuits. Figure 3-16 shows an ac voltage source across two inductors. Because of the parallel connection, the instantaneous voltage is the same across all parts of the circuit. The values of this voltage can be calculated at any instant in time with this equation:

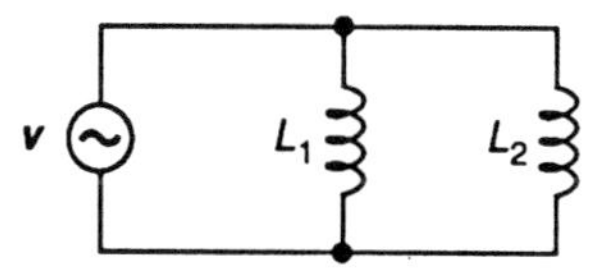

Fig. 3-16.

$$v = V_p \cos 2\pi ft$$

Figure 3-17 shows a sketch of this sinusoidal voltage.

Because the instantaneous voltage is the same across both inductors, each inductor current *lags* its voltage by 90°. Therefore, the inductor currents are given by

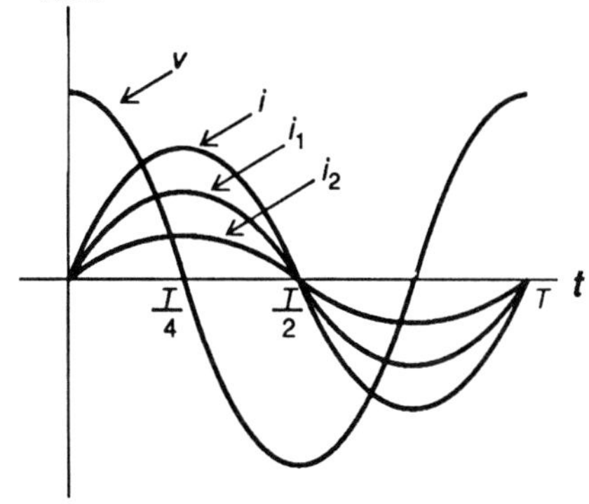

Fig. 3-17.

$$i_1 = I_{p1} \sin 2\pi ft$$
$$i_2 = I_{p2} \sin 2\pi ft$$

We can add these two currents to get the total current:

$$i = i_1 + i_2 = I_{p1} \sin 2\pi ft + I_{p2} \sin 2\pi ft$$

or

$$i = (I_{p1} + I_{p2}) \sin 2\pi ft$$

This says the total current is a sine wave with a peak value equal to the sum of the two peak values.

Figure 3-17 shows the inductor currents and the total current. As you can see, i_1 and i_2 lag v by a quarter cycle. Notice that the total current i is the sum of the individual currents i_1 and i_2. Because of this, the total current i also lags the voltage v by a quarter cycle. Make sure you understand the waveforms in this figure. The key ideas are

1. *The same voltage is across all components.*
2. *The individual currents lag the voltage by a quarter cycle.*
3. *The total current is the sum of the individual currents.*

Fig. 3-18.

EXAMPLE 3-10

In Fig. 3-18, an rms voltage of 4 V is across two inductors in series. Sketch the waveforms.

SOLUTION

The inductors are in series. The total inductance is

$$L = 5 \text{ mH} + 3 \text{ mH} = 8 \text{ mH}$$

The total inductive reactance is

$$X_L = 2\pi(1 \text{ kHz})(8 \text{ mH}) = 50.3\ \Omega$$

An alternative way to find the total inductive reactance is to add the individual inductive reactances. The first inductor has an inductive reactance of

$$X_{L1} = 2\pi(1 \text{ kHz})(5 \text{ mH}) = 31.4\ \Omega$$

The second inductor has an inductive reactance of

$$X_{L2} = 2\pi(1 \text{ kHz})(3 \text{ mH}) = 18.8\ \Omega$$

The sum of these is

$$X_L = 31.4\ \Omega + 18.8\ \Omega = 50.2\ \Omega$$

which is the approximately the same as before.

Next, use Ohm's law to get the rms current:

$$I = \frac{4 \text{ V}}{50.3\ \Omega} = 79.5 \text{ mA}$$

The peak current is

$$I_p = 1.414(79.5 \text{ mA}) = 112 \text{ mA}$$

and the peak voltage is

$$V_p = 1.414(4 \text{ V}) = 5.66 \text{ V}$$

Since the frequency is 1 kHz, the period is

$$T = \frac{1}{1 \text{ kHz}} = 1 \text{ ms}$$

Here is how to get the individual voltages. As previously calculated, the first inductor has an inductive reactance of 31.4 Ω. The peak current through this inductor is 112 mA. Therefore, the peak voltage across the first inductor is

$$V_{p2} = (112 \text{ mA})(31.4\ \Omega) = 3.52 \text{ V}$$

Since the second inductor has an inductive reactance of 18.8 Ω, the peak voltage across this inductor is

$$V_{p1} = (112 \text{ mA})(18.8\ \Omega) = 2.11 \text{ V}$$

When we add these two peak voltages, we get the total peak voltage:

$$V_p = 3.52 \text{ V} + 2.11 \text{ V} = 5.63 \text{ V}$$

This is the approximately the same peak voltage calculated earlier by a different method.

Figure 3-19 shows the waveforms. There is a single current in both inductors. The first inductor has a peak voltage of 3.52 V and the second inductor has a peak voltage of 2.11 V. The sum of the two peak voltages equals the total peak voltage of 5.63 V. Notice that all three voltage waveforms are in phase with each other. Furthermore, they lead the current by a quarter cycle.

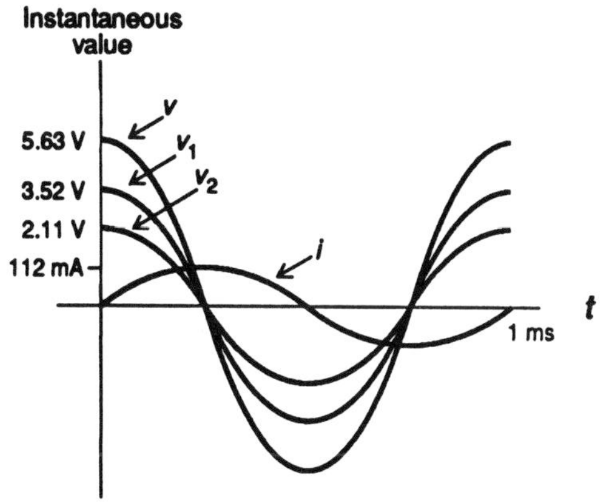

Fig. 3-19.

EXAMPLE 3-11

In Fig. 3-20, an rms voltage of 4 V is across two inductors in parallel. Sketch the waveforms.

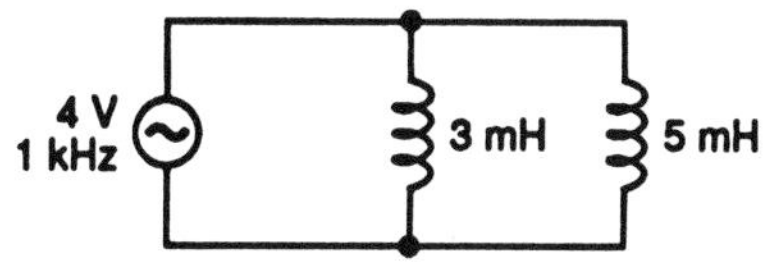

Fig. 3-20.

SOLUTION

The inductors are in parallel. The total inductance is

$$L = \frac{(3 \text{ mH})(5 \text{ mH})}{3\text{mH} + 5 \text{ mH}} = 1.88 \text{ mH}$$

The total inductive reactance is

$$X_L = 2\pi(1 \text{ kHz})(1.88 \text{ mH}) = 11.8\ \Omega$$

Here is an alternative way to find the total inductance reactance. The first inductor has an inductive reactance of

$$X_{L1} = 2\pi(1 \text{ kHz})(3 \text{ mH}) = 18.8\ \Omega$$

The second inductor has an inductive reactance of

$$X_{L2} = 2\pi(1 \text{ kHz})(5 \text{ mH}) = 31.4\ \Omega$$

The sum of these is

$$X_L = \frac{(18.8\ \Omega)(31.4\ \Omega)}{18.8\ \Omega + 31.4\ \Omega} = 11.8\ \Omega$$

which is the same as before.

Next, use Ohm's law to get the total rms current:

$$I = \frac{4 \text{ V}}{11.8\ \Omega} = 339 \text{ mA}$$

The peak current is

$$I_p = 1.414(339 \text{ mA}) = 479 \text{ mA}$$

and the peak voltage is

$$V_p = 1.414(4 \text{ V}) = 5.66 \text{ V}$$

Since the frequency is 1 kHz, the period is

$$T = \frac{1}{1 \text{ kHz}} = 1 \text{ ms}$$

Here is how get the individual currents. Since the first inductor has an inductive reactance of 18.8 Ω, the peak current through this inductor is

$$I_{p1} = \frac{5.66 \text{ V}}{18.8\ \Omega} = 301 \text{ mA}$$

Since the second inductor has an inductive reactance of 31.4 Ω, the peak current through this inductor is

$$I_{p2} = \frac{5.66 \text{ V}}{31.4\ \Omega} = 180 \text{ mA}$$

When we add these two peak currents, we get the total peak current:

I_p = 301 mA +180 mA = 481 mA

This is approximately the same peak current calculated earlier by a different method.

Figure 3-21 shows the waveforms. There is a single voltage across both inductors. The first inductor has a peak current of 301 mA and the second inductor has a peak current of 180 mA. The sum of the two peak currents equals the total peak current of 481 mA. Notice that all three current waveforms are in phase with each other. Furthermore, they lag the voltage by a quarter cycle.

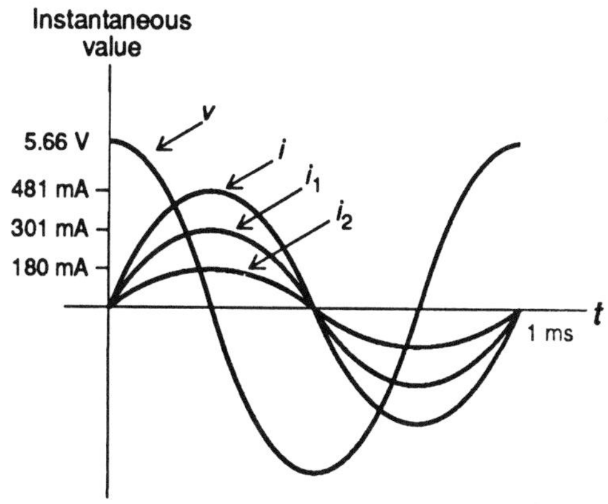

Fig. 3-21.

3-5. PHASORS

You can see from the preceding discussions that ac circuits are much more complicated than dc circuits. Because the ac source voltage varies sinusoidally throughout the cycle, all currents and voltages in an ac circuit vary sinusoidally. When resistors and inductors are used in the same circuit, the analysis becomes very difficult.

When Charles P. Steinmetz began working with ac circuits, he realized that shortcuts were needed to simplify the analysis. One of the shortcuts he invented was the *phasor*. With phasors, the analysis of ac circuits is greatly simplified because we avoid working directly with sinusoidal quantities.

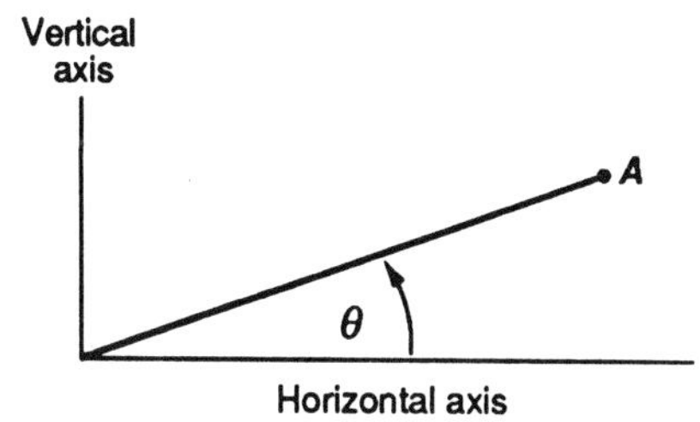

Fig. 3-22.

Rotating Conductor

Recall the ac generator. As the conductor rotates, it forms an angle with the horizontal axis as shown in Fig. 3-22. The rotation angle determines the amount of flux being cut. Assume the magnetic field is vertical. Then, no flux is cut when the angle is 0° and maximum flux is cut when the angle is 90°. Furthermore, the instantaneous voltage is given by

$$v = V_p \sin \theta$$

This voltage is sinusoidal because the amount of flux being cut is proportional to the sine of the rotation angle.

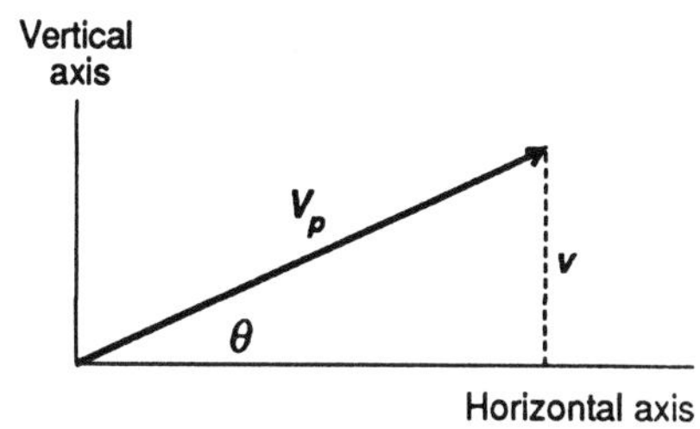

Fig. 3-23.

Definition of Phasor

The concept of a rotating conductor producing a sinusoidal voltage led Steinmetz to a new metaphor, an equivalent way of visualizing sinusoidal quantities. To begin with, he proposed using a rotating arrow instead of a conductor as shown in Fig. 3-23. He called this rotating arrow a phasor. The length or magnitude of the phasor is V_p and the angle it makes with the horizontal axis is theta.

This phasor acts the same as the conductor. As it rotates between 0° and 360°, it generates a sinusoidal voltage given by

$$v = V_p \sin \theta$$

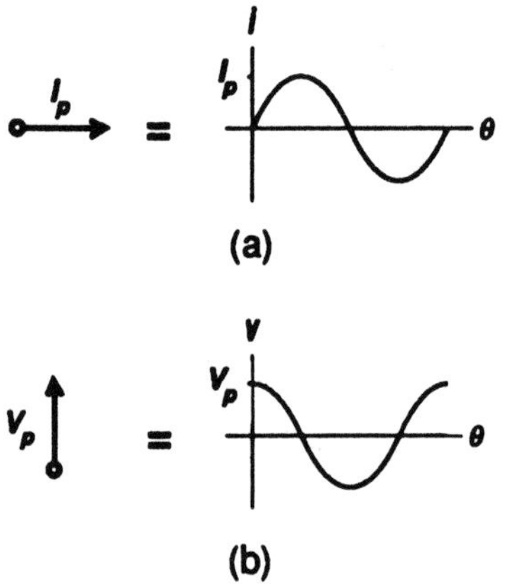

Fig. 3-24.

If we know the length of the phasor and its angle, we can calculate the instantaneous value of induced voltage. For instance, if the phasor has a magnitude of 10 V and the angle is 30°, the instantaneous voltage is

$$v = (10\ \text{V}) \sin 30° = 5\ \text{V}$$

Phasor Code for Sinusoids

A sinusoid always has the same shape because it is a trigonometric function. The starting point, however, may differ from one sinusoid to another. For instance, the voltage across an inductor leads the current by 90°. Steinmetz decided to use phasors to represent these sinusoidal quantities. To get a unique code, he drew each phasor as it appeared at the beginning of the cycle. For instance, Fig. 3-24a shows the phasor for a sinusoidal current through an inductor, and Fig. 3-24b shows the phasor for the sinusoidal voltage.

As you see, the phasor is drawn at whatever angle is necessary to produce the initial voltage. For instance, the sinusoidal current of Fig. 3-24a has an initial value of zero; therefore, the phasor is drawn at an angle of 0°. The sinusoidal voltage of Fig. 3-24b has an initial voltage of V_p; therefore, the phasor is drawn at an angle of 90°.

In Fig. 3-24, the equal signs mean we can replace the sine waves by their phasors. This will simplify the drawings for ac circuits. Instead of having to sketch a sine wave, we can draw its phasor. This is valid because the phasor contains the necessary information about the sine wave, including its peak value and initial value.

Any Sinusoidal Quantity

The current through an inductor is given by

$$i = I_p \sin \theta$$

The induced voltage across an inductor is given by

$$v = V_p \sin (\theta + 90°)$$

In this equation, the 90° is called the *phase angle* of the sine wave.

In a purely resistive circuit, the total voltage has a phase angle of 0° because it is in phase with the current. In a purely inductive circuit, the total voltage has a phase angle of 90°. In a circuit that combines resistors and inductors, the total voltage leads the current by a phase angle that is between 0° and 90°. In other words, the instantaneous voltage in some ac circuits may look like Fig. 3-25.

Fig. 3-25.

This voltage has a peak voltage V_p and a phase angle ϕ (Greek letter Phi, pronounced *fee*). Given the values of V_p and ϕ, there is only one sine wave we can draw. Because of this, we can replace any sinusoidal quantity by its phasor. In the next chapter, we will continue the discussion of phasors with any phase angle.

EXAMPLE 3-12

Show the phasors for current and voltage in Fig. 3-26.

SOLUTION

The inductive reactance is

$$X_L = 2\pi(3\text{ kHz})(5\text{ mH}) = 94.2\ \Omega$$

The rms current is

$$I = \frac{12\text{ V}}{94.2\ \Omega} = 127\text{ mA}$$

The peak current is

$$I_p = 1.414(127\text{ mA}) = 180\text{ mA}$$

and the peak voltage is

$$V_p = 1.414(12\text{ V}) = 17\text{ V}$$

Figure 3-26b shows the phasors. Look at how neat this is. This phasor diagram gives us almost all of the essential information about the sinusoidal current and voltage. We can see that the sinusoidal current has a peak value of 180 mA, while the sinusoidal voltage has a peak value of 17 V and a phase angle of 90°.

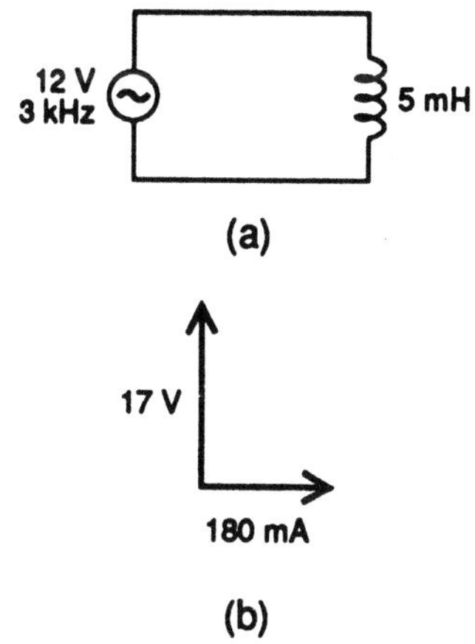

Fig. 3-26.

EXAMPLE 3-13

In Fig. 3-27, an rms current of 32 mA flows through an inductance of 27 μH. If the frequency is 450 kHz, draw the phasor diagram for the sinusoidal current and voltage.

SOLUTION

The inductive reactance is

$$X_L = 2\pi(450\text{ kHz})(27\ \mu\text{H}) = 76.3\ \Omega$$

The rms voltage across the inductor is

$$V = (32\text{ mA})(76.3\ \Omega) = 2.44\text{ V}$$

Now, calculate the peak current and peak voltage:

$$I_p = 1.414(32\text{ mA}) = 45.2\text{ mA}$$

$$V_p = 1.414(2.44\text{ V}) = 3.45\text{ V}$$

Figure 3-27b shows the phasor diagram.

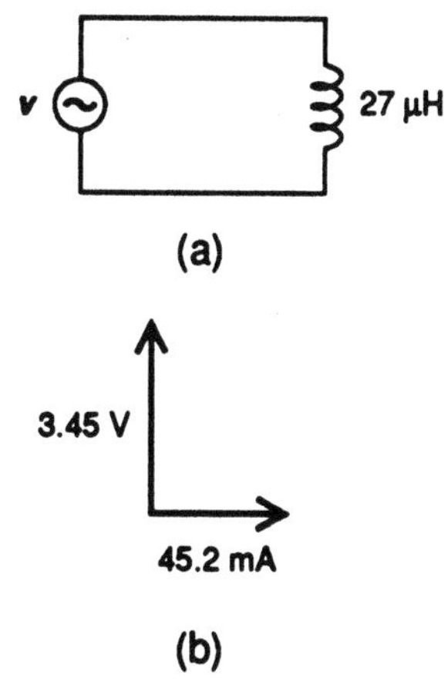

Fig. 3-27.

MULTIPLE-CHOICE TEST

The following questions may have more than one right answer. Select the best answer. This is the one that is always true, or covers more situations, or agrees with the material of this chapter, gives more information, etc.

1. Inductive reactance equals the rms voltage divided by the
   a. instantaneous current
   b. rms current
   c. peak current
   d. average current

2. When frequency increases, inductive reactance
   a. decreases
   b. is unaffected
   c. increases
   d. none of the above

3. When inductance increases, inductive reactance
   a. decreases
   b. is unaffected
   c. increases
   d. none of the above

4. When two inductors are in series, the total inductive reactance is
   a. less than the smaller
   b. less than the larger
   c. equal to the sum of the two inductive reactances
   d. equal to the product over the sum of the inductive reactances

5. When two inductors are in parallel, the total inductive reactance is
   a. less than the smaller
   b. less than the larger
   c. equal to the sum of the two inductive reactances
   d. equal to the product over the sum of the inductive reactances

6. The sum of two in-phase sine waves is a
   a. sine wave
   b. cosine wave
   c. square wave
   d. none of the above

7. The instantaneous current is the same in all parts of a
   a. series ac circuit
   b. parallel ac circuit
   c. parallel dc circuit
   d. none of the above

8. The length of a phasor equals the
   a. flux
   b. peak value
   c. rotation angle
   d. frequency

9. When the current through an inductor has a phase angle of 0°, the voltage has a phase angle of
   a. 0°
   b. 90°
   c. 180°
   d. 270°

10. If you divide the peak voltage across an inductor by the peak current, you get the
    a. resistance
    b. frequency
    c. period
    d. inductive reactance

11. When the current through an inductor has a phase angle of 90°, the voltage has a phase angle of
    a. 0°
    b. 90°
    c. 180°
    d. 270°

12. The rms voltage across an inductor divided by the inductive reactance gives the
    a. peak voltage
    b. rms voltage
    c. peak current
    d. rms current

13. The rms current through an inductor times the inductive reactance gives the
    a. peak voltage
    b. rms voltage
    c. peak current
    d. rms current

14. When the frequency doubles, the inductive reactance
    a. drops in half
    b. stays the same
    c. doubles
    d. quadruples

15. If you divide the inductive reactance by $2\pi f$, you get the
    a. peak voltage
    b. peak current
    c. rms current
    d. inductance

16. Two series inductors have reactances of 30 Ω and 60 Ω. The total inductive reactance is
    a. 20 Ω
    b. 30 Ω
    c. 60 Ω
    d. 90 Ω

17. Two in-phase sinusoidal voltages have peak values of 10 V and 15 V. If these sinusoidal voltages are added, the total peak voltage is
    a. 0 V
    b. 10 V
    c. 15 V
    d. 25 V

18. A sinusoidal voltage has a phase angle of 90°. Its phasor points
    a. to the left
    b. straight up
    c. to the right
    d. straight down

19. Two parallel inductors have reactances of 30 Ω and 60 Ω. The total inductive reactance is
    a. 20 Ω
    b. 30 Ω
    c. 60 Ω
    d. 90 Ω

20. A sinusoidal current has a peak value of 10 mA and a phase angle of 0°. The phasor has a
    a. length of 7.07 mA and an angle of 0°
    b. length of 7.07 mA and angle of 90°
    c. length of 10 mA and angle of 0°
    d. length of 10 mA and angle of 90°

21. A sinusoidal voltage has a peak value of 10 V and a phase angle of 90°. The phasor has a
    a. length of 7.07 V and an angle of 0°
    b. length of 7.07 V and angle of 90°
    c. length of 10 V and angle of 0°
    d. length of 10 V and angle of 90°

PROBLEMS

Sec. 3-1: Basic Ideas

3-1. What is the rms current in Fig. 3-28?
3-2. If the applied voltage doubles in Fig. 3-28, what is the rms current?
3-3. If the inductive reactance doubles in Fig. 3-28, what is the rms current?
3-4. If the inductive reactance drops in half in Fig. 3-28, what is the rms current?
3-5. 50 μA flows through an inductive reactance of 24 kΩ. What is the rms voltage across the inductor?
3-6. A sinusoidal voltage with a peak value of 15 V is across an inductive reactance of 2 kΩ. What is the rms current?

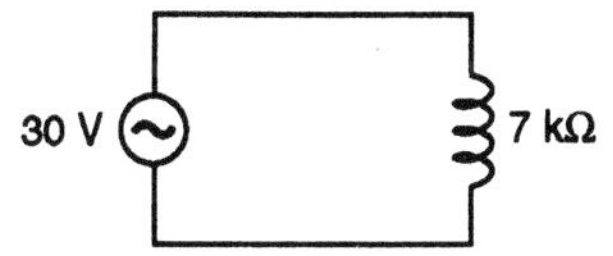

Fig. 3-28.

Sec. 3-2: The Effect of F and L

3-7. What is the rms current in Fig. 3-29?
3-8. If the frequency changes to 500 Hz in Fig. 3-29, what is the rms current?
3-9. If the frequency changes to 5 kHz in Fig. 3-29, what is the rms current?
3-10. 15 V rms is across an inductance of 100 mH. What is the rms current if the frequency is 10 kHz?
3-11. 250 μA flows through an inductance of 800 μH. What is the rms voltage across the inductor if the frequency is 1 MHz?
3-12. A sinusoidal voltage with a peak value of 15 V is across an inductance of 35 mH. If the frequency is 20 kHz, what is the rms current?

Fig. 3-29.

Sec. 3-3: Series and Parallel Inductors

3-13. What is the rms current in Fig. 3-30?
3-14. If both inductances are doubled in Fig. 3-30, what is the rms current?
3-15. If the frequency decreases to 1 kHz, what is the rms current in Fig. 3-30?
3-16. What is the total rms current in Fig. 3-31?
3-17. In Fig. 3-31, what is the rms current through the 200 μH? Through the 400 μH?
3-18. If the frequency doubles in Fig. 3-31, what is the total rms current? The rms current through the left inductor? Through the right inductor?

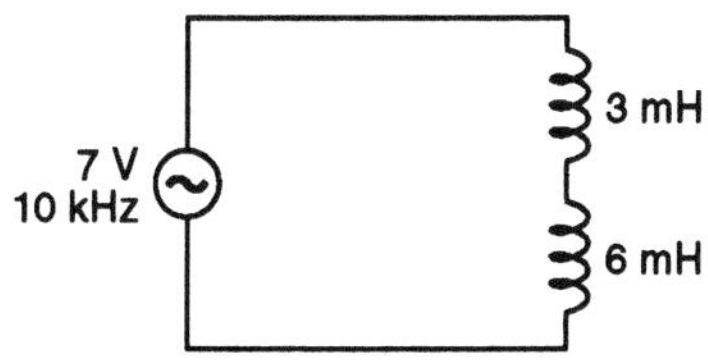

Fig. 3-30.

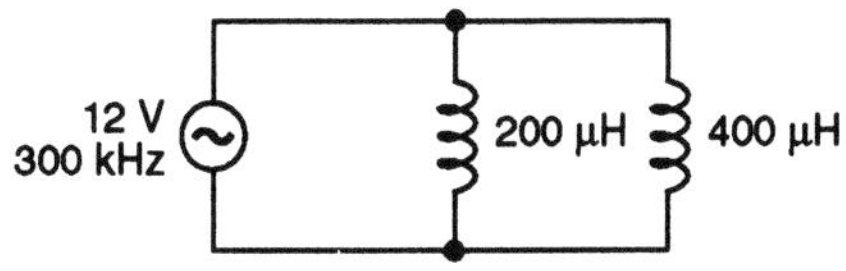

Fig. 3-31.

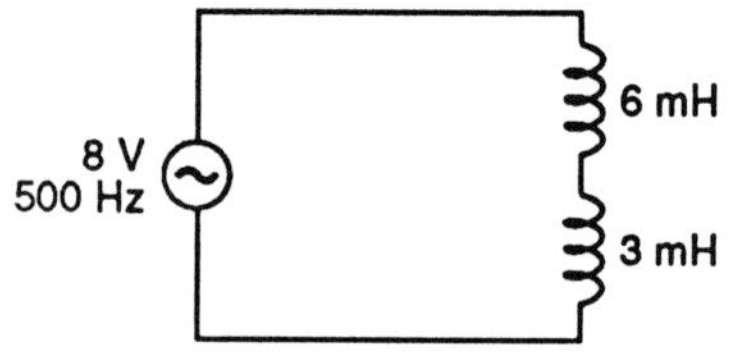

Fig. 3-32.

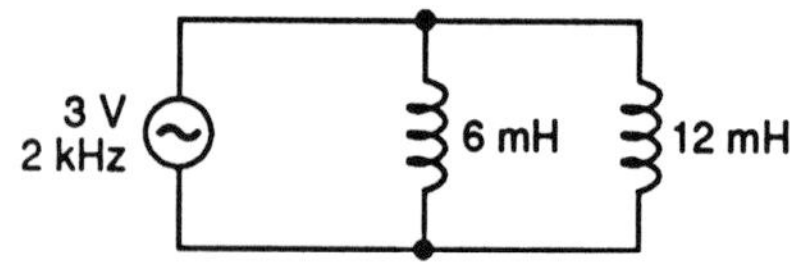

Fig. 3-33.

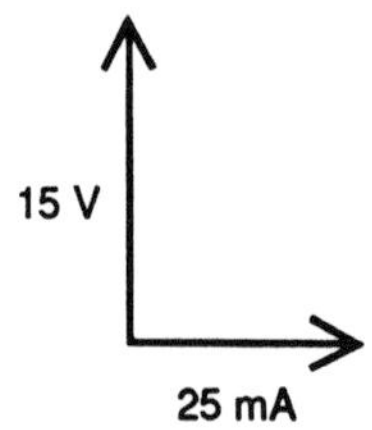

Fig. 3-34.

Sec. 3-4. Instantaneous Analysis

3-19. What are the inductive reactances for the inductors of Fig. 3-32? The total inductive reactance? The peak current? The peak voltages across the inductors?

3-20. Sketch the current and voltage waveforms for Fig. 3-32.

3-21. If the applied voltage doubles in Fig. 3-32, what is the peak current? The peak voltages across the inductors? Sketch the current and voltage waveforms.

3-22. If the frequency doubles in Fig. 3-32, what is the peak current? The peak voltages across the inductors? Sketch the current and voltage waveforms.

3-23. What are the inductive reactances for the inductors of Fig. 3-33? The total inductive reactance? The peak current? The peak voltages across the inductors?

3-24. Sketch the current and voltage waveforms for Fig. 3-33.

3-25. If the applied voltage doubles in Fig. 3-33, what is the peak current? The peak voltages across the inductors? Sketch the current and voltage waveforms.

3-26. If the frequency doubles in Fig. 3-33, what is the peak current? The peak voltages across the inductors? Sketch the current and voltage waveforms.

Sec. 3-5. Phasors

3-27. Figure 3-34 shows the phasor diagram for an inductor. What is the peak voltage across the inductor? The rms voltage? The peak current? The rms current?

3-28. An inductor has the phasor diagram of Fig. 3-34. If the voltage remains at 15 V but the frequency doubles, what happens to the current phasor? The voltage phasor?

3-29. Draw the phasor diagram for Fig. 3-28. Your drawing should look like Fig. 3-34 but the numbers should be different.

3-30. Draw the phasor diagram for Fig. 3-29.

CHAPTER 4
INDUCTIVE CIRCUITS

This chapter shows you how to analyze ac circuits that have resistors and inductors. Since the analysis uses phasors, one of the first things to learn is how to add two phasors. Then, you will be ready understand how to apply Ohm's and Kirchhoff's laws to ac circuits.

4-1. PHASORS

Let's begin by reviewing the phasors for ideal resistors and ideal inductors. This will be followed by a discussion of how to add phasors.

Phasor Current and Voltage for a Resistor

Ohm's law applies at each instant if a component is *purely resistive.* This means we can use the formula

$$i = \frac{v}{R} \tag{4-1}$$

at any instant in an ac circuit that contains only resistances. If the voltage is sinusoidal, it is given by

$$v = V_p \sin \theta$$

Substituting this into Eq. (4-1), we get

$$i = \frac{V_p \sin \theta}{R}$$

or

$$i = I_p \sin \theta$$

where $I_p = V_p/R$. This means a sinusoidal voltage across a resistance produces a sinusoidal current of the same phase angle. In terms of phasors, it means the phasor voltage and phasor current have a zero angle between them, as shown in Fig. 4-1.

Fig. 4-1.

Fig. 4-1.

When you look at a phasor diagram like Fig. 4-1, you can mentally convert the phasors into equivalent sine waves. When you see the I phasor of Fig. 4-1, you should also see in another part of your mind a sine wave with a phase angle of 0°. The V phasor should also bring to mind a sine wave with a phase angle of 0°.

Phasor Current and Voltage for an Inductor

When a component is purely inductive, the instantaneous voltage across the inductance is given by

$$v = L\frac{di}{dt} \qquad (4\text{-}2)$$

If the current is sinusoidal, it is given by

$$i = I_p \sin \theta$$

As shown in Chap. 2, the voltage is sinusoidal and given by

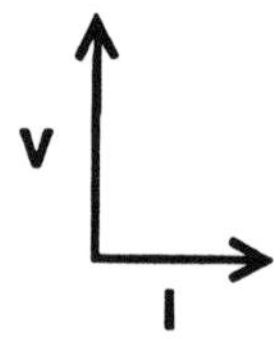

Fig. 4-2.

$$v = V_p \sin(\theta + 90°)$$

where $V_p = I_pX_L$. This means a sinusoidal current through an inductance produces a sinusoidal voltage with a phase angle of 90°. In terms of phasors, it means the phasor voltage leads the phasor current by 90°, as shown in Fig. 4-2.

EXAMPLE 4-1

Figure 4-3a shows a resistive circuit. If the source voltage has a peak value of 60 V, what is the current? The voltage across each resistor?

SOLUTION

To begin with, visualize the phasor of the source voltage as shown in Fig. 4-3b. As you see, the phasor has a magnitude of 60 V and a phase angle of 0°. This voltage produces an in-phase current with a peak value of

$$I_p = \frac{V_p}{R} = \frac{60\text{ V}}{3\text{ k}\Omega} = 20\text{ mA}$$

Figure 4-3c shows the phasor for this current. Notice that it is in phase with the applied voltage.

Since the peak current is 20 mA, the first resistor has a peak voltage of

$$V_{p1} = I_pR_1 = (20\text{ mA})(1\text{ k}\Omega) = 20\text{ V}$$

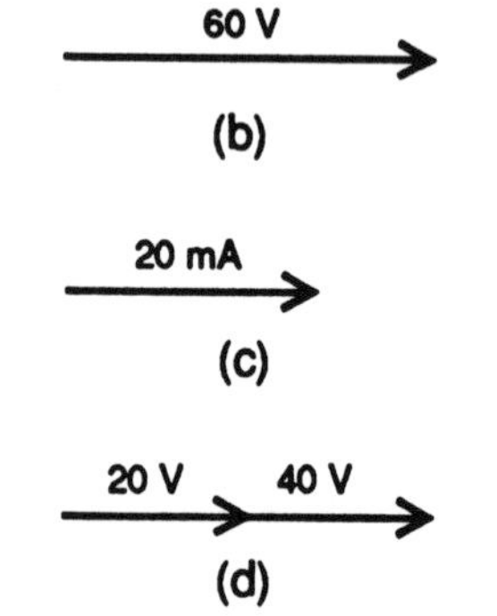

Fig. 4-3.

The second resistor has a peak voltage of

$$V_{p2} = I_p R_2 = (20 \text{ mA})(2 \text{ k}\Omega) = 40 \text{ V}$$

Figure 4-3d shows the phasors for these two voltages. Notice that these voltages are in phase and add to a total of 60 V, the magnitude of the applied voltage.

EXAMPLE 4-2

Figure 4-4a shows an inductive circuit. The first inductor has an inductive reactance of 1 kΩ, and the second has an inductive reactance of 2 kΩ. If the source voltage has a peak value of 60 V, what is the current? The voltage across each resistor?

SOLUTION

Again, visualize the phasor of the source voltage as shown in Fig. 4-4b. This voltage produces a current with a peak value of

$$I_p = \frac{60 \text{ V}}{3 \text{ k}\Omega} = 20 \text{ mA}$$

Figure 4-4c shows the phasor for this current. Notice that it lags the applied voltage by 90°. This is equivalent to saying the applied voltage leads the current by 90°.

Since the peak current is 20 mA, the first inductor has a peak voltage of

$$V_{p1} = I_p X_{L1} = (20 \text{ mA})(1 \text{ k}\Omega) = 20 \text{ V}$$

The second inductor has a peak voltage of

$$V_{p2} = I_p X_{L2} = (20 \text{ mA})(2 \text{ k}\Omega) = 40 \text{ V}$$

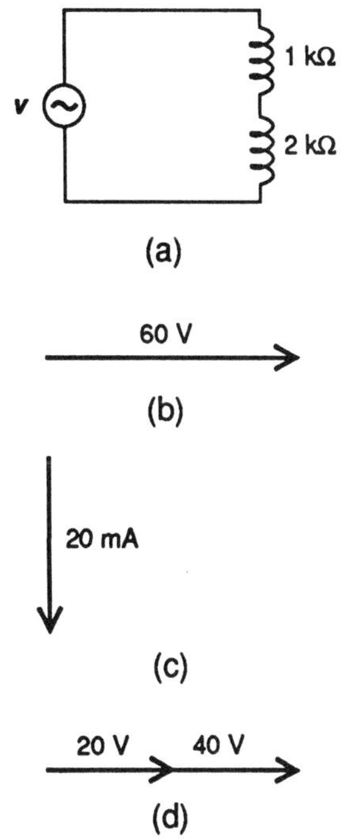

Fig. 4-4.

Figure 4-4d shows the phasors for these two voltages. Notice that these voltages are in phase and add to a total of 60 V, the magnitude of the applied voltage.

As you have seen, a purely resistive circuit and a purely inductive circuit have voltages that are in phase with the applied voltage. The difference is this: the current of the resistive circuit is in phase with the applied voltage, but the current of the inductive circuit lags the applied voltage by 90°.

4-2. VECTOR ADDITION

When an ac circuit contains only resistances, it is easy to analyze because all voltages and currents are in phase. In this case, we can apply Ohm's law to the peak values of the circuit. Similarly, when an ac circuit contains only inductances, it is still easy to analyze because all voltages lead the currents by 90°. Again, Ohm's law is valid for the peak values, because we use inductive reactance instead of resistance.

But how do we analyze an ac circuit that has both resistance and inductance? Before we answer this question, we will discuss *vectors*. Normally, vectors are covered in trigonometry and physics courses. What follows is a brief discussion.

Vectors

The most common type of quantity is called a *scalar*. You have been working with scalar quantities most of your life. A scalar is a quantity that has magnitude only. Examples of scalar quantities are dc voltage, dc current, temperature, time, etc. Scalar quantities are usually symbolized by italic notation. For example, scalar voltage is V, scalar current is I, and so on. When you add scalar quantities, you use ordinary addition. For instance, in a series dc circuit with two resistors, the total voltage is given by

$$V_T = V_1 + V_2$$

A *vector* is a quantity that has magnitude and direction. Vectors are usually symbolized by boldface notation. For example, vector voltage is **V**, vector current is **I**, and so on. When you add vector quantities, you have to use a new kind of addition called *vector addition*. This type of addition takes the direction of the quantity into account as well as the magnitude.

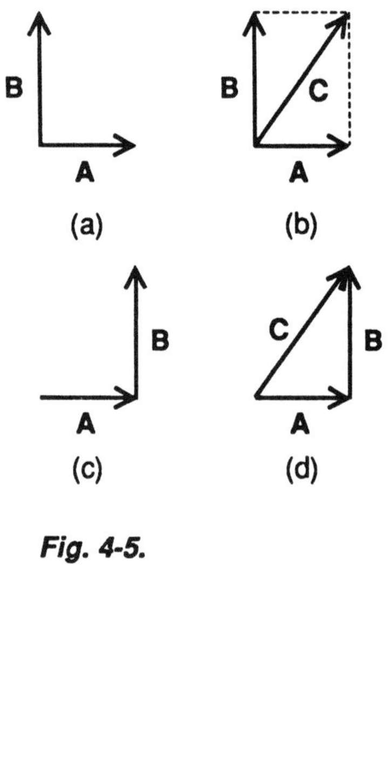

Fig. 4-5.

Vector Addition

Figure 4-5a shows two vectors, **A** and **B**. Notice that **B** leads **A** by 90°. Here is how you add the vectors. Form a parallelogram using the **A** and **B** sides as shown in Fig. 4-5b. Then, draw a new vector from the origin to the opposite vertex. This new vector is called the *resultant* vector. It represents the vector sum of **A** and **B**.

Figure 4-5c shows an alternative way to think about the vector sum. Here is how it works. To get the vector sum of **A** and **B**, move the tail of **B** on top of the head of **A** as shown. Then, draw a new vector from the tail of **A** to the head of **B** as indicated in Fig. 4-5d. The new vector is the vector sum of **A** and **B**.

The first method is called the *parallelogram* method (Fig. 4-5b), and the second method is the *tail-to-head* method (Fig. 4-5d). Both methods give the same resultant vector, so we can use either method.

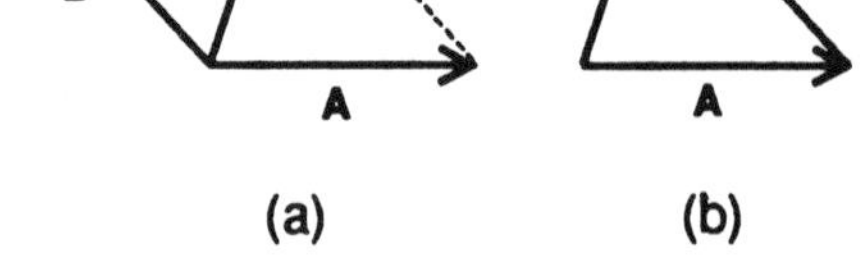

Fig. 4-6.

Any Angle

The most important case is two vectors with 90° between them. But the angle is not restricted to 90°. For instance, the angle is greater than 90° for the **A** and **B** vectors of Fig. 4-6a. When the parallelogram method is used, the resultant vector **C** appears as shown here. The same result can be obtained with the tail-to-head method shown in Fig. 4-6b.

Figure 4-7a shows two vectors with an angle of less than 90° between them. In this case, the parallelogram method produces a C vector whose magnitude is greater than the magnitude of A or B. Figure 4-7b shows the same result with the tail-to-head method.

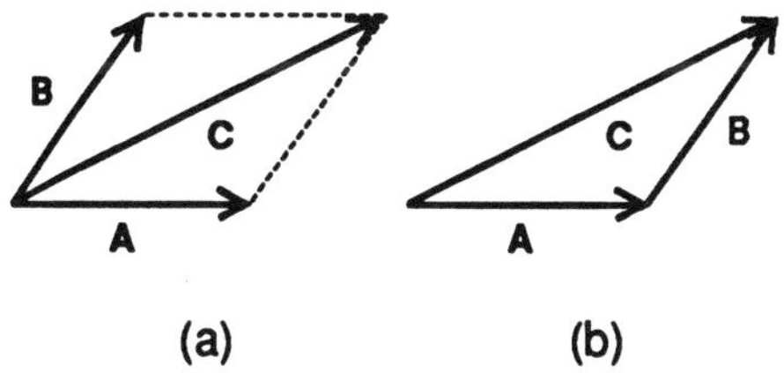

Fig. 4-7.

EXAMPLE 4-3

What is the vector sum in Fig. 4-8a?

SOLUTION

Here you see one vector pointing up and another pointing down. With the tail-to-head method, the resultant vector has a magnitude of 2 and points upward as shown in Fig. 4-8b.

In general, whenever the angle between two vectors is 180°, you can subtract the smaller from the larger to get the magnitude of the resultant vector. The direction of the resultant vector is the same as the larger vector.

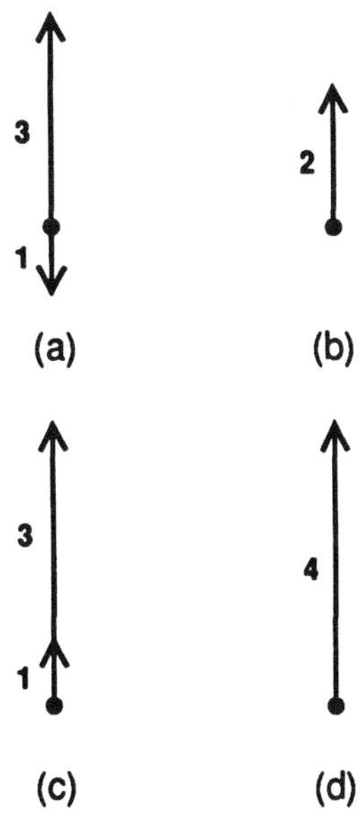

Fig. 4-8.

EXAMPLE 4-4

What is the vector sum in Fig. 4-8c?

SOLUTION

Here you see both vectors pointing up. With the tail-to-head method, the resultant vector has a magnitude of 4 and points upward as shown in Fig. 4-8d.

In general, whenever the angle between two vectors is 0°, you can add the magnitudes to get the magnitude of the resultant vector. The direction of the resultant vector is the same as either vector.

4-3. SERIES CIRCUITS

A phasor is a special kind of vector because it represents a sinusoidal quantity. Since they are vectors, the phasors of ac circuits must be added using the rules of vector addition. For instance, the phasor voltages in a series ac circuit may be added by the tail-to-head method to get the total voltage.

RL Circuit

Figure 4-9a shows a series *RL* circuit where an ac source is connected to a resistor and inductor in series. Because it is a series circuit, the same instantaneous current exists in the resistor and the inductor. This means a sinusoidal source voltage produces a sinusoidal current in all parts of the circuit. At any instant in time, the current has the same value in the resistor and the inductor. Stated another way, the current is in phase in both the resistor and inductor.

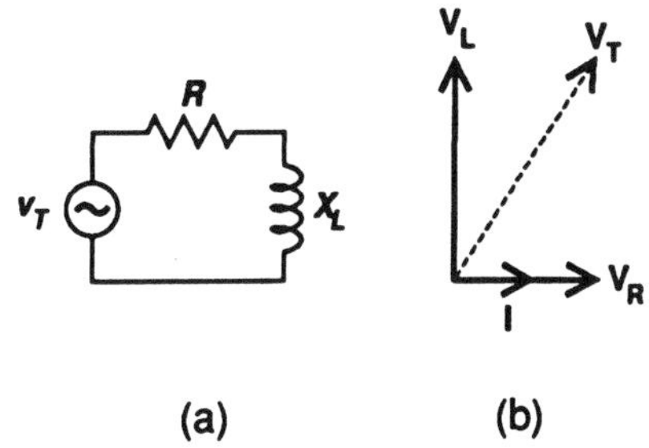

Fig. 4-9.

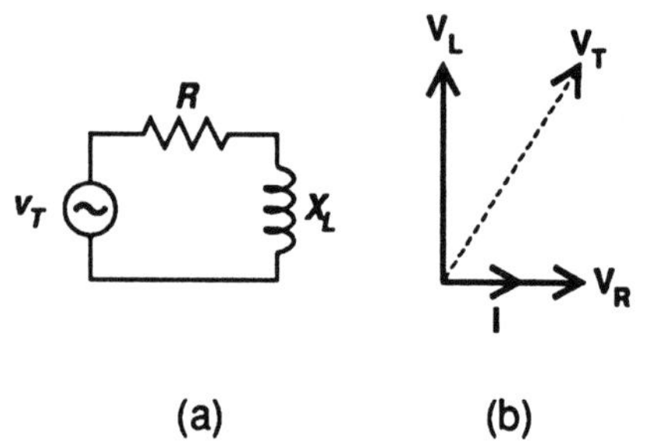

Fig. 4-9.

To simplify the analysis, we will assume the instantaneous current has a phase angle of 0°. Therefore, it is given by

$$i = I_p \sin \theta$$

Figure 4-9b shows the phasor current. As you see, it is a horizontal vector pointing to the right. This current is the same for the resistor and the inductor.

From earlier discussions, we know the voltage across the resistance is always in phase with the current through the resistance. This is why the phasor voltage across the resistor is drawn with a phase angle of 0° in Fig. 4-9b. On the other hand, the voltage across an inductance leads the current by 90°. This is why the phasor voltage across the inductance is shown with a phase angle of 90°.

Instantaneous Quantities

Kirchhoff's voltage law is always true for instantaneous voltages. In Fig. 4-9a, the sum of the instantaneous voltages is

$$v_T = V_R \sin \theta + V_L \cos \theta$$

which can be written as

$$v_T = V_R \sin 2\pi ft + V_L \cos 2\pi ft \qquad (4\text{-}3)$$

This is the equation to use if you need to calculate the total voltage at a particular instant.

Phasor Quantities

Most of the time, we don't need to know the instantaneous voltages in a circuit. All we usually need to know are the peak or rms voltages. In this case, it is lot easier and faster to use phasor voltages. Kirchhoff's voltage law is valid for phasor voltages, provided we use vector addition. When phasor voltages are used, the sum of the phasor voltages is

$$\mathbf{V_T} = \mathbf{V_R} + \mathbf{V_L} \qquad (4\text{-}4)$$

This says the total phasor voltage of the circuit equals the sum of the individual phasor voltages. In this equation, the plus sign stands for vector addition because we are dealing with vector quantities.

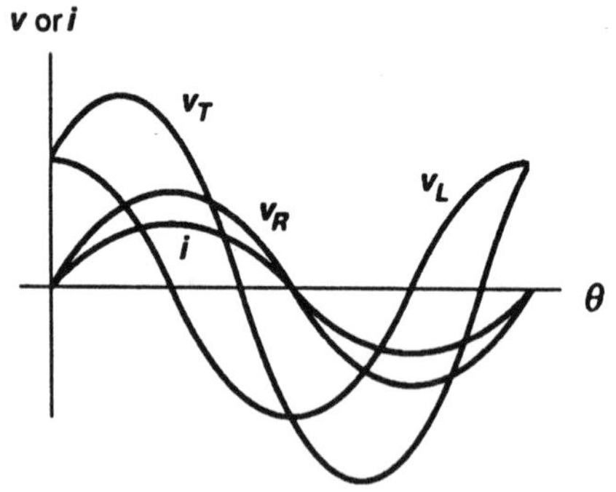

Fig. 4-10.

Why Phasor Quantities are Preferred

When you convert each phasor voltage of Fig. 4-9b into its equivalent sine wave, you get Fig. 4-10. Notice that the total voltage has a phase angle between 0° and 90°. Figure 4-10 shows the peak value and phase angle of each voltage. But Fig. 4-9b contains the same information. It is also a lot easier to draw Fig. 4-9b than Fig. 4-10. For this reason, most ac analysis uses phasor diagrams like Fig. 4-9b rather than complicated sinusoidal diagrams like Fig. 4-10.

From now on, we will emphasize phasors rather than sine waves. But remember this. Sine waves of voltage and current are what actually exist in time; they are what you will see on an oscilloscope. But sine waves are too hard to work with when analyzing circuits. Therefore, we use phasors in their place. These phasors do not exist except in our minds as mental equivalents of sine waves. Phasors are to sine waves as shorthand notation is to words and sentences.

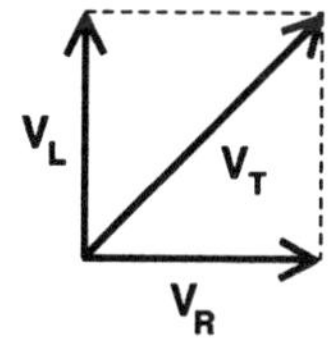

Fig. 4-11.

Pythagorean Theorem

When a resistance is in series with an inductance, the phasor voltages are 90° apart as shown in Fig. 4-11. Because of the right triangle, we can use the Pythagorean theorem to calculate the total voltage:

$$V_T = \sqrt{V_R^2 + V_L^2} \qquad (4\text{-}5)$$

For instance, if the peak voltage across the resistor is 3 V and the peak voltage across the inductance is 4 V, the total voltage is

$$V_T = \sqrt{(3\ \text{V})^2 + (4\ \text{V})^2} = 5\ \text{V}$$

Because rms voltages are directly proportional to peak voltages, we can also use rms voltages in Eq. (4-5). As an example, if the rms voltage across the resistor is 9 V and the rms voltage across the inductance is 14 V, the total voltage is

$$V_T = \sqrt{(9\ \text{V})^2 + (14\ \text{V})^2} = 16.6\ \text{V}$$

Quadrature Sum

QUADRATURE SUM

$$V_T = \sqrt{V_R^2 + V_L^2}$$

Equation (4-5) makes one thing clear. The total voltage in the *RL* circuit is not the ordinary sum of the two voltages. Instead, it is the *quadrature sum*, defined as the square root of the sum of the squares of the quantities being combined. You have to make an effort to remember this. Whenever you build a series *RL* circuit, the individual voltages you measure cannot be added to get the total voltage. Because the sinusoidal voltages are 90° apart, you have to use the Pythagorean theorem to calculate the total voltage.

EXAMPLE 4-5

In Fig. 4-12, you measure an rms voltage of 3.75 V across the resistor and an rms voltage of 2.86 V across the inductor. What is the total rms voltage?

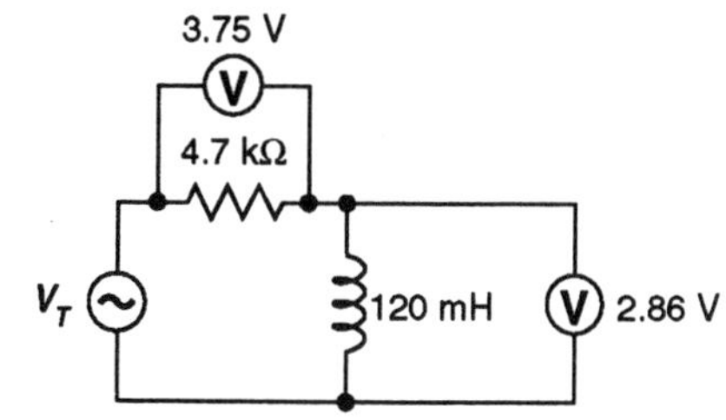

Fig. 4-12.

SOLUTION

At first, you may be tempted to add the voltages to get the total voltage. But you must use the quadrature sum, the square root of the sum of the squares:

$$V_T = \sqrt{(3.75\ \text{V})^2 + (2.86\ \text{V})^2} = 4.72\ \text{V}$$

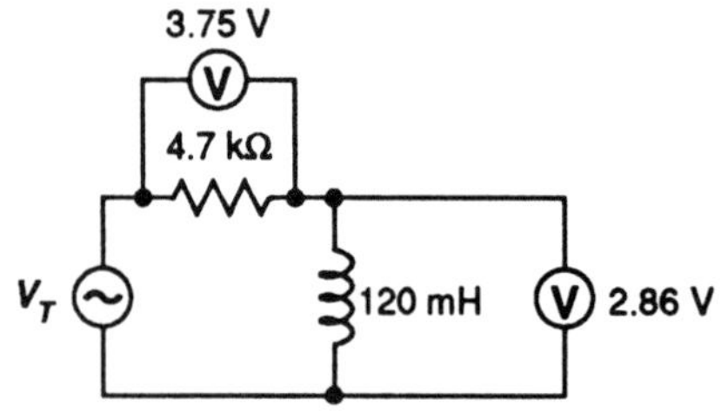

Fig. 4-12.

This is the total rms voltage. This is what you would measure if you connect an ac voltmeter across the voltage source. The reason this voltage is less than the sum is because the two voltages being measured are 90° out of phase.

EXAMPLE 4-6

In Fig. 4-13a, the rms current is 2 mA. Calculate the rms value of the applied voltage.

SOLUTION

The inductive reactance is

$$X_L = 2\pi fL = 2\pi(1\text{ kHz})(500\text{ mH}) = 3.14\text{ k}\Omega$$

The current is the same in all parts of a series circuit. Therefore, an rms current of 2 mA flows through the resistor and the inductor. The rms voltage across the resistance is

$$V_R = (2\text{ mA})(4.7\text{ k}\Omega) = 9.4\text{ V}$$

and the rms voltage across the inductance is

$$V_L = (2\text{ mA})(3.14\text{ k}\Omega) = 6.28\text{ V}$$

With the Pythagorean theorem, the total voltage is

$$V_T = \sqrt{(9.4\text{ V})^2 + (6.28\text{ V})^2} = 11.3\text{ V}$$

Figure 4-13b shows the phasor voltages.

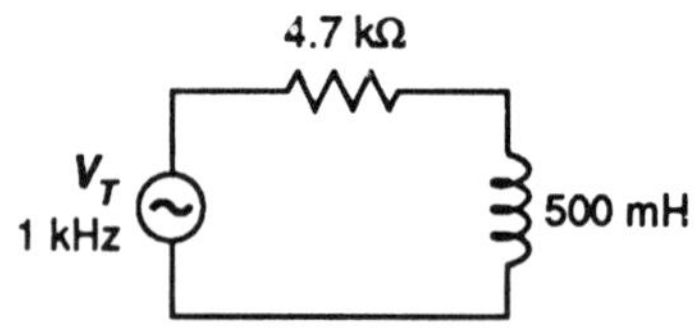

(a)

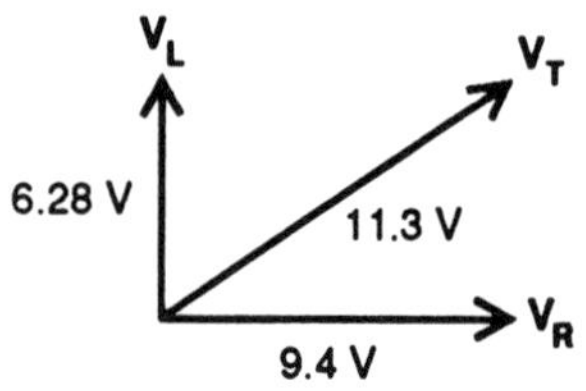

(b)

Fig. 4-13.

EXAMPLE 4-7

In Fig. 4-14a, the rms voltage across the inductance is 80 V. What is the resistance?

SOLUTION

With the Pythagorean theorem,

$$120\text{ V} = \sqrt{V_R^2 + (80\text{ V})^2}$$

Solve for V_R as follows. Square both sides to get

$$14400 = V_R^2 + 6400$$

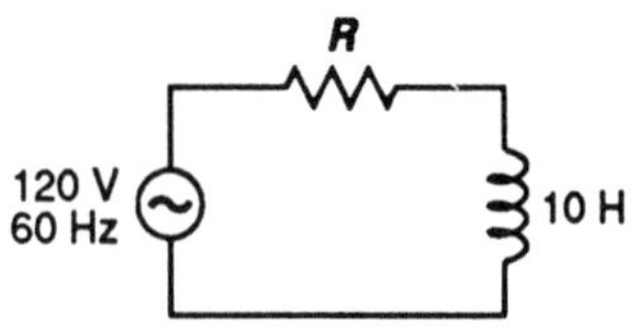

(a)

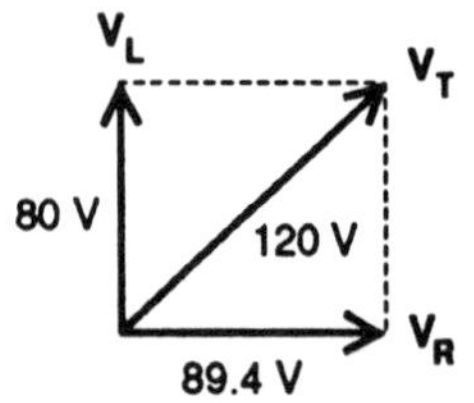

(b)

Fig. 4-14.

$V_R^2 = 14400 - 6400 = 8000$

Take the square root of both sides to get

$V_R = 89.4$ V

Figure 4-14b shows the phasor voltages.

The inductive reactance is

$X_L = 2\pi(60 \text{ Hz})(10 \text{ H}) = 3.77 \text{ k}\Omega$

The rms current through the inductance is

$$I = \frac{80 \text{ V}}{3.77 \text{ k}\Omega} = 21.2 \text{ mA}$$

Since we have the rms voltage across the resistance and the current through it, we can calculate the resistance:

$$R = \frac{89.4 \text{ V}}{21.2 \text{ mA}} = 4.22 \text{ k}\Omega$$

EXAMPLE 4-8

In Fig. 4-15a, the rms current is 4 mA. What is the inductance?

SOLUTION

The rms voltage across the resistance is

$V_R = (4 \text{ mA})(3 \text{ k}\Omega) = 12$ V

With the Pythagorean theorem,

$15 \text{ V} = \sqrt{(12 \text{ V})^2 + V_L^2}$

Solve for V_L as follows. Square both sides to get

$225 = 144 + V_L^2$

Rearrange to get

$V_L^2 = 225 - 144 = 81$

Take the square root of both sides to get

$V_L = 9$ V

Figure 4-15b shows the phasor voltages.

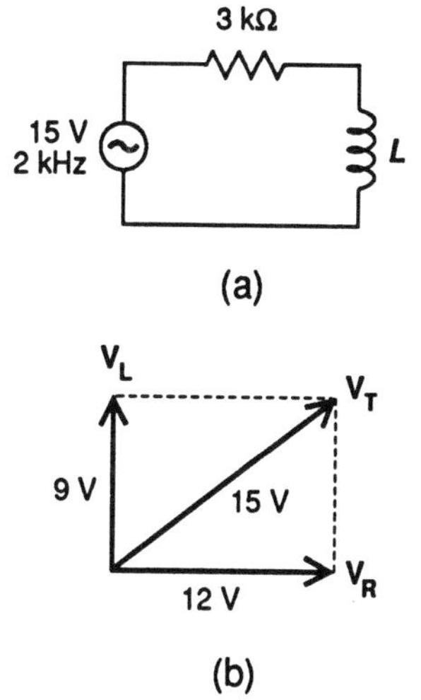

Fig. 4-15.

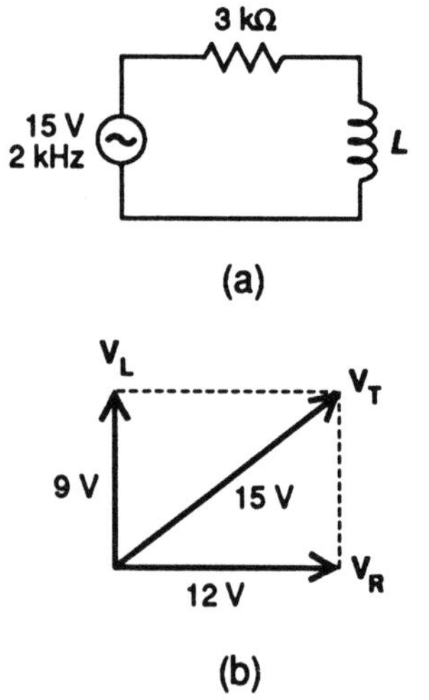

Fig. 4-15.

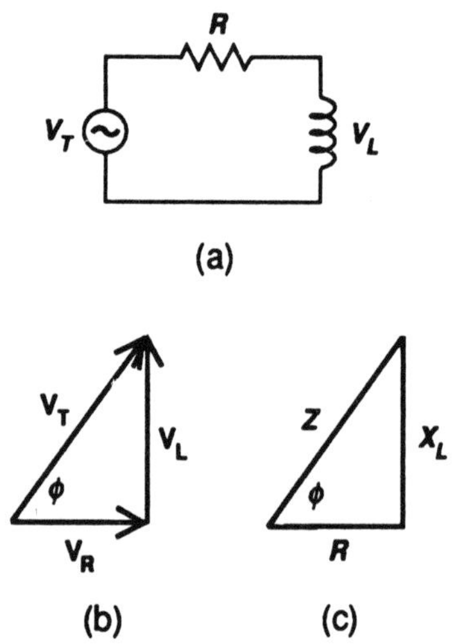

Fig. 4-16.

MEMORIZE

$$Z = \sqrt{R^2 + X_L^2}$$

Since we have the rms voltage across the inductance and the current through it, we can calculate the inductive reactance:

$$X_L = \frac{9\text{ V}}{4\text{ mA}} = 2.25\text{ k}\Omega$$

Now, solve for the inductance as follows:

$$L = \frac{X_L}{2\pi f} = \frac{2.25\text{ k}\Omega}{2\pi(2\text{ kHz})} = 179\text{ mH}$$

4-4. IMPEDANCE

Figure 4-16a shows a series *RL* circuit. The phasor voltages for the resistor and the inductor are 90° out of phase as shown in Fig. 4-16b. As you see, the inductive voltage leads the resistive voltage by 90°. The impedance of the circuit is defined as the total opposition to current. This opposition is a combined effect produced by the resistance and inductance acting together. Mathematically, the impedance is defined as

$$Z = \frac{V_T}{I}$$

This says the impedance equals the total voltage divided by the current.

Since the total voltage is the quadrature sum of the individual voltages, we can substitute to get

$$Z = \frac{\sqrt{V_R^2 + V_L^2}}{I}$$

Next, we can divide numerator and denominator by the current to get

$$Z = \sqrt{(V_R/I)^2 + (V_L/I)^2}$$

Since $R = V_R/I$ and $X_L = V_L/I$, the foregoing equation becomes

$$Z = \sqrt{R^2 + X_L^2} \qquad (4\text{-}6)$$

This says the impedance of the circuit equals the square root of the sum of the square of resistance and the square of inductive reactance.

Equation (4-6) is often visualized by drawing the impedance triangle shown in Fig. 4-16c. The impedance represents the total opposition to current, the combined effect of the resistance and inductive reactance. Because of this, we can calculate the current with this equation:

$$I = \frac{V_T}{Z} \qquad (4\text{-}7)$$

Given the total voltage, we can divide by the impedance to find the current. An alternative form of Eq. (4-7) is

$$Z = \frac{V_T}{I} \qquad (4\text{-}8)$$

You can use this to calculate the impedance when you know the total voltage and current.

EXAMPLE 4-9

What is the current in Fig. 4-17a? Draw the phasor diagram for the voltages.

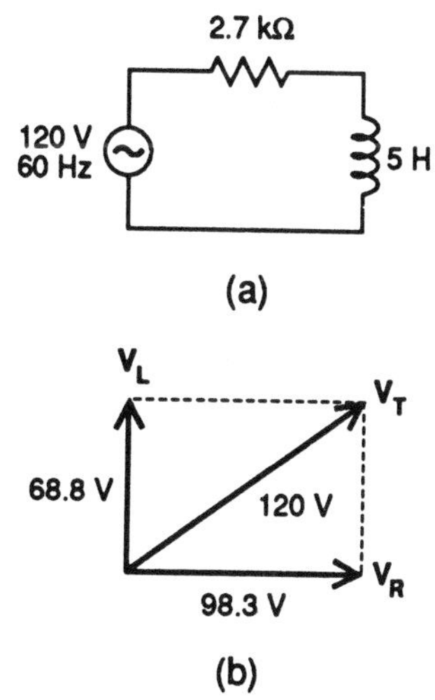

Fig. 4-17.

SOLUTION

Start by calculating the inductive reactance:

$$X_L = 2\pi(60\text{ Hz})(5\text{ H}) = 1.89\text{ k}\Omega$$

Next, calculate the impedance:

$$Z = \sqrt{(2.7\text{ k}\Omega)^2 + (1.89\text{ k}\Omega)^2} = 3.3\text{ k}\Omega$$

Now, calculate the current:

$$I = \frac{120\text{ V}}{3.3\text{ k}\Omega} = 36.4\text{ mA}$$

After you have the current, you can use Ohm's law to calculate the individual voltages as follows:

$$V_R = (36.4\text{ mA})(2.7\text{ k}\Omega) = 98.3\text{ V}$$

$$V_L = (36.4\text{ mA})(1.89\text{ k}\Omega) = 68.8\text{ V}$$

Figure 4-17b shows the phasor diagram.

EXAMPLE 4-10

What is the current in Fig. 4-18a? Draw the phasor diagram for the voltages.

SOLUTION

Start by calculating the inductive reactance:

$X_L = 2\pi(5\ \text{kHz})(120\ \text{mH}) = 3.77\ \text{k}\Omega$

Next, calculate the impedance:

$$Z = \sqrt{(4.7\ \text{k}\Omega)^2 + (3.77\ \text{k}\Omega)^2} = 6.03\ \text{k}\Omega$$

Next, calculate the current:

$$I = \frac{10\ \text{V}}{6.03\ \text{k}\Omega} = 1.66\ \text{mA}$$

The individual voltages are

$V_R = (1.66\ \text{mA})(4.7\ \text{k}\Omega) = 7.8\ \text{V}$

$V_L = (1.66\ \text{mA})(3.77\ \text{k}\Omega) = 6.26\ \text{V}$

Figure 4-18b shows the phasor diagram.

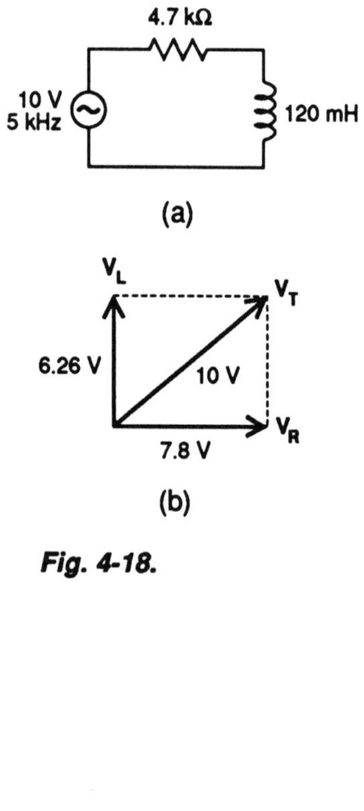

Fig. 4-18.

EXAMPLE 4-11

Draw the phasor diagram for Fig. 4-18a for each of these conditions:
a. Frequency decreases to 500 Hz
b. Frequency increases to 50 kHz

SOLUTION

a. When the frequency is 500 Hz, the inductive reactance is

$X_L = 2\pi(500\ \text{Hz})(120\ \text{mH}) = 377\ \Omega$

The impedance is

$$Z = \sqrt{(4.7\ \text{k}\Omega)^2 + (377\ \Omega)^2} = 4.72\ \text{k}\Omega$$

The remaining calculations are

$$I = \frac{10\ \text{V}}{4.72\ \text{k}\Omega} = 2.12\ \text{mA}$$

$V_R = (2.12\ \text{mA})(4.7\ \text{k}\Omega) = 9.96\ \text{V}$

$V_L = (2.12\ \text{mA})(377\ \Omega) = 0.8\ \text{V}$

Figure 4-19a shows the phasor diagram. Notice that the voltage across the resistance is almost equal to the total voltage.

b. When the frequency increases to 50 kHz, the inductive reactance is

$$X_L = 2\pi(50\text{ kHz})(120\text{ mH}) = 37.7\text{ k}\Omega$$

The impedance is

$$Z = \sqrt{(4.7\text{ k}\Omega)^2 + (37.7\text{ k}\Omega)^2} = 38\text{ k}\Omega$$

The remaining calculations are

$$I = \frac{10\text{ V}}{38\text{ k}\Omega} = 0.263\text{ mA}$$

$$V_R = (0.263\text{ mA})(4.7\text{ k}\Omega) = 1.24\text{ V}$$

$$V_L = (0.263\text{ mA})(37.7\text{ k}\Omega) = 9.92\text{ V}$$

Figure 4-19b shows the phasor diagram. Notice that the voltage across the inductance is almost equal to the total voltage.

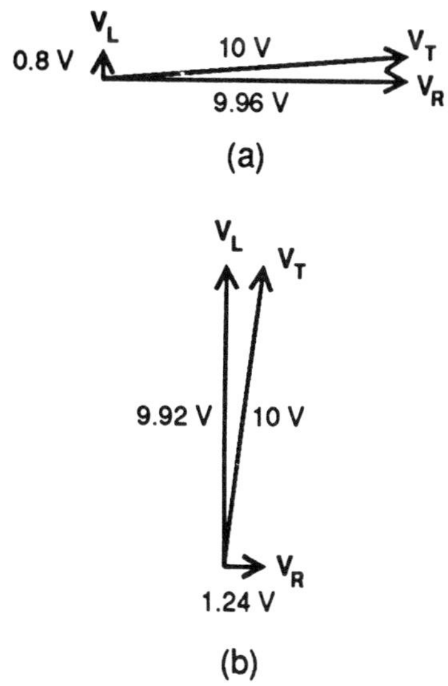

Fig. 4-19.

4-5. ANGLE OF THE IMPEDANCE

The preceding section showed you how to calculate the impedance of a series *RL* circuit. It is also important to be able to calculate the angle between the total voltage and current.

Finding the Angle

Figure 4-20a shows the phasor voltages for a series *RL* circuit. Figure 4-20b shows the impedance triangle for the same circuit. In either case, notice the angle between the horizontal side and the hypotenuse. The tangent of this angle equals the ratio of X_L to R:

$$\tan \phi = \frac{X_L}{R} \tag{4-9}$$

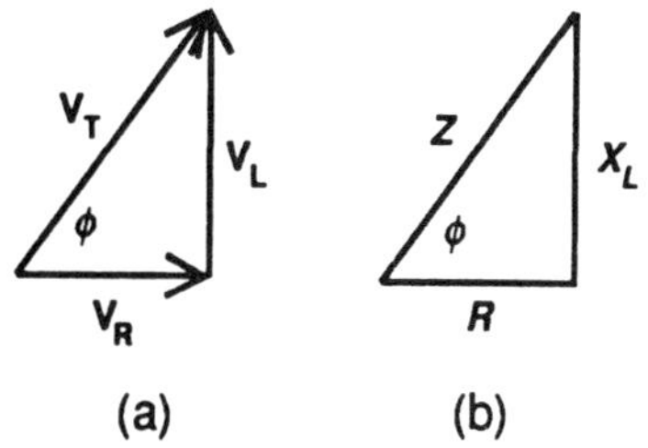

Fig. 4-20.

The angle is called the angle of the impedance. It may vary between 0° and 90°. After you calculate the tangent of this angle with Eq. (4-9), you can look up the angle in a table of tangents or you can use a calculator to get the angle.

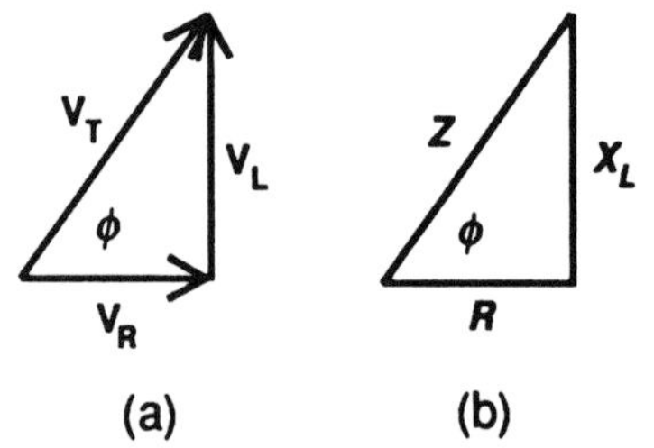

Fig. 4-20.

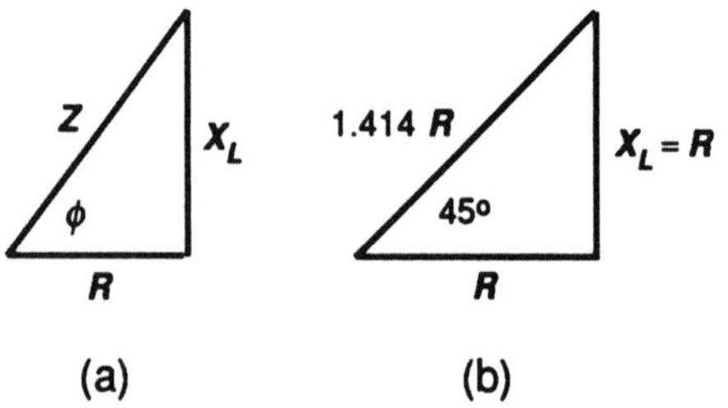

Fig. 4-21.

The tangent of the angle can also be calculated with

$$\tan \phi = \frac{V_L}{V_R} \qquad (4\text{-}10)$$

Figure 4-20a shows that the angle of impedance represents the angle between the total voltage and the current.

Critical Frequency

The impedance triangle of Fig. 4-21a will change when the frequency changes. When the frequency is very low, the angle is near 0°. On the other hand, when the frequency is very high, the angle approaches 90°.

There is an important frequency called the *critical* frequency, symbolized f_c. This is the frequency where X_L equals R. At this frequency,

$$X_L = 2\pi f_c L = R$$

Solving for the critical frequency gives

$$f_c = \frac{R}{2\pi L} \qquad (4\text{-}11)$$

Figure 4-21b shows the impedance triangle for the critical frequency. In this case, the impedance is

$$Z = \sqrt{R^2 + X_L^2} = \sqrt{R^2 + R^2} = \sqrt{2R^2} = 1.414R$$

Also, the angle of impedance is

$$\tan \phi = \frac{X_L}{R} = \frac{R}{R} = 1$$

or

$$\phi = 45°$$

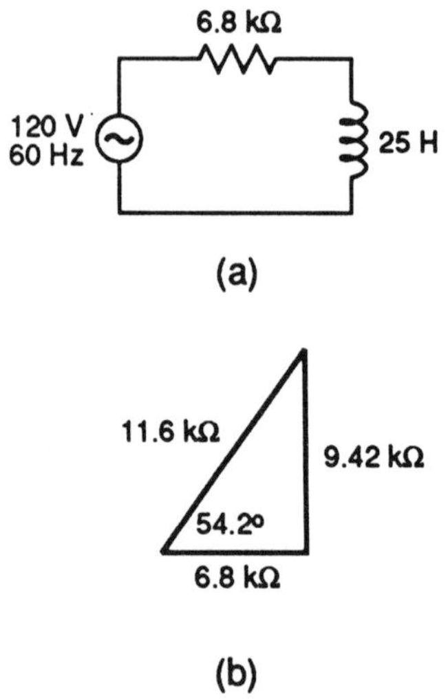

Fig. 4-22.

EXAMPLE 4-12

What is the angle of the impedance in Fig. 4-22a? Draw the impedance triangle.

SOLUTION

Start by calculating the inductive reactance:

$$X_L = 2\pi(60\text{ Hz})(25\text{ H}) = 9.42\text{ k}\Omega$$

Next, get the impedance:

$$Z = \sqrt{(6.8\ k\Omega)^2 + (9.42\ k\Omega)^2} = 11.6\ k\Omega$$

The tangent of the angle is

$$\tan \phi = \frac{9.42\ k\Omega}{6.8\ k\Omega} = 1.39$$

Solving for ϕ, we get

$$\phi = 54.2^\circ$$

Figure 4-22b shows the impedance triangle.

EXAMPLE 4-13

Draw the phasor diagram for Fig. 4-22a.

SOLUTION

In the preceding example, we calculated the impedance and the angle:

$$Z = 11.6\ k\Omega$$

$$\phi = 54.2^\circ$$

Now, calculate the current in Fig. 4-22a:

$$I = \frac{120\ V}{11.6\ k\Omega} = 10.3\ mA$$

This current flows through the resistor and inductor to produce voltages of

$$V_R = (10.3\ mA)(6.8\ k\Omega) = 70\ V$$

$$V_L = (10.3\ mA)(9.42\ k\Omega) = 97\ V$$

Figure 4-23a shows the phasor diagram of the voltages. Since the current is in phase with the resistor voltage, Fig. 4-23b shows the phasor diagram of total voltage and current. As you can see, the total voltage leads the current by 54.2°.

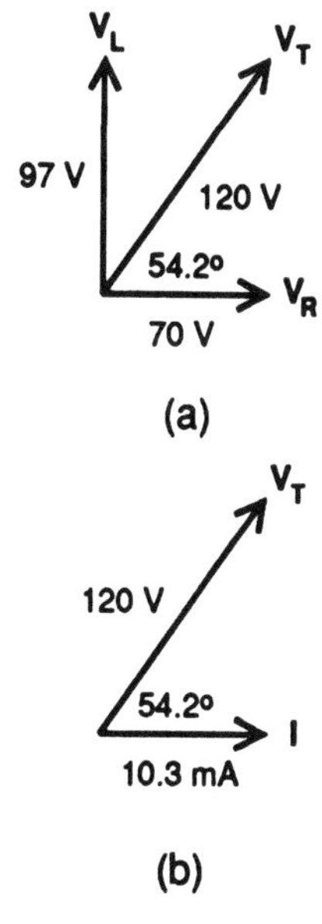

Fig. 4-23.

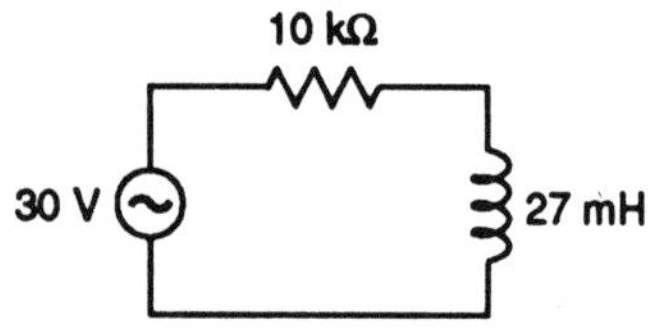

Fig. 4-24.

EXAMPLE 4-14

What is the critical frequency in Fig. 4-24?

SOLUTION

At the critical frequency, the inductive reactance equals the resistance, which means

$$2\pi f_c(27\ \text{mH}) = 10\ \text{k}\Omega$$

Solving for the critical frequency, we get

$$f_c = \frac{10\ \text{k}\Omega}{2\pi(27\ \text{mH})} = 58.9\ \text{kHz}$$

At this frequency, the inductive reactance equals the resistance. You can check the work by calculating the inductive reactance:

$$X_L = 2\pi(58.9\ \text{kHz})(27\ \text{mH}) = 10\ \text{k}\Omega$$

As you see, the inductive reactance equals 10 kΩ, the same value as the resistance. If the frequency were less than 58.9 kHz, the inductive reactance would be less than 10 kΩ. If the frequency were more than 58.9 kHz, the inductive reactance would be greater than 10 kΩ.

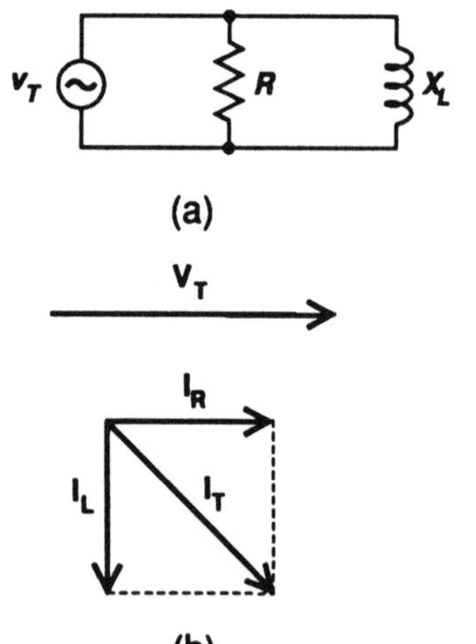

Fig. 4-25.

4-6. PARALLEL CIRCUITS

With parallel circuits, the applied voltage has the same value across each branch. But the branch currents are usually different, and their phasor sum equals the total phasor current. Again, phasors are used because they are easier to work with than sine waves.

Figure 4-25a shows an ac source across a resistor and inductor in parallel. Because of the parallel connection, the same instantaneous voltage appears across the resistor and the inductor simultaneously. Visualize the applied voltage as a sine wave. Then, the corresponding phasor voltage V_T is horizontal as shown in Fig. 4-25b.

Since the voltage and current are in phase in a resistor, the phasor current I_R in the resistor has an angle of 0° as shown in Fig. 4-25b. Since the voltage leads the current by 90° in an ideal inductor, the phasor current I_L has an angle of −90° as shown. The total phasor current is the phasor sum of the individual phasor currents. As before, the Pythagorean theorem gives the length of the hypotenuse, equivalent to the magnitude of the total phasor current.

When analyzing parallel *RL* circuits, the magnitude of each phasor is either an rms value or a peak value, whichever is convenient in the particular circuit you are analyzing. If you use rms values for the phasors, then all phasors represent rms values. If you use peak values, then all are peak values.

Notice how parallel circuits differ from series circuits. In a series circuit, the current is constant in all parts of the circuit, but the voltages are usually different. Furthermore, you use phasor addition of the individual voltages to find the total voltage. In a parallel circuit, the situation is reversed. The voltage is constant across all branches, and the branch currents are usually different. In this case, you use phasor addition of the currents to find the total current.

PARALLEL CIRCUIT

$$I_T = \sqrt{I_R^2 + I_L^2}$$

Here are the equations used in analyzing parallel circũits. The current in the resistive branch is

$$I_R = \frac{V_T}{R} \qquad (4\text{-}12)$$

The current in the inductive branch is

$$I_L = \frac{V_T}{X_L} \qquad (4\text{-}13)$$

The formula for the total current is

$$I_T = \sqrt{I_R^2 + I_L^2} \qquad (4\text{-}14)$$

Again, the quadrature sum refers to taking the square root of the sum of the squares of individual quantities to find the total quantity.

EXAMPLE 4-15

In Fig. 4-26a, the rms voltage is 8 V. What is the total rms current?

SOLUTION

The inductive reactance is

$$X_L = 2\pi fL = 2\pi(1\text{ kHz})(500\text{ mH}) = 3.14\text{ k}\Omega$$

The current through the resistive branch is

$$I_R = \frac{8\text{ V}}{4.7\text{ k}\Omega} = 1.7\text{ mA}$$

The current through the inductive branch is

$$I_L = \frac{8\text{ V}}{3.14\text{ k}\Omega} = 2.55\text{ mA}$$

With the Pythagorean theorem, the total current is

$$I_T = \sqrt{(1.7\text{ mA})^2 + (2.55\text{ mA})^2} = 3.06\text{ mA}$$

Figure 4-26b shows the phasor currents.

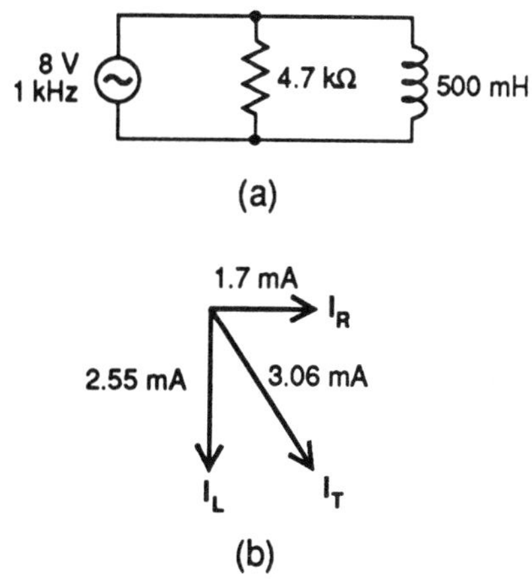

Fig. 4-26.

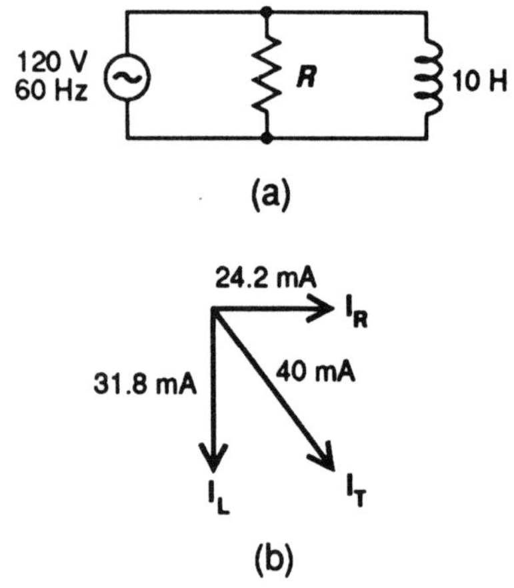

Fig. 4-27.

EXAMPLE 4-16

In Fig. 4-27a, the total rms current is 40 mA. What is the resistance?

SOLUTION

The inductive reactance is

$$X_L = 2\pi(60 \text{ Hz})(10 \text{ H}) = 3.77 \text{ k}\Omega$$

The rms current through the inductance is

$$I_L = \frac{120 \text{ V}}{3.77 \text{ k}\Omega} = 31.8 \text{ mA}$$

With the Pythagorean theorem,

$$(40 \text{ mA})^2 = \sqrt{I_R^2 + (31.8 \text{ mA})^2}$$

Solve for I_R to get

$$I_R = 24.2 \text{ mA}$$

Figure 4-27b shows the phasor currents.

Since we have the rms voltage across the resistance and the current through it, we can calculate the resistance:

$$R = \frac{120 \text{ V}}{24.2 \text{ mA}} = 4.96 \text{ k}\Omega$$

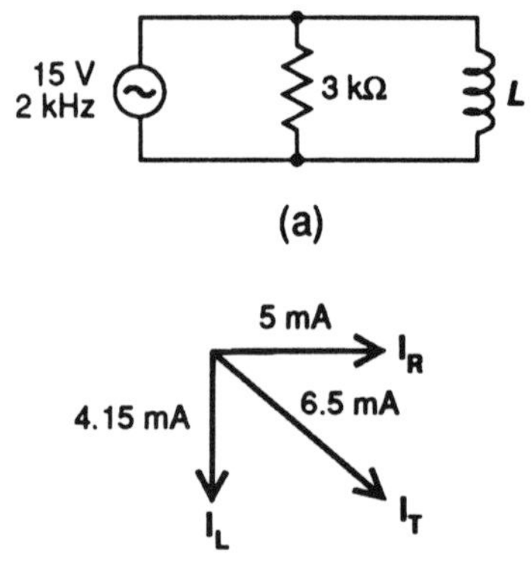

Fig. 4-28.

EXAMPLE 4-17

In Fig. 4-28a, the total rms current is 6.5 mA. What is the inductance?

SOLUTION

The rms current through the resistance is

$$I_R = \frac{15 \text{ V}}{3 \text{ k}\Omega} = 5 \text{ mA}$$

With the Pythagorean theorem,

$$(6.5 \text{ mA})^2 = \sqrt{(5 \text{ mA})^2 + I_L^2}$$

Solve for I_L to get

$I_L = 4.15$ mA

Figure 4-28b shows the phasor currents.

Since we have the rms voltage across the inductance and the current through it, we can calculate the inductive reactance:

$$X_L = \frac{15\ \text{V}}{4.15\ \text{mA}} = 3.61\ \text{k}\Omega$$

Now, solve for the inductance as follows:

$$L = \frac{X_L}{2\pi f} = \frac{3.61\ \text{k}\Omega}{2\pi(2\ \text{kHz})} = 287\ \text{mH}$$

4-7. PARALLEL IMPEDANCE AND ANGLE

The impedance of a parallel circuit equals the total voltage divided by the total current. The formula for impedance is

$$Z = \frac{V_T}{I_T} \qquad (4\text{-}15)$$

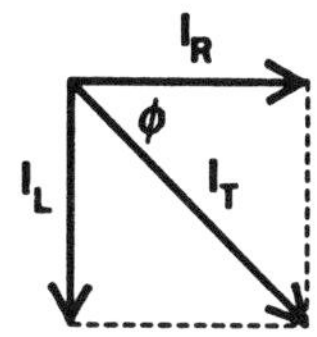

Fig. 4-29.

where the voltage and current are expressed in rms or peak values. The easiest way to find the impedance of a parallel *RL* circuit is to calculate each branch current. Then, apply the Pythagorean theorem to find the total current (Fig. 4-29). After you find the total current, you can calculate the impedance of the circuit with Eq. (4-15).

In Fig. 4-29, the total current lags the resistive current by an angle of *phi*. The tangent of this angle equals the opposite side over the adjacent side:

$$\tan\phi = \frac{I_L}{I_R} \qquad (4\text{-}16)$$

Because $I_L = V/X_L$ and $I_R = V/R$, the foregoing equation can be rewritten as

$$\tan\phi = \frac{R}{X_L} \qquad (4\text{-}17)$$

When you know the values of I_L and I_R, or the values of R and X_L, you can calculate the tangent of the angle. Then, you can use trig tables or a calculator to find the angle.

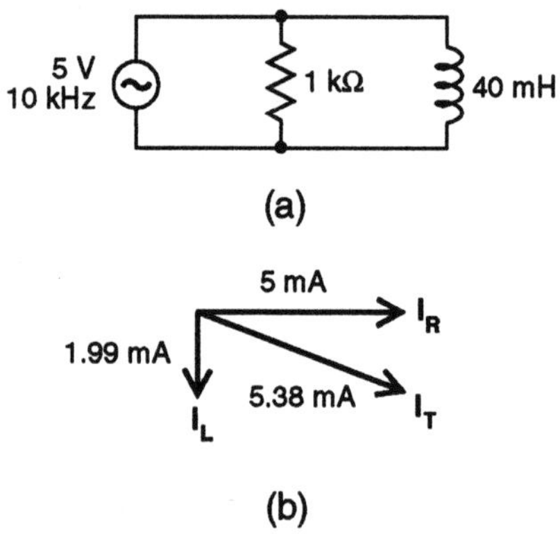

Fig. 4-30.

EXAMPLE 4-18

What is the impedance and angle in Fig. 4-30?

SOLUTION

The approach is to calculate each branch current, the total current, the impedance, and the angle. Here is how it looks. The current through the resistor equals the total voltage divided by the resistance:

$$I_R = \frac{5\text{ V}}{1\text{ k}\Omega} = 5\text{ mA}$$

The inductive reactance is

$$X_L = 2\pi(10\text{ kHz})(40\text{ mH}) = 2.51\text{ k}\Omega$$

The current through the inductive reactance is the total voltage divided by the inductive reactance:

$$I_L = \frac{5\text{ V}}{2.51\text{ k}\Omega} = 1.99\text{ mA}$$

With the Pythagorean theorem, the quadrature sum is

$$I_T = \sqrt{(5\text{ mA})^2 + (1.99\text{ mA})^2} = 5.38\text{ mA}$$

Figure 4-30b shows the phasor currents.

Now, divide the total voltage by the total current to get the impedance:

$$Z = \frac{5\text{ V}}{5.38\text{ mA}} = 929\ \Omega$$

The tangent of the angle equals the inductive current divided by the resistive current:

$$\tan\phi = \frac{1.99\text{ mA}}{5\text{ mA}} = 0.398$$

Solving for the angle, we get

$$\phi = 21.7^\circ$$

We can get the same angle as follows. The tangent of the angle equals the resistance divided by the inductive reactance:

$$\tan \phi = \frac{1\ \text{k}\Omega}{2.51\ \text{k}\Omega} = 0.398$$

which gives the same angle, 21.7°. Therefore, we can find the angle by either method.

In summary, we have calculated an impedance of 929 Ω and an angle of 21.7°. In Fig. 4-30b, we can see that the total current lags the resistive current by 21.7°. An equivalent statement is that the resistive current leads the total current by 21.7°.

4-8. MORE ABOUT PARALLEL IMPEDANCE

The total impedance of a parallel *RL* circuit equals the total voltage divided by the total current. As shown in the preceding section, we can calculate each branch current, the total current, and the impedance. This is one way to find the impedance.

Finding Impedance with Formulas

There is another way to calculate the impedance of a parallel *RL* circuit. The total impedance is given by

$$Z = \frac{V_T}{I_T}$$

We can substitute the quadrature sum for the total current to get

$$Z = \frac{V_T}{\sqrt{I_R^2 + I_L^2}}$$

Now, divide the numerator and the denominator by V_T to get

$$Z = \frac{1}{\sqrt{(I_R/V_T)^2 + (I_L/V_T)^2}}$$

Since $R = V_T/I_R$ and $X_L = V_T/I_L$, we can rewrite the foregoing equation as

$$Z = \frac{1}{\sqrt{(1/R)^2 + (1/X_L)^2}} \qquad (4\text{-}18)$$

This equation allows us to calculate the impedance of a parallel *RL* circuit. The equation is not as difficult as it first appears. With a scientific calculator, you can easily work out the impedance of a parallel circuit.

PRODUCT OVER QUADRATURE SUM

$$Z = \frac{RX_L}{\sqrt{R^2 + X_L^2}}$$

By applying some algebra to Eq. (4-18), we can expand the denominator and rearrange the equation to get this equivalent form:

$$Z = \frac{RX_L}{\sqrt{(R^2 + X_L^2}} \tag{4-19}$$

This is easy to remember. Why? Because it is similar to the product-over-sum rule you learned in dc circuits. The equation says the impedance equals the product over the quadrature sum. You can use this equation instead of Eq. (4-18) if you prefer.

Equation (4-18) is faster on a calculator, but you may not remember it. In this case, you can fall back on Eq. (4-19). Equation (4-19) is slower on a calculator, but the chances are that you will remember it because it is the product over the quadrature sum.

Calculating the Branch Currents

After you calculate the impedance with Eq. (4-18) or (4-19), you can calculate the total current with

$$I_T = \frac{V_T}{Z} \tag{4-20}$$

One way to get the branch currents is like this. The resistive current equals the total voltage divided by the resistance:

$$I_R = \frac{V_T}{R}$$

The inductive current equals the total voltage divided by the inductive reactance:

$$I_L = \frac{V_T}{X_L}$$

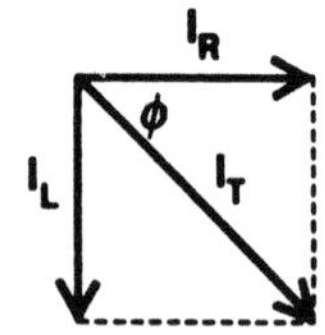

Fig. 4-31.

But there is another way to find the branch currents. In Fig. 4-31, we can use trigonometry to calculate the resistive and inductive currents. Since the cosine of the angle is given by the adjacent side over the hypotenuse, we can calculate the resistive current with

$$I_R = I_T \cos \phi \tag{4-21}$$

Since the sine of the angle is given by the opposite side over the hypotenuse, we can calculate the inductive current with

$$I_L = I_T \sin \phi \tag{4-22}$$

EXAMPLE 4-19

In Example 4-18, we calculated an impedance of 929 Ω. Recalculate the impedance using the new formulas.

SOLUTION

In Example 4-18, $R = 1$ kΩ and $X_L = 2.51$ kΩ. With Eq. (4-18),

$$Z = \frac{1}{\sqrt{(1/1\ \text{k}\Omega)^2 + (1/2.51\ \text{k}\Omega)^2}} = 929\ \Omega$$

With Eq. (14-19), the product over the quadrature sum gives

$$Z = \frac{(1\ \text{k}\Omega)(2.51\ \text{k}\Omega)}{\sqrt{(1\ \text{k}\Omega)^2 + (2.51\ \text{k}\Omega)^2}} = 929\ \Omega$$

You can see that we get the same answer using the formulas as we did by direct calculations in Example 4-18.

EXAMPLE 4-20

What is the total current in Fig. 4-32? The branch currents?

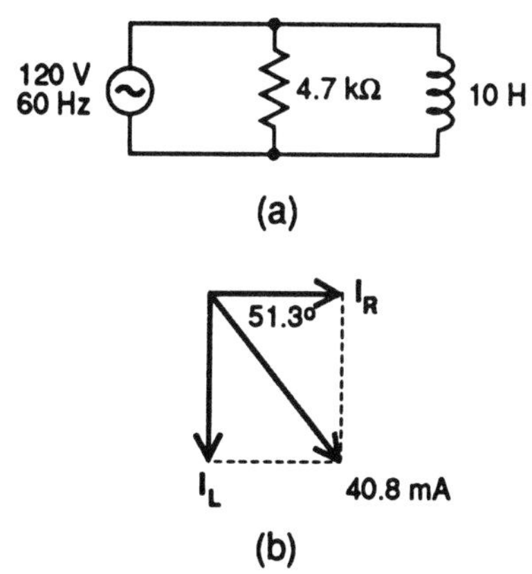

Fig. 4-32.

SOLUTION

First, get the inductive reactance:

$$X_L = 2\pi(60\ \text{Hz})(10\ \text{H}) = 3.77\ \text{k}\Omega$$

Next, calculate the impedance with Eq. (4-18):

$$Z = \frac{1}{\sqrt{(1/4.7\ \text{k}\Omega)^2 + (1/3.77\ \text{k}\Omega)^2}} = 2.94\ \text{k}\Omega$$

The total current equals the total voltage divided by the impedance:

$$I_T = \frac{120\ \text{V}}{2.94\ \text{k}\Omega} = 40.8\ \text{mA}$$

The tangent of the angle equals the resistance divided by the inductive reactance:

$$\tan\phi = \frac{4.7\ \text{k}\Omega}{3.77\ \text{k}\Omega} = 1.25$$

which gives an angle of

$$\phi = 51.3^\circ$$

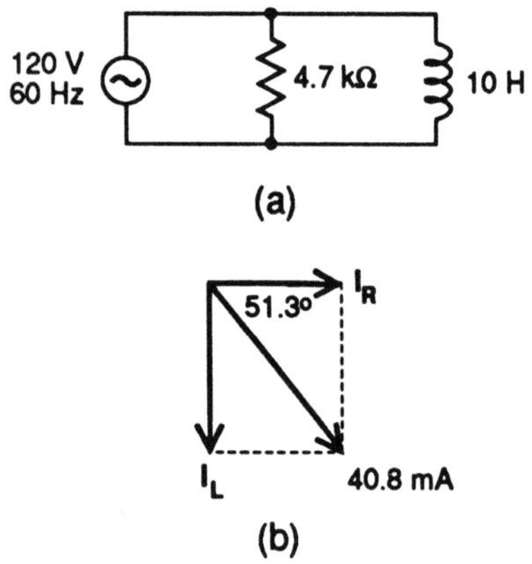

Fig. 4-32.

Figure 4-32b shows the phasor currents. Here you see that the total current lags the resistive current by 51.3°.

Now, we can use trigonometry to calculate the branch currents. With Fig. 4-32b,

$$I_R = (40.8 \text{ mA}) \cos 51.3° = 25.5 \text{ mA}$$

and

$$I_L = (40.8 \text{ mA}) \sin 51.3° = 31.8 \text{ mA}$$

4-9. THE *Q* OF INDUCTIVE CIRCUITS

The figure of merit, symbolized by *Q*, tells you how inductive an *RL* circuit is. When *Q* is much greater than 1, the circuit is highly inductive. When *Q* is much less than 1, the circuit is highly resistive. Let's find out what this means.

Q of Series *RL* Circuits

The *Q* of a series *RL* circuit is defined as the ratio of inductive voltage to resistive voltage:

$$Q = \frac{V_L}{V_R} \tag{4-23}$$

When the inductive voltage equals the resistive voltage, *Q* equals 1. When the inductive voltage is larger than the resistive voltage, *Q* is greater than 1. When the inductive voltage is less than the resistive voltage, *Q* is less than 1. Because of this, *Q* indicates whether is a circuit is primarily inductive or resistive. When *Q* is greater than 1, the circuit is primarily inductive because more voltage is dropped across the inductor. When *Q* is less than 1, the circuit is primarily resistive because more voltage is dropped across the resistor.

Since the current is the same throughout a series circuit, Eq. (4-23) can be rewritten as

$$Q = \frac{V_L}{V_R} = \frac{IX_L}{IR}$$

$$Q = \frac{X_L}{R} \tag{4-24}$$

SERIES CIRCUIT

$$Q = \frac{X_L}{R}$$

This says the *Q* of a series *RL* circuit equals the inductive reactance divided by the resistance.

In fact, the Q of a series RL circuit is equal to the tangent of the angle. You can see this by looking at Fig. 4-33. In symbols,

$$Q = \tan \phi \qquad (4\text{-}25)$$

When the angle is 45°, the Q of the circuit is 1. When the angle is greater than 45°, the Q of the circuit is greater than 1 and the circuit is more inductive than resistive. When the angle is less than 45°, the Q of the circuit is less than 1 and the circuit is more resistive than inductive.

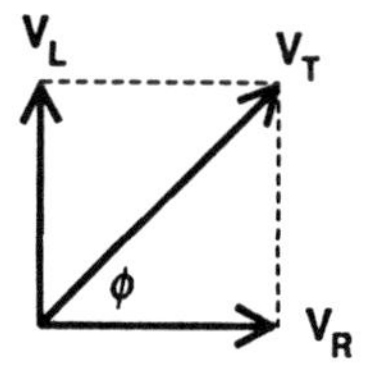

Fig. 4-33.

Q of Parallel RL Circuits

The Q of a parallel RL circuit is defined as the ratio of inductive current to resistive current:

$$Q = \frac{I_L}{I_R} \qquad (4\text{-}26)$$

When the inductive current equals the resistive current, Q equals 1. When the inductive current is larger than the resistive current, Q is greater than 1. When the inductive current is less than the resistive current, Q is less than 1. When Q is greater than 1, the circuit is primarily inductive because more current flows the inductor. When Q is less than 1, the circuit is primarily resistive because more current flows through the resistor.

Since the voltage is the same across the branches of a parallel circuit, Eq. (4-26) can be rewritten as

$$Q = \frac{I_L}{I_R} = \frac{V_T/X_L}{V_T/R}$$

or

$$Q = \frac{R}{X_L} \qquad (4\text{-}27)$$

PARALLEL CIRCUIT

$$Q = \frac{R}{X_L}$$

This says the Q of a parallel RL circuit equals the resistance divided by inductive reactance. This is exactly the opposite to the Q of a series RL circuit. In a parallel RL circuit, the Q increases when the resistance increases or when the inductive reactance decreases.

Again, the Q of a series RL circuit is equal to the tangent of the angle. You can see this by looking at Fig. 4-34. In symbols,

$$Q = \tan \phi \qquad (4\text{-}28)$$

When the angle is 45°, the Q of the circuit is 1. When the angle is greater than 45°, the Q of the circuit is greater than 1 and the circuit is more inductive than resistive. When the angle is less than 45°, the Q of the circuit is less than 1 and the circuit is more resistive than inductive.

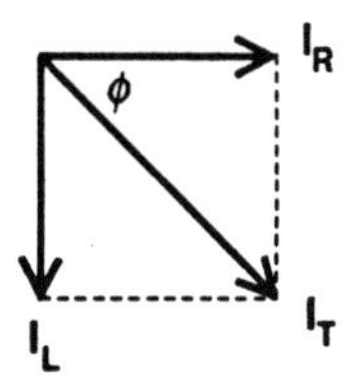

Fig. 4-34.

EXAMPLE 4-21

A series *RL* circuit has $R = 6.8\ \text{k}\Omega$ and $X_L = 15\ \text{k}\Omega$. What is the Q of the circuit?

SOLUTION

Since this is a series circuit, Q equals the inductive reactance divided by the resistance:

$$Q = \frac{15\ \text{k}\Omega}{6.8\ \text{k}\Omega} = 2.2$$

EXAMPLE 4-22

A parallel *RL* circuit has $R = 10\ \text{k}\Omega$ and $X_L = 200\ \Omega$. What is the Q of the circuit?

SOLUTION

Since this is a parallel circuit, Q equals the resistance divided by the inductive reactance:

$$Q = \frac{10\ \text{k}\Omega}{200\ \Omega} = 50$$

EXAMPLE 4-23

The circuit of Fig. 4-35 has a Q of 10. What is the inductance?

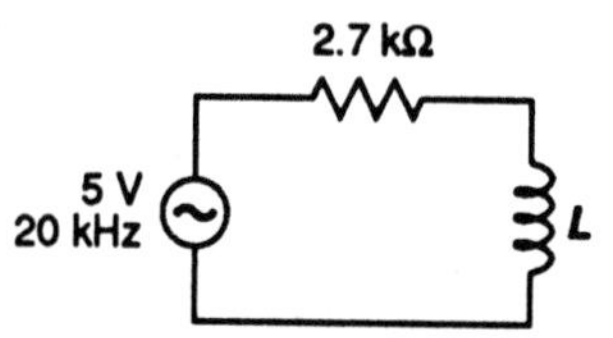

Fig. 4-35.

SOLUTION

This is a series circuit. Substitute into the Q equation to get

$$10 = \frac{X_L}{2.7\ \text{k}\Omega}$$

Solve for the inductive reactance to get

$$X_L = 10(2.7\ \text{k}\Omega) = 27\ \text{k}\Omega$$

Now, solve the inductance:

$$L = \frac{X_L}{2\pi f} = \frac{27\ \text{k}\Omega}{2\pi(20\ \text{kHz})} = 215\ \text{mH}$$

EXAMPLE 4-24

The circuit of Fig. 4-36 has a Q of 80. What is the inductance?

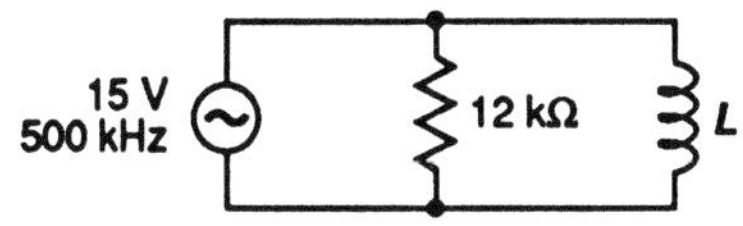

Fig. 4-36.

SOLUTION

This is a parallel circuit. Substitute into the Q equation to get

$$80 = \frac{12\ \text{k}\Omega}{X_L}$$

Solve for the inductive reactance to get

$$X_L = \frac{12\ \text{k}\Omega}{80} = 150\ \Omega$$

Now, solve the inductance:

$$L = \frac{X_L}{2\pi f} = \frac{150\ \Omega}{2\pi(500\ \text{kHz})} = 47.7\ \mu\text{H}$$

MULTIPLE-CHOICE TEST

The following questions may have more than one right answer. Select the best answer. This is the one that is always true, or covers more situations, or agrees with the material of this chapter, etc.

1. In a resistive circuit, the voltage and current are
   a. equal
   b. in phase
   c. 90° out of phase
   d. 180° out of phase

2. In a series *RL* circuit the total voltage times the cosine of the angle is the
   a. resistive current
   b. inductive current
   c. resistive voltage
   d. inductive voltage

3. In an inductive circuit, the current
   a. leads the voltage
   b. lags the voltage
   c. is in phase with the voltage
   d. none of the above

4. Quantities that have both magnitude and direction are called
   a. scalars
   b. vectors
   c. currents
   d. voltages

5. A phasor is equivalent to
   a. a scalar
   b. a sinusoidal quantity
   c. a phase angle
   d. a quadrature sum

6. When two vectors are separated by 90°, the resultant vector has a length that is given by the
   a. algebraic sum
   b. tangent of the angle
   c. quadrature sum
   d. critical frequency

7. The impedance of a circuit is the
   a. tangent of the angle between current and voltage
   b. angle between the current and the voltage
   c. quadrature sum of the current and voltage
   d. total opposition to current

8. In a series *RL* circuit, the impedance equals the quadrature sum of
   a. current and voltage
   b. resistance and inductance
   c. frequency and inductive reactance
   d. inductive reactance and resistance

9. When the inductive voltage increases in a series *RL* circuit, the angle
   a. decreases
   b. stays the same
   c. increases
   d. none of the above

10. When the frequency decreases in a series *RL* circuit, the angle
    a. decreases
    b. stays the same
    c. increases
    d. none of the above

11. When the frequency increases in a series *RL* circuit, the current
    a. decreases
    b. stays the same
    c. increases
    d. none of the above

12. At the critical frequency, how is the inductive reactance related to the resistance?
    a. less than
    b. equal
    c. greater than
    d. none of the above

13. In a parallel *RL* circuit, the total current equals the quadrature sum of the
    a. branch currents
    b. branch voltages
    c. resistance and inductance
    d. inductive reactance and resistance

14. When the frequency doubles, the total current in a parallel *R L* circuit
    a. drops in half
    b. stays the same
    c. doubles
    d. decreases

15. The impedance of an ac circuit is defined as the
    a. total voltage divided by the total current
    b. quadrature sum of the resistance and inductive reactance
    c. product over the quadrature sum
    d. total power divided by the total voltage

16. The *Q* of an a parallel circuit equals the
    a. inductive reactance divided by the resistance
    b. resistance divided by the inductive reactance
    c. tangent of the angle
    d. angle

17. The *Q* of a parallel *RL* circuit equals the
    a. inductive reactance divided by the resistance
    b. resistance divided by the inductive reactance
    c. impedance times the sine of the angle
    d. impedance divided by the angle

18. The *Q* of a series *RL* circuit equals the
    a. inductive reactance divided by the resistance
    b. resistance divided by the inductive reactance
    c. impedance divided by the angle
    d. impedance times the sine of the angle

19. In a parallel *RL* circuit the total current times the sine of the angle is the
    a. resistive current
    b. inductive current
    c. resistive voltage
    d. inductive voltage

Fig. 4-37.

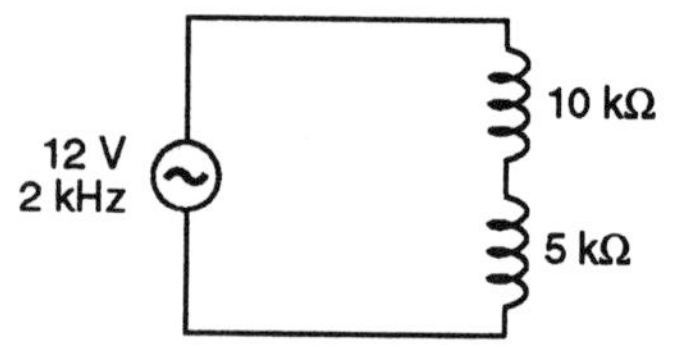

Fig. 4-38.

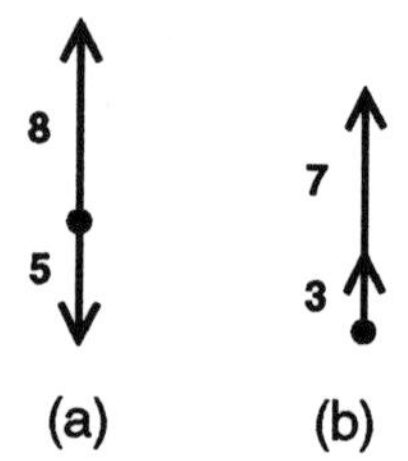

Fig. 4-39.

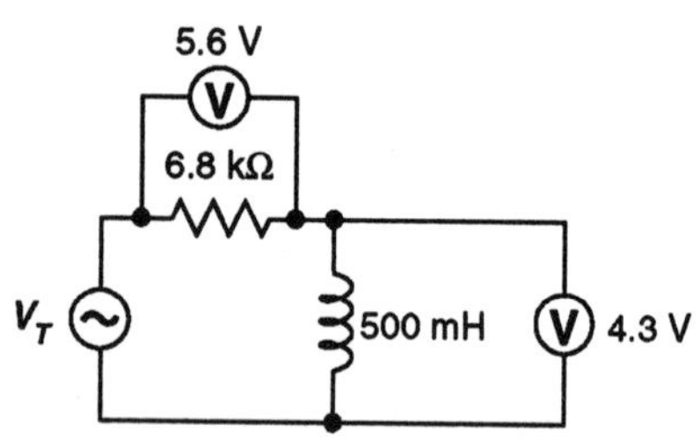

Fig. 4-40.

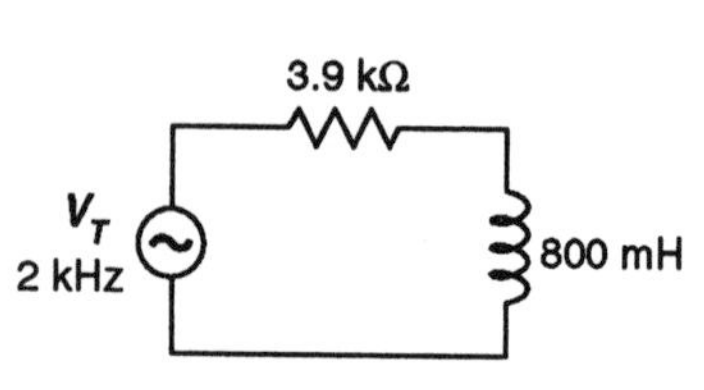

Fig. 4-41.

PROBLEMS

Sec. 4-1: Phasors

4-1. Figure 4-37 shows a resistive circuit. What is the current? The voltage across each resistor? Draw the phasors.

4-2. Figure 4-38 shows an inductive circuit. What is the current? The voltage across each inductor? Draw the phasors.

4-3. If the inductances are doubled in Fig. 4-38, what is the current? The voltage across each inductor? Draw the phasors.

4-4. If the frequency is reduced by a factor of 2 in Fig. 4-38, what is the current? The voltage across each inductor? Draw the phasors.

Sec. 4-2: Vector Addition

4-5. What is the vector sum in Fig. 4-39a?

4-6. What is the vector sum in Fig. 4-39b?

Sec. 4-3: Series Circuits

4-7. In Fig. 4-40, you measure an rms voltage of 5.6 V across the resistor and an rms voltage of 4.3 V across the inductor. What is the total rms voltage?

4-8. In Fig. 4-41, the rms current is 3 mA. What is the rms value of the applied voltage? The rms voltage across the resistor? The rms voltage across the inductor?

4-9. The inductance of Fig. 4-41 is doubled and the applied voltage is increased to get an rms current of 5 mA. What is the applied voltage?

4-10. The frequency is doubled in Fig. 4-41 and the applied voltage is increased to get an rms current of 4 mA. What is the applied voltage?

4-11. In Fig. 4-42, the rms voltage across the inductance is 30 V. What is the voltage across the resistance?

4-12. In Fig. 4-42, the rms voltage across the inductance is 35.4 V. What is the resistance?

4-13. In Fig. 4-43, the rms current is 3 mA. What is the voltage across the inductance?

4-14. In Fig. 4-43, the applied voltage is increased to get a current of 5 mA. What is the inductance?

Sec. 4-4. Impedance

4-15. The total voltage is 15 V rms in Fig. 4-41. What is the current? The voltage across the resistor? The voltage across the inductor?

4-16. The frequency is doubled in Fig. 4-41. What is the current if the total voltage is 5 V rms? The voltage across the resistor? The voltage across the inductor?

4-17. The inductance is reduced to 400 mH in Fig. 4-41. If the applied voltage is 20 V, what is the current? The voltage across the resistor? The voltage across the inductor?

4-18. What is the current in Fig. 4-42 if the resistance is 15 kΩ?

4-19. What is the voltage across the resistor of Fig. 4-42 if the resistance is 8 kΩ?

4-20. If the resistance is 8 kΩ in Fig. 4-42, what is the voltage across the inductor?

4-21. What is the current in Fig. 4-44? The voltage across the resistor? The voltage across the inductor?

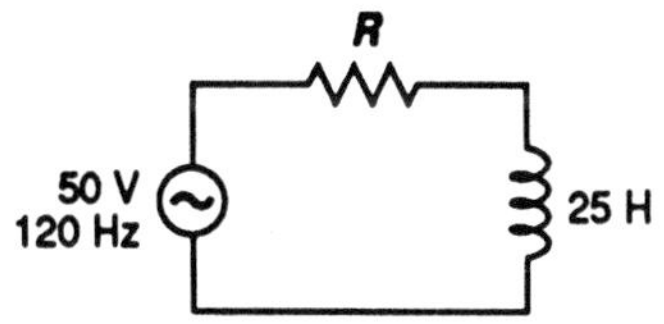

Fig. 4-42.

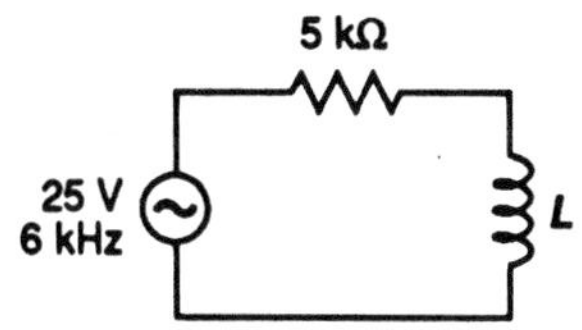

Fig. 4-43.

Sec. 4-5. Angle of Impedance

4-22. What is the angle of the impedance in Fig. 4-41?

4-23. What is the angle of impedance in Fig. 4-42 if the resistance is half the inductive reactance?

4-24. What is the angle of the impedance in Fig. 4-43 if the voltage across the inductor is 18 V?

4-25. What is the critical frequency in Fig. 4-41?

4-26. What is the critical frequency in Fig. 4-44?

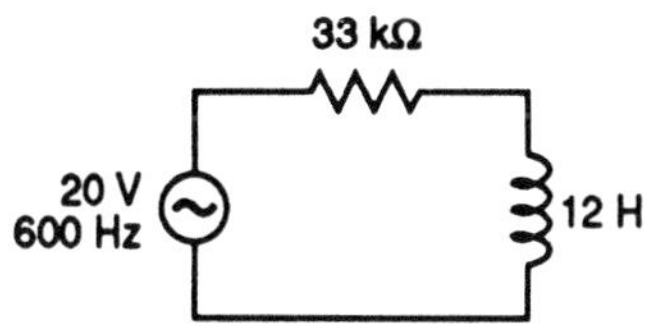

Fig. 4-44.

Sec. 4-6. Parallel Circuits

4-27. In Fig. 4-45, the rms voltage is 7 V. What is the total rms current?

4-28. If the rms voltage of Fig. 4-45 is increased to 14 V, what is the total rms current?

4-29. If the frequency is doubled in Fig. 4-45, what is the total rms current?

4-30. If the inductance is tripled in Fig. 4-45, what is the total rms current?

4-31. In Fig. 4-46, the total rms current is 30 mA. What is the resistance?

4-32. In Fig. 4-47, the total rms current is 35 mA. What is the inductance?

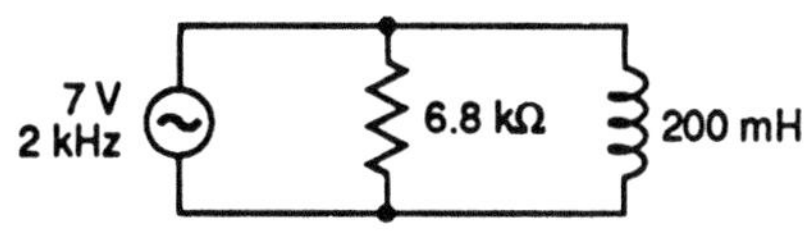

Fig. 4-45.

Sec. 4-7. Parallel Impedance and Angle

4-33. What is the impedance and angle in Fig. 4-45?

4-34. If the frequency doubles in Fig. 4-45, what is the impedance and the angle?

4-35. If the inductance triples in Fig. 4-45, what is the impedance and the angle?

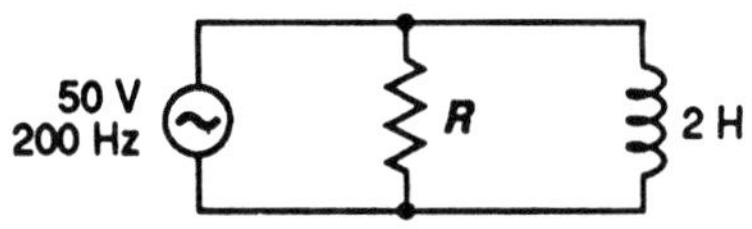

Fig. 4-46.

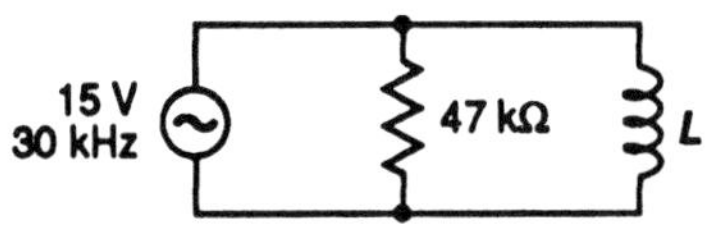

Fig. 4-47.

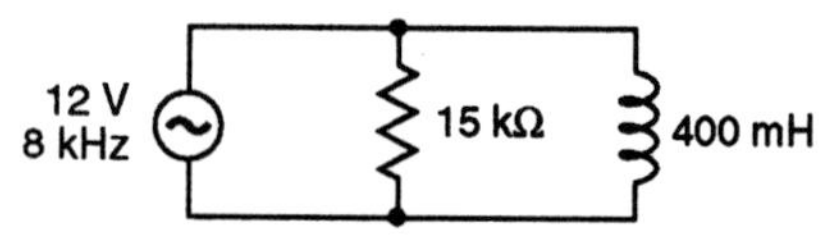

Fig. 4-48.

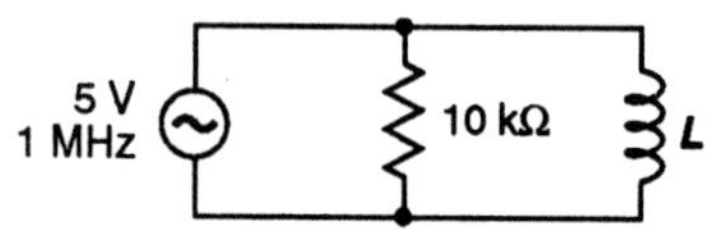

Fig. 4-49.

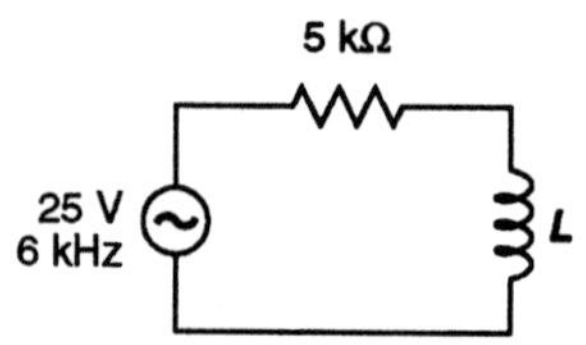

Fig. 4-50.

Sec. 4-8. More About Parallel Impedance

4-36. Calculate the impedance of Fig. 4-48 using Eq. (4-18).

4-37. Calculate the impedance of Fig. 4-48 using Eq. (4-19).

4-38. Use Eq. (4-18) to calculate the impedance of Fig. 4-49 for an inductance of 1 mH. Then, calculate the total current and the branch currents.

Sec. 4-9. The *Q* of Inductive Circuits

4-39. A series *RL* circuit has $R = 2.7$ kΩ and $X_L = 18$ kΩ. What is the *Q* of the circuit?

4-40. A parallel *RL* circuit has $R = 33$ kΩ and $X_L = 425$ Ω. What is the *Q* of the circuit?

4-41. The circuit of Fig. 4-50 has a *Q* of 20. What is the inductance?

4-42. The circuit of Fig. 4-49 has a *Q* of 120. What is the inductance?

CHAPTER 5
CAPACITANCE

There are three basic components: the resistor, the inductor, and the *capacitor*. Previous chapters discussed the resistor and the inductor. Now, you are ready to study the capacitor. The capacitor is one of the most important components in electronics.

5-1. HOW A CAPACITOR WORKS

A capacitor is a device that can store electric charge indefinitely. Here is the basic idea of how it does this. Figure 5-1a shows a pair of metal plates. These parallel plates have no net charge because every atom inside them is neutral. With the switch open as shown, the voltage between the plates is zero.

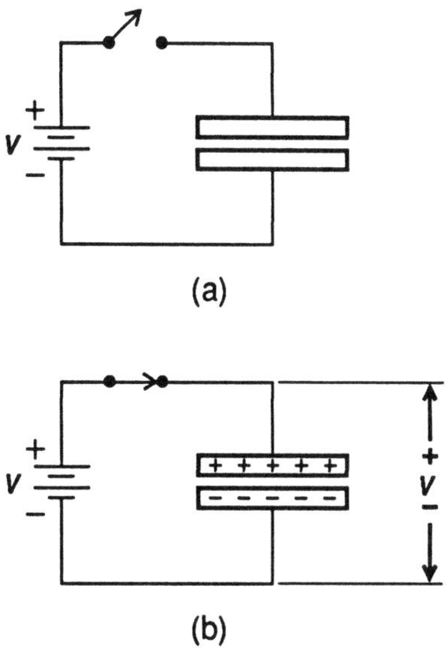

Fig. 5-1.

Charging a Capacitor

When the switch closes, free electrons will flow from the negative battery terminal (excess of free electrons) to the lower plate. Simultaneously, free electrons will flow from the upper plate to the positive battery terminal (deficiency of free electrons).

When free electrons flow into the lower plate, it becomes negatively charged. When free electrons leave the upper plate, it becomes positively charged. The flow of free electrons will continue until the voltage between the plates equals the applied voltage as shown in Fig. 5-1b.

The charging of the plates is almost but not quite instantaneous. Because of the small resistance in the connecting wires, the current is limited. Therefore, it takes a time to transfer free electrons between the plates. The point is that the plates become fully charged in a short time.

If the switch is opened as shown in Fig. 5-2, the parallel plates can store their charges indefinitely because these charges are trapped on the plates. Any device that can store an electric charge like this is called a *capacitor*. Theoretically, the capacitor will remain fully charged as long as the switch is open.

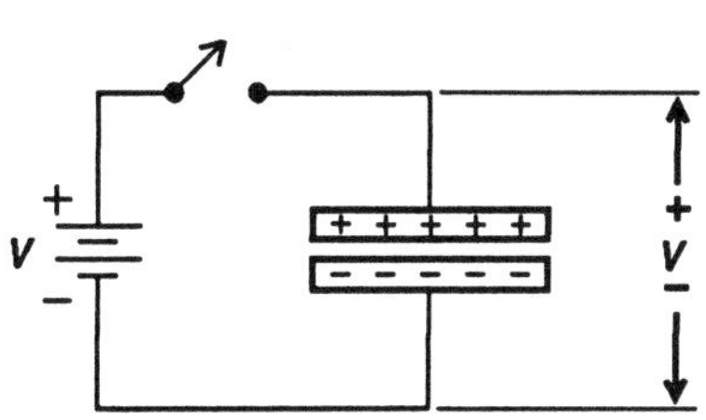

Fig. 5-2.

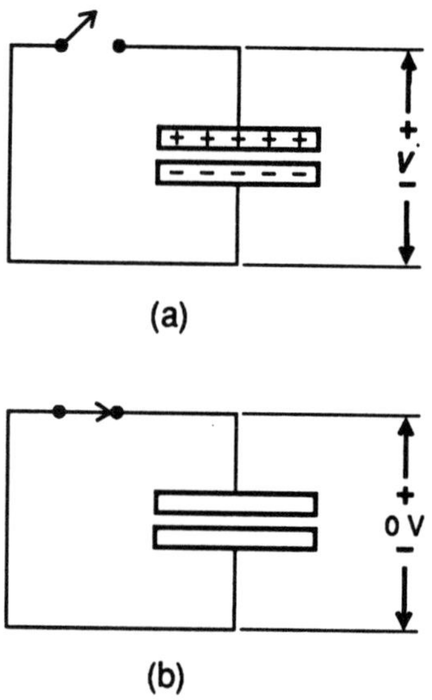

Fig. 5-3.

Discharging a Capacitor

Assume that we replace the battery of Fig. 5-2 by a short as shown in Fig. 5-3a. The capacitor voltage equals V, the applied voltage when it was charged earlier. With the switch open, the capacitor remains charged indefinitely.

If the switch is closed in Fig. 5-3a, a complete path exists between the lower and upper plates. The mutual repulsion of the free electrons on the lower plate forces some of these electrons to flow out of the lower plate. At the same time, the positive charge on the upper plate attracts free electrons. These two factors produce an electron flow from the lower to the upper plate. The flow continues until the capacitor is fully discharged. Fully discharged means the voltage across the capacitor has *decreased to zero* as shown in Fig. 5-3b.

5-2. INCREASING THE STORED CHARGE

Certain factors can increase the stored charge. For instance, a larger plate area increases the charge, a smaller distance between the plates increases the charge, and an insulator between the plates increases the charge. This section explains why.

Effect of Plate Area

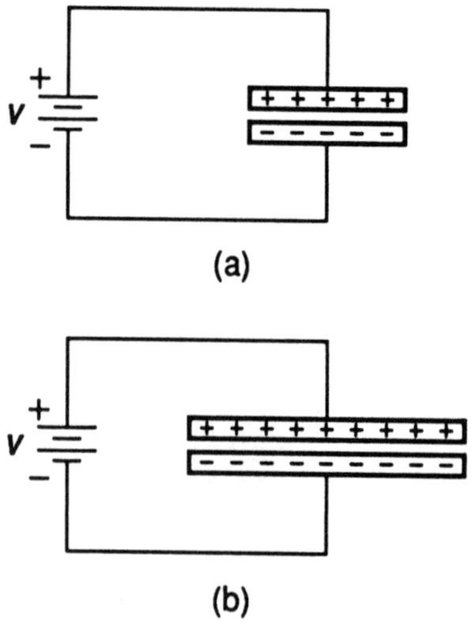

Fig. 5-4.

Figure 5-4a shows a fully charged capacitor. It is fully charged when its voltage equals the applied voltage. Figure 5-4b shows a capacitor whose plates are twice as long as those of Fig. 5-4a. This larger capacitor can store more charge as shown in Fig. 5-4b. The extra plate area means that more atoms in the upper plate will lose electrons and more atoms in the lower plate will gain electrons.

The stored charge is *directly proportional to the plate area.* If you double the plate area, the stored charge doubles. Triple the plate area and you get three times as much stored charge.

Distance between Plates

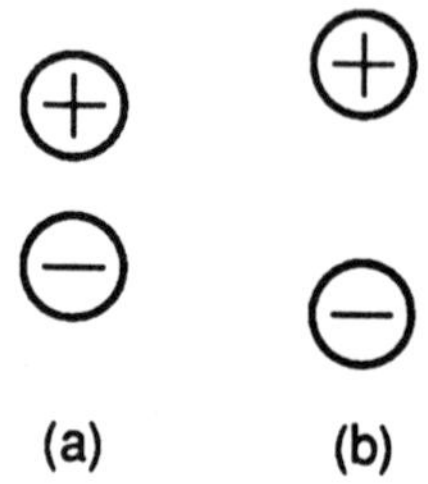

Fig. 5-5.

Figure 5-5a shows a positive and a negative charge. Assume an outside force pulls these charges apart to the new position shown in Fig. 5-5b. Because the charges have been separated, they have more potential energy. If these charges are allowed to move, they can do work as they move back toward each other.

Voltage is defined as the work that one coulomb of charges can do as the charges move back together. The further the charges are apart initially, the more work they can do as they move toward each other. This means that voltage increases as the separation of charges increases. Conversely, voltage decreases when the charges are closer together.

The same concept applies to charged plates. When the distance between the plates of Fig. 5-6a is increased to get Fig. 5-6b, the voltage between the plates increases. Notice that the charge is the same in Fig. 5-6b, but the increased separation of the charges means the charges have more potential energy than before.

On the other hand, if you start with the plates well separated as shown in Fig. 5-6b, the voltage is high. As you move the plates together as shown in Fig. 5-6a, potential energy is lost and the voltage decreases.

When a capacitor is connected across a constant voltage source, the final capacitor voltage equals the applied voltage. Because of this, more charge will be stored on the capacitor whose plates are closer together. The additional charges are necessary to raise the capacitor voltage enough to equal the applied voltage. In other words, the stored charge is *inversely proportional to the distance between the plates.* As you reduce the distance between the plates, you get more stored charge.

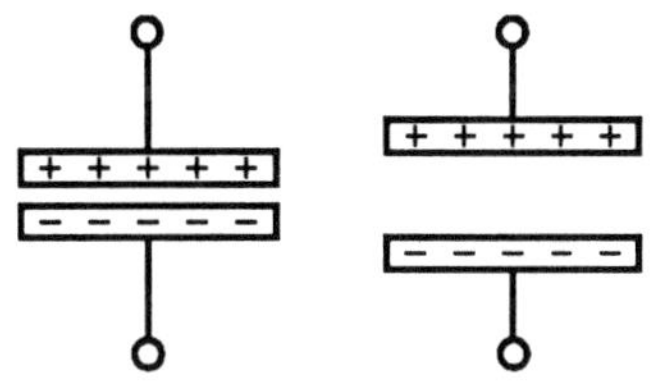

Fig. 5-6.

Electric Field

Figure 5-7a shows a stationary positive charge. This positive charge will repel another positive charge in the direction shown by the outward arrows. For instance, if you place a small movable positive charge just to the right of the large stationary charge shown, the movable charge moves to the right in the direction shown by the 3 o'clock arrow. If the small movable positive charge is placed below the large positive charge, it moves straight down along the 6 o'clock arrow.

The space around the positive charge of Fig. 5-7a is called a *field.* Because of the force this charge exerts on another positive charge, we can say a field of force surrounds the large stationary positive charge. To distinguish this field of force from other types of fields, we call it an electric field of force, or simply an *electric field.* (Other kinds of fields are gravitational and magnetic fields.)

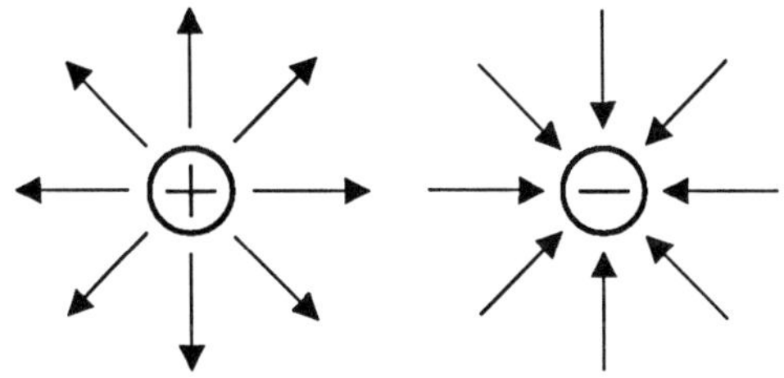

Fig. 5-7.

Negative charges also have electric fields. It is customary to draw the arrows in the direction of force *felt by a positive charge.* This is why the electric field around a negative charge points inward as shown in Fig. 5-7b.

The use of arrows for the electric field is a visual aid. Because the electric field is invisible, we have to draw lines on paper to see it. By looking at these lines of force, we can tell at a glance how other charges are affected. Positive charges move in the direction of the arrows, and negative charges move in the opposite direction.

The electric field is a visual aid, similar to phasors. It is another viewpoint, another angle of seeing, another metaphor. If you forgot what a metaphor means in science, review Sec. 12-8 in your Dc Circuits textbook.

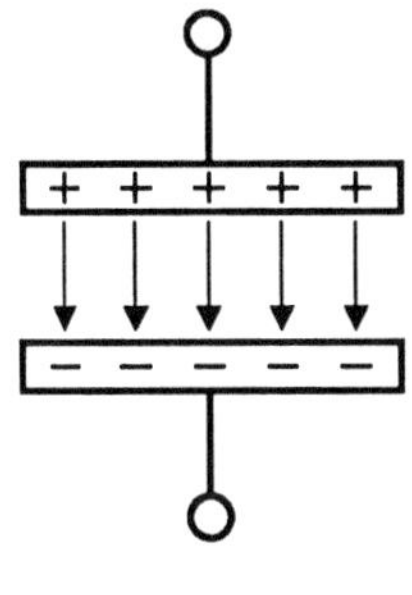
Fig. 5-8.

Effect of Dielectric

Figure 5-8 shows a parallel-plate capacitor with a downward electric field between the plates. If a positive charge were in the space between the plates, it would feel a downward force. On the other hand, an electron would feel an upward force.

Figure 5-9a shows a single atom between a pair of charged plates. The valence electron is repelled by the lower plate and attracted to the upper plate. This distorts the orbit of the valence electron from a circular orbit to an elliptical orbit. Because the center of the ellipse is above the core of the atom, the electron spends more time above the core than below. The elliptical distortion of the orbit is called *polarization.*

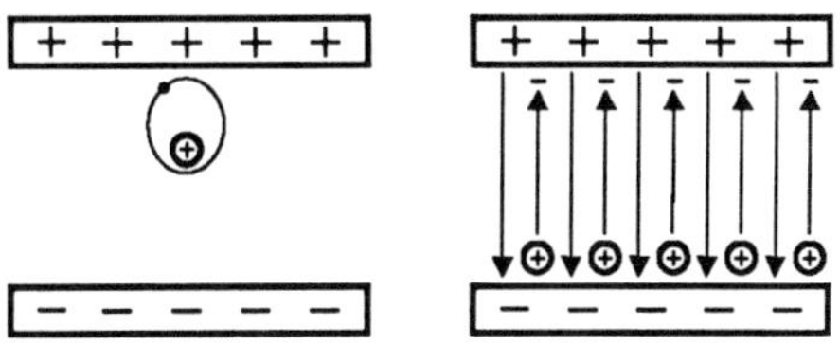

Fig. 5-9.

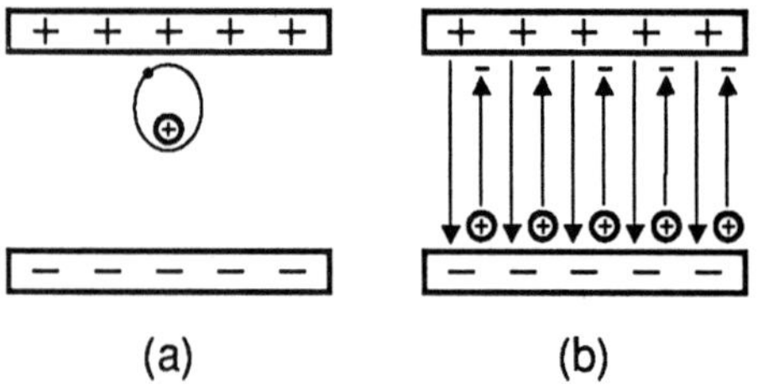

Fig. 5-9.

When a *dielectric* (another name for an insulator) is used between the plates, the orbits of the dielectric are polarized. The effect of this polarization is the same as having small electric *dipoles* (pairs of positive and negative charges) as shown in Fig. 5-9b. Each dipole has an electric field pointing upward. These dipole fields oppose the original field produced by the parallel plates. Therefore, the net electric field is less with a dielectric than without.

Since the net electric field is smaller with a dielectric, the applied voltage can transfer more free electrons from the upper plate to the lower plate. In other words, a dielectric between the plates increases the stored charge for the same value of applied voltage.

The *dielectric constant* K_ε of a material indicates the ease of polarizing its atoms. A vacuum has a dielectric constant of 1, which means no polarization. A material with a dielectric constant of 10 will produce 10 times as much stored charge as a vacuum. This means that the stored charge in a capacitor is directly proportional to the dielectric constant. The higher the dielectric constant, the larger the stored charge.

Table 5-1 shows some materials and their typical dielectric constants. Air or a vacuum has no effect on stored charge. Other materials, however, can have a pronounced effect on the stored charge. For instance, aluminum oxide produces 10 times as much charge as air.

TABLE 5-1. DIELECTRICS

Material	**Dielectric constant K_ε**
Air or vacuum	1
Teflon	2
Plastic film	2.5
Polystyrene	2.6
Polyester	3
Oil	4
Paper	4
Glass	4.2
Porcelain	6.2
Mica	7.5
Aluminum oxide	10
Tantulum oxide	11
Ceramic	1000

Dielectric Strength

The greater the voltage between the plates, the greater the polarization of the dielectric atoms. If the voltage is too large, electrons can be pulled away from the dielectric atoms. When this happens, the dielectric changes from an insulator to a conductor. Dislodging electrons like this is called *voltage breakdown.*

The dielectric strength of a material is its ability to resist voltage breakdown. The defining formula is

$$\text{Dielectric strength} = \frac{V_{max}}{d} \qquad (5\text{-}1)$$

where V_{max} is the breakdown voltage and d is the thickness of the dielectric.

Engineering handbooks list the dielectric strength of different materials. Each dielectric can take only so much voltage before it breaks down. Because of this, a manufactured capacitor has a *voltage rating* which you cannot exceed without risking breakdown.

EXAMPLE 5-1

A capacitor has a voltage rating of 80 V. If the distance between capacitor plates is 0.0001 m, what is the dielectric strength of the material between the plates?

SOLUTION

Divide 80 V by 0.0001 m to get

$$\text{Dielectric strength} = \frac{80\text{ V}}{0.0001\text{ m}} = 800\text{ kV/m}$$

5-3. CAPACITOR FORMULAS

Capacitance is defined as the ability of a capacitor to store charge. This is the general idea, but we need something mathematical to work with. We need a defining formula for capacitance.

Definition of capacitance

Capacitance is mathematically defined as the stored charged divided by the voltage between the plates. In symbols,

$$C = \frac{Q}{V} \qquad (5\text{-}2)$$

MEMORIZE

$$C = \frac{Q}{V}$$

A *farad* (abbreviated F) is the metric unit of measure for capacitance. It is defined as

1 farad = 1 coulomb per volt

or

1 F = 1 C/V

A farad is the amount of capacitance that stores 1 coulomb for each volt between the charged plates.

Calculating Stored Charge

Sometimes, you know the values of capacitance and voltage, but not the stored charge. In this case, you can rearrange Eq. (5-2) and use the following formula to calculate the stored charge:

$$Q = CV \tag{5-3}$$

This says the stored charge equals the capacitance times the voltage.

Calculating Capacitor Voltage

Occasionally, you are given the values of stored charge and capacitance, but not the voltage. When this happens, you can rearrange Eq. (5-2) and use this formula to calculate the voltage:

$$V = \frac{Q}{C} \tag{5-4}$$

This says the voltage equals the stored charged divided by the capacitance.

Practical Prefixes

In the electronics industry, people use picofarads (pF) and microfarads (μF) for all capacitors, even though theoretically any prefix can be used. The reason for using only pF and μF is because the largest commercial capacitors used to be less than 100 μF. Although much larger capacitances are now being manufactured, the custom of using only pF and μF persists.

EXAMPLE 5-2

What is the capacitance if the stored charge is 5 μC and the voltage is 10 V?

SOLUTION

Capacitance equals stored charge divided by voltage:

$$C = \frac{Q}{V} = \frac{5\ \mu\text{C}}{10\ \text{V}} = 0.5\ \mu\text{F}$$

This means that the capacitor stores 0.5 μC for each volt between its plates.

EXAMPLE 5-3

What is the stored charge if $C = 10\ \mu\text{F}$ and $V = 2$ V?

SOLUTION

The stored charge equals the capacitance times the voltage:

$$Q = CV = (10\ \mu F)(2\ V) = 20\ \mu C$$

EXAMPLE 5-4

What is the voltage if $Q = 30\ \mu C$ and $C = 1.5\ \mu F$?

SOLUTION

The voltage equals the stored charge divided by the capacitance:

$$V = \frac{Q}{C} = \frac{30\ \mu C}{1.5\ \mu F} = 20\ V$$

5-4. ANOTHER FORMULA FOR CAPACITANCE

Recall that the stored charged is directly proportional to the plate area and inversely proportional to the distance between the plates. Because of this, we can write the following formula for capacitance:

$$C = \frac{KA}{d} \tag{5-5}$$

where K = the constant of proportionality
A = area of either plate
d = distance between plates

The constant K depends on the units of measure and the dielectric.

In the metric system, the plate area is measured in square meters and the distance between the plates in meters. In this case, Eq. (5-5) may be written like this:

$$C = \frac{8.85(10^{-12})K_\varepsilon A}{d} \tag{5-6}$$

where $8.85(10^{-12})$ is constant of proportionality for air or a vacuum, and K_ε is the dielectric constant of Table 5-1.

EXAMPLE 5-5

A capacitor has a plate area of 0.1 m^2 and a distance between the plates of 0.001 m. If the dielectric constant is 4, what is the capacitance?

SOLUTION

Use Eq. (5-6) to get

$$C = \frac{8.85(10^{-12})(4)(0.1)}{0.001} = 3540 \text{ pF}$$

5-5. CURRENT AND CAPACITANCE

IMPORTANT

$$i = C\frac{dv}{dt}$$

In an earlier chapter, we derived this important equation for an inductance:

$$v = L\frac{di}{dt}$$

This says the induced voltage equals the inductance times the instantaneous rate of current change with time. There is a similar equation for a capacitor, given by

$$i = C\frac{dv}{dt} \tag{5-7}$$

This says the charging current into a capacitor equals the capacitance times the instantaneous rate of voltage change with time.

Equation (5-7) is of major importance and should be memorized. With Eq. (5-7), we can calculate the charging current using the method of small changes. By working out the small changes between two close points on a waveform, we can get an accurate value for the instantaneous rate of change of voltage with time. Then, multiplying by the capacitance gives the charging current.

Incidentally, when the instantaneous rate of voltage change is positive in Eq. (5-7), the charging current is positive. This means the capacitor is charging because its voltage is increasing. On the other hand, when the instantaneous rate of voltage change is negative, the current is negative. This means the capacitor is discharging because its voltage is decreasing.

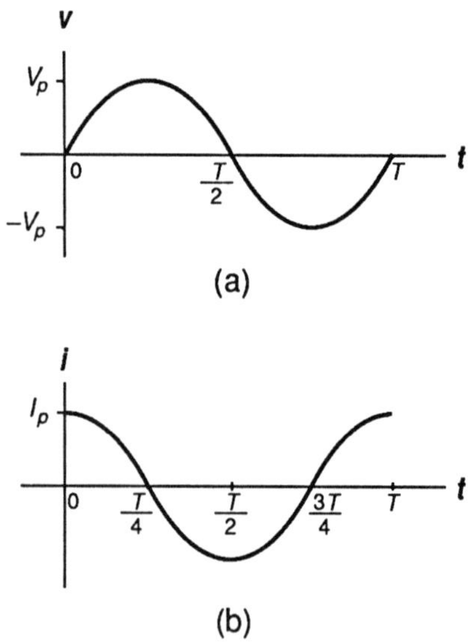

Fig. 5-10.

CRITICAL POINTS

t	dv/dt
0	Max +
$T/4$	Zero
$T/2$	Max –
$3T/4$	Zero
$2T$	Max +

Current Leads Voltage

When the voltage is sinusoidal, the instantaneous value is given by

$$v = V_p \sin 2\pi ft$$

Figure 5-10a shows the voltage waveform. The instantaneous rate of change dv/dt equals the slope of a tangent line drawn to any point on this waveform. From earlier discussions, we know that the instantaneous rate of change is maximum at the beginning of the sine wave, zero at the positive peak, maximum negative at the middle of the sine wave, and zero at the negative peak. (See Critical Points at left.)

Equation (5-7) says the charging current equals the capacitance multiplied by the instantaneous rate of change of voltage with time. By the same logic used in our discussion of inductors, we can see that the charging current is sinusoidal but shifted by a quarter of cycle. Because of this shift, the charging current leads the voltage as shown in Fig. 5-10b. Furthermore, we can write the charging current in either of two forms:

$$i = I_p \sin(2\pi ft + 90^\circ)$$

or

$$i = I_p \cos 2\pi ft$$

In terms of phasors, the phasor current leads the phasor voltage by 90° as shown in Fig. 5-11.

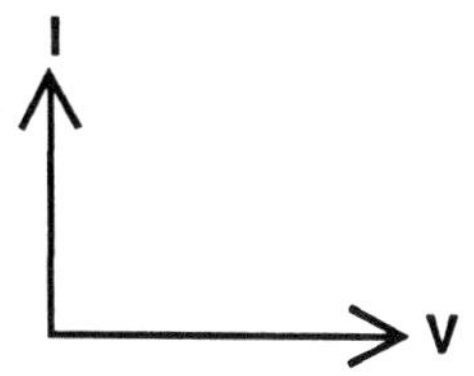

Fig. 5-11.

The Derivation

Where does Eq. (5-7) come from? The proof follows. If you are not interested in seeing the mathematical proof, you may skip to Example 5-6.

The capacitance of a capacitor is given by

$$C = \frac{Q}{V}$$

When the voltage is time-varying, it is customary to use lower-case letters for the charge and voltage as follows:

$$C = \frac{q}{v}$$

By cross-multiplying, this becomes

$$q = Cv$$

Because of the direct proportion between charge and voltage, the equation can be written in terms of small changes as follows:

$$\Delta q = C\Delta v$$

We can divide both sides by Δt to get

$$\frac{\Delta q}{\Delta t} = \frac{C\Delta v}{\Delta t}$$

When the changes are infinitesmally small, the average rates of change become instantaneous rates of change:

$$\frac{dq}{dt} = \frac{Cdv}{dt}$$

The left side of this equation equals the instantaneous current, so the equation may be written like this:

$$i = C \frac{dv}{dt}$$

This final equation says the charging current into a capacitor equals the capacitance times the instantaneous rate of voltage change with time.

EXAMPLE 5-6

A sinusoidal voltage is across a capacitance of 5 μF. If the instantaneous rate of voltage change is 157 V/s at $t = 0$, what is charging current at this instant?

SOLUTION

Multiply the capacitance by the instantaneous rate of change of voltage to get the charging current:

$$i = (5\ \mu\text{F})(157\ \text{V/s}) = 0.785\ \text{mA}$$

EXAMPLE 5-7

A capacitance of 1000 pF has $dv/dt = -1000$ V/s. What is the charging current?

SOLUTION

The charging current is

$$i = (1000\ \text{pF})(-1000\ \text{V/s}) = -1\ \mu\text{A}$$

Because the answer is negative, the capacitor is actually discharging rather than charging.

5-6. CAPACITOR OPPOSES CHANGE IN VOLTAGE

The current into a capacitor is given by

$$i = C \frac{dv}{dt}$$

We can rearrange this basic equation to get

$$C = \frac{i}{dv/dt} \qquad (5\text{-}8)$$

This says that the capacitance equals the charging current divided by the rate of voltage change with respect to time. This is useful when you have the value of charging current and the instantaneous rate of voltage change.

Equation (5-8) can be rearranged to get this equation:

$$\frac{dv}{dt} = \frac{i}{C} \tag{5-9}$$

Notice the following. As the capacitance approaches infinity, the rate of voltage change approaches zero. Stated another way, a capacitor opposes changes in voltage. The larger the capacitance, the greater the opposition to voltage change.

Since it always takes time to charge a capacitor, we can state the following important fact: the *voltage across a capacitor cannot change instantaneously*. Remember this important idea.

EXAMPLE 5-8

What is the capacitance if $i = 1\ \mu A$ and $dv/dt = 4\ V/s$?

SOLUTION

Divide the current by the instantaneous rate of voltage change as follows:

$$C = \frac{i}{dv/dt} = \frac{1\ \mu A}{4\ V/s} = 0.25\ \mu F$$

EXAMPLE 5-9

What does dv/dt equal if $i = 1\ \mu A$ and $C = 1\ \mu F$? If $C = 100\ \mu F$?

SOLUTION

A current of 1 μA into a capacitance of 1 μF produces an instantaneous rate of voltage change of

$$\frac{dv}{dt} = \frac{1\ \mu A}{1\ \mu F} = 1\ V/s$$

When the capacitance is increased to 100 μF, the instantaneous rate of voltage change decreases to

$$\frac{dv}{dt} = \frac{1\ \mu A}{100\ \mu F} = 0.01\ V/s$$

As you see, the larger the capacitance, the smaller the rate of voltage change. This is why we say a capacitor opposes the change in voltage. As the capacitance approaches infinity, the rate of voltage change approaches zero.

5-7. SERIES AND PARALLEL CAPACITORS

Two inductors in series are added to get the total inductance. Two inductors in parallel are combined by the product over the sum to get the total inductance. Capacitors are again different from inductors. They combine in exactly the opposite way.

PARALLEL

$C_T = C_1 + C_2$

Parallel Capacitances

When two capacitances are in parallel, they are added to get the total capacitance:

$$C_T = C_1 + C_2 \qquad (5\text{-}10)$$

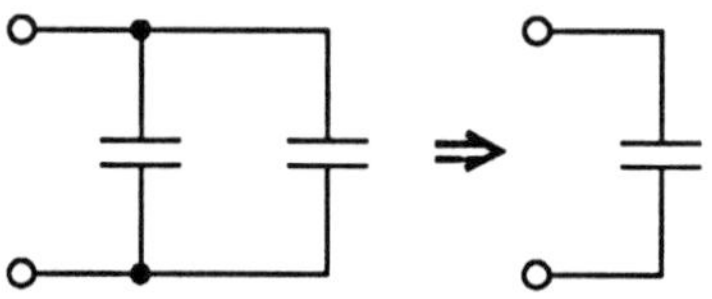

Fig. 5-12.

This formula makes sense when you look at Fig. 5-12. Two capacitors in parallel have the effect of increasing the plate area. This permits more charge to be stored, so that the total capacitance is greater.

Here is the mathematical proof. The total capacitance equals

$$C_T = \frac{Q_T}{V_T} = \frac{Q_1 + Q_2}{V_T}$$

The total charge Q_T equals the sum of the individual charges because currents add in a parallel circuit. The foregoing equation can be written as

$$C_T = \frac{Q_1}{V_T} + \frac{Q_2}{V_T}$$

But all voltages are equal in a parallel circuit. Therefore,

$$C_T = \frac{Q_1}{V_1} + \frac{Q_2}{V_2} = C_1 + C_2$$

The same idea applies to any number of capacitances in parallel. You can add them to get the total capacitance.

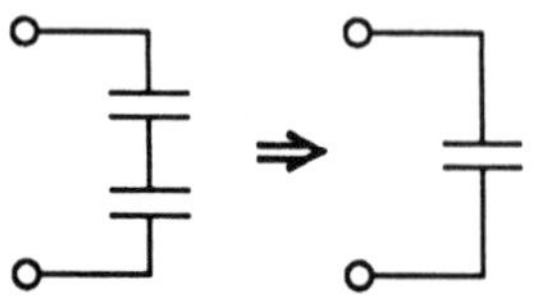

Fig. 5-13.

Capacitances in Series

When two capacitances are in series, the total capacitance is less than either capacitance. In Fig. 5-13, the two capacitors in series have the same effect as increasing the distance between the plates of a capacitor. This implies less stored charge, so that the total capacitance is smaller.

Here is the mathematical proof. The total capacitance equals

$$C_T = \frac{Q_T}{V_T} = \frac{Q_T}{V_1 + V_2}$$

The total voltage V equals the sum of the individual voltages because voltages add in a series circuit. If we divide the numerator and the denominator by Q_T, we get

$$C_T = \frac{1}{V_1/Q_T + V_2/Q_T}$$

Since the current is the same in all parts of a series circuit, all stored charges are equal. Therefore, the equation reduces as follows:

$$C_T = \frac{1}{V_1/Q_1 + V_2/Q_2}$$

or

$$C_T = \frac{1}{1/C_1 + 1/C_2} \qquad (5\text{-}11)$$

An alternative form of this equation is the product over the sum:

$$C_T = \frac{C_1 C_2}{C_1 + C_2} \qquad (5\text{-}12)$$

SERIES

$$C_T = \frac{C_1 C_2}{C_1 + C_2}$$

By a similar proof, the total capacitance of n series capacitances is

$$C_T = \frac{1}{1/C_1 + 1/C_2 + 1/C_3 + \ldots + 1/C_n} \qquad (5\text{-}13)$$

You can see from these equations that capacitors in series combine like resistors in parallel. For instance, two capacitances in series have a total capacitance equal to the product of the capacitance divided by the sum of the capacitances. As a check on your calculations, remember that the total capacitance of several capacitors in series has to be less than smallest capacitance.

EXAMPLE 5-10

If C_1 = 3 pF, C_2 = 6 pF, and C_3 = 2 pF are in parallel, what is the total capacitance?

SOLUTION

When capacitors are in parallel, you add the capacitances to get the total capacitance:

$$C_T = 3\text{ pF} + 6\text{ pF} + 2\text{ pF} = 11\text{ pF}$$

EXAMPLE 5-11

If $C_1 = 3$ pF, $C_2 = 6$ pF, and $C_3 = 2$ pF are in series, what is the total capacitance?

SOLUTION

$$C_T = \frac{1}{1/3 \text{ pF} + 1/6 \text{ pF} + 1/2 \text{ pF}} = 1 \text{ pF}$$

As a check, notice that the total capacitance is less than the smallest capacitance.

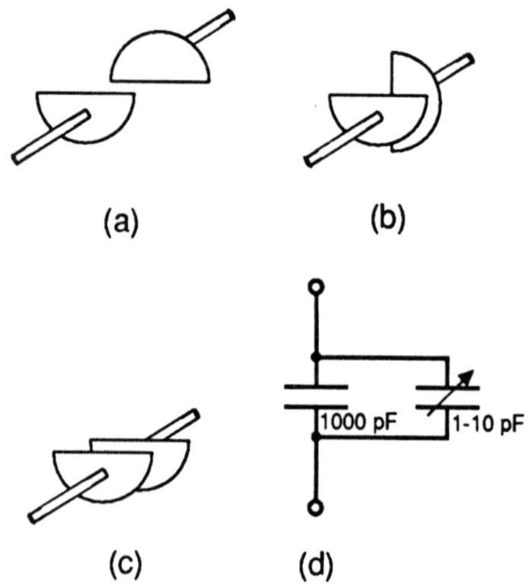

Fig. 5-14.

5-8. MANUFACTURED CAPACITORS

By using different dielectrics, manufacturers can produce capacitors with capacitances from less than 1 pF to more than 10,000 μF.

Air Capacitors

A *variable capacitor* can change capacitance over a 10:1 range (typical) by changing the plate area. For instance, Fig. 5-14a shows the capacitor plates adjusted for minimum capacitance, Fig. 5-14b shows the plates adjusted for an intermediate value of capacitance, and Fig. 5-14c shows the plates adjusted for maximum capacitance.

A *trimmer capacitor* is a small variable capacitor that changes the distance between the plates. The typical way to use a trimmer capacitor is in parallel with another capacitor (see Fig. 5-14d). The arrow through the trimmer capacitor indicates that it is adjustable. By varying the trimmer capacitor, we can make fine adjustments to the total capacitance. Since the capacitances are in parallel, the total capacitance of Fig. 5-14d varies from 1001 pF to 1010 pF.

Air capacitors have values from less than 1 pF to more than 500 pF, with about a 10-to-1 range for variable capacitors and somewhat less for trimmer capacitors.

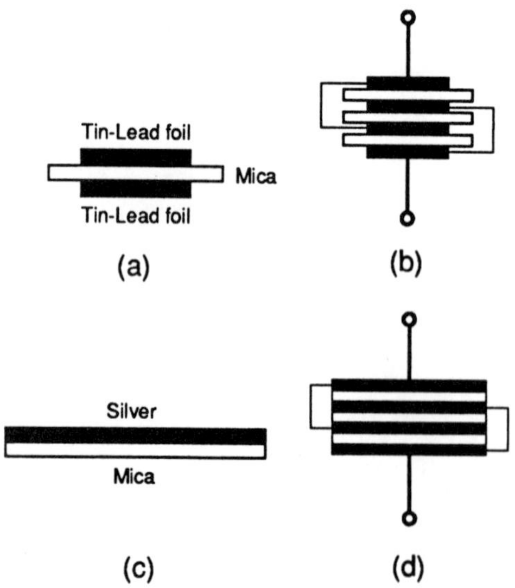

Fig. 5-15.

Mica Capacitors

Mica is a chemically inert mineral whose insulating properties change only slightly with time, temperature, moisture, etc. It is easily cut into sheets as thin as 0.0001 inch. Because of this, mica was one of the first materials used as a dielectric.

Some mica capacitors look like Fig. 5-15a, a thin sheet of mica between two pieces of tin-lead foil. This sandwich acts like a parallel-plate capacitor. For added strength and capacitance, we can stack and parallel several sheets of mica and foil as shown in Fig. 5-15b. This the way mica capacitors were originally made.

The modern way is better. A very thin layer of silver is bonded to the mica by furnace heating. The resulting *silvered-mica* sheet (Fig. 5-15c) has no air gaps, is more reliable, and is easier to work with. By stacking and paralleling these silvered mica sheets (Fig. 5-15d), we get a more reliable capacitor than the mica-foil capacitor previously described. Silvered mica capacitors have values from less than 1 pF to more than 10,000 pF.

Glass Capacitors

Structurally, a *glass capacitor* is like the mica-foil capacitor of Fig. 5-15b, except sheets of glass (the dielectric) and aluminum foil (the plates) are used. Glass capacitors cost more than mica capacitors, but they have an advantage: they are moisture resistant. Used in high-humidity environments, glass capacitors show less change than any other type of capacitor. Glass capacitors have values from less than 1 pF to more than 10,000 pF (same as mica).

Paper Capacitors

This type of capacitor uses sheets of *paper dielectric* and metal foil (Fig. 5-16a). Because paper is flexible, the sheets can be rolled into a cylindrical structure like Fig. 5-16b. This paper-foil cylinder gives larger capacitance values than are possible with mica or glass capacitors.

Some modern paper capacitors use *metallized paper*, made by depositing metal vapor on paper (Fig. 5-16c). The paper and its metal coating eliminate air gaps, improve reliability, etc. By rolling up sheets of metallized paper, we get a paper capacitor with greater reliability and higher capacitance than the paper-foil type.

Paper capacitors have a band or stripe on the end connected to the outside foil (Fig. 5-16d). When a schematic diagram shows one side of a paper capacitor grounded, most people ground the banded end. This is not necessary, but it has this advantage: it grounds the outside foil. The effect is to shield the inner foil from stray electric fields, unwanted signals, noise, etc. Paper capacitors have values from less than 0.001 μF to more than 1 μF.

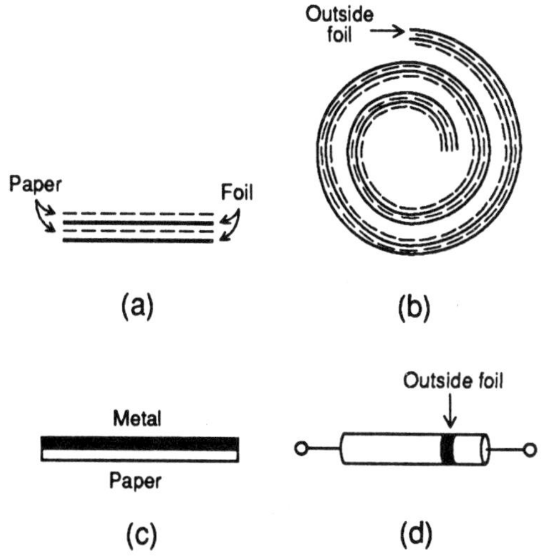

Fig. 5-16.

Plastic-film Capacitors

Plastics are excellent dielectrics. The commonly used plastic dielectrics are polyester and polystyrene. These materials have high dielectric strength and can be produced in very thin sheets like film.

Polyester-film capacitors use separate sheets of metal foil and polyester like Fig. 5-17a, or they use metallized polyester film like Fig. 5-17b. In either case, sheets of the flexible film can be rolled into a cylindrical structure to get large capacitance values (similar to Fig. 5-16b). Polyester film has a higher dielectric strength than paper. For the same voltage rating, thinner sheets can be used to get higher capacitance values than with paper.

The main disadvantage of polyester-film capacitors is their temperature sensitivity; large changes in capacitor occur when the temperature changes. Polystyrene is much less temperature sensitive than polyester. This is why

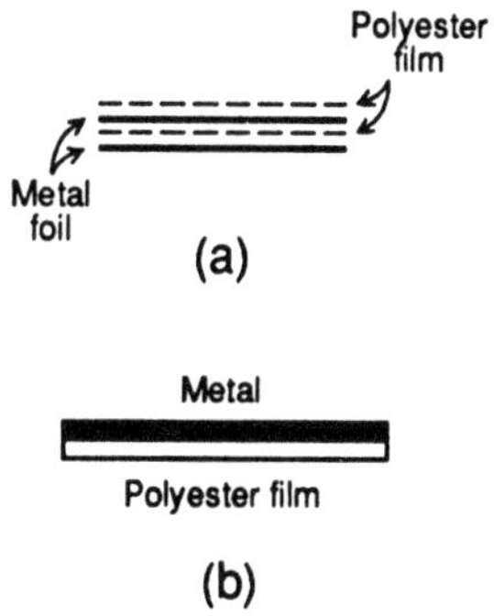

Fig. 5-17.

polystyrene-film capacitors are preferred when large temperature variations are expected.

Polyester-film capacitors have values from less than 0.001 μF to more than 10 μF. Polystyrene-film capacitors have values from less than 5 pF to more than 0.5 μF.

Ceramic Capacitors

Clay is transformed by extreme heat. At temperatures near 1000°F, chemically combined water is driven off and the clay undergoes an irreversible chemical change. The resulting product is called a *ceramic*. Examples of ceramic are porcelain, china, brick, etc.

The main advantage of ceramic materials is their tremendous range in dielectric properties. For instance, porcelain has a dielectric constant of 6 (approximate); barium-strontium-titanate (another ceramic) has a dielectric constant of 7500. By mixing ceramic materials, we can get a wide range in dielectric constant and other properties. Using ceramic dielectrics, manufacturers can produce capacitors with positive, negative, or zero temperature coefficients. Ceramic capacitors have values from less than 10 pF to more than 1 μF.

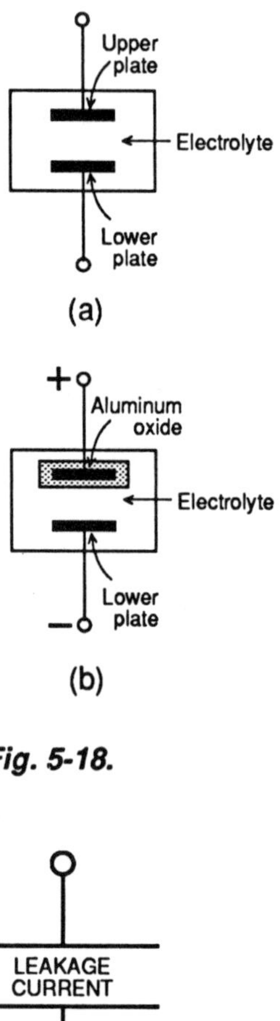

Fig. 5-18.

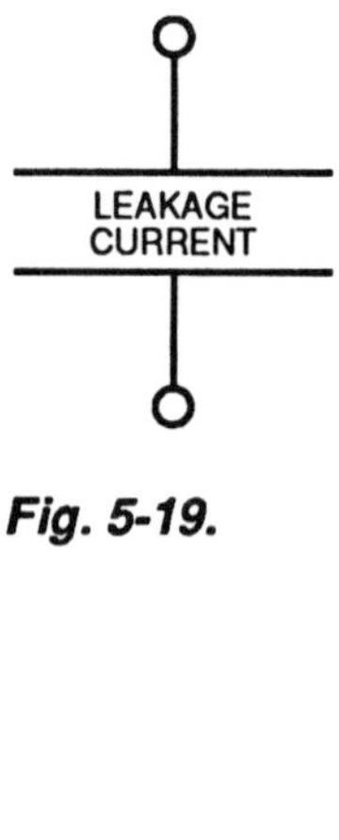

Fig. 5-19.

Electrolytic Capacitors

Electrolytic capacitors have more capacitance than any other type, because the dielectric is an extremely thin layer of oxide grown on one of the plates. Here is how an electrolytic capacitor is formed. Initially, two aluminum plates and an *electrolyte* (a conducting paste or liquid) are in a container (Fig. 5-18a). After voltage has been applied for a while, electrochemical action produces a layer of aluminum oxide on the positive plate (Fig. 5-18b). The process of growing this layer is a slow one, and is referred to as *forming the capacitor.*

After the capacitor has been formed, the oxide layer acts like a dielectric between the upper aluminum plate and the electrolyte. The lower aluminum plate merely makes contact with the electrolyte, which functions as the negative plate.

The electrolytic capacitor of Fig. 5-18b works fine as long as the upper plate is positive and the lower one is negative. But if you try to reverse the applied voltage, the oxide layer no longer acts like an insulator. Instead, you get a large current and a loss of capacitor action. For this reason, most electrolytic capacitors are *polarized.* They work properly only with one polarity of voltage. This is the reason a manufacturer puts a plus sign (or other marking) on the positive end of the electrolytic capacitor.

The thin oxide layer is the key to why electrolytic capacitors have more capacitance than other types. But the thin oxide layer has a disadvantage: its resistance is relatively low compared with other dielectrics. Instead of acting like an open, the oxide layer passes a small current (see Fig. 5-19). This unwanted current is called the *leakage current.* This may or may not be a problem, depending on the circuit. Electrolytic capacitors have values from less than 1 μF to more than 10,000 μF.

Summary

We have discussed the basic types of capacitors. Table 5-2 shows the type, approximate range, and main advantage of each. As shown, mica capacitors are stable (little change with time, temperature, etc.). The ceramic capacitors offer the greatest range in capacitance values and temperature coefficients. The electrolytic capacitors offer the largest available capacitance values.

TABLE 5-2. CAPACITOR TYPES

Type	Range	Advantage
Air	1 to 500 pF	Variable
Mica	1 to 10,000 pF	Stable values
Glass	1 to 10,000 pF	Moisture resistant
Paper	0.001 to 1 μF	Inexpensive
Polyester	0.001 to 10 μF	Physically small
Polystyrene	5 pF to 0.5 μF	Temperature stable
Ceramic	10 pF to 1 μF	Variety
Electrolytic	1 to 10,000 μF	Large capacitance values

MULTIPLE-CHOICE TEST

The following questions may have more than one right answer. Select the best answer. This is the one that is always true, or covers more situations, or agrees with the material of this chapter, etc.

1. When the plate area increases, the stored charge
   a. decreases
   b. stays the same
   c. increases
   d. none of the above

2. If the voltage across a capacitor is sinusoidal, the charging current is
   a. constant
   b. time-varying
   c. rectangular
   d. sinusoidal

3. The charging current into capacitor is directly proportional to the
   a. capacitance
   b. instantaneous voltage
   c. distance between the plates
   d. power

4. If the voltage is sinusoidal, the charging current into a capacitor is
   a. in phase with the voltage
   b. 90° ahead of the voltage
   c. 180° behind the voltage
   d. 90° behind the voltage

5. The instantaneous rate of change of a sine wave is maximum positive at
   a. 0°
   b. 90°
   c. 180°
   d. 270°

6. Capacitance equals charge divided by
   a. time
   b. voltage
   c. charging current
   d. number of plates

7. The material used between the plates of a capacitor is called a
   a. filler
   b. dielectric
   c. dipole
   d. polarization

8. The charging current into a capacitor is directly proportional to the
   a. resistance
   b. voltage
   c. current
   d. rate of voltage change

9. The voltage across an capacitor
   a. is in phase with the current
   b. leads the current by 90°
   c. leads the current by 180°
   d. lags the current by 90°

10. The total capacitance of two parallel capacitances is
    a. less than either capacitance
    b. the sum of the capacitances
    c. the product over the sum of the capacitances
    d. none of the above

11. If the instantaneous rate of voltage change doubles, the current into a capacitor will
    a. drop in half
    b. stay the same
    c. double
    d. quadruple

12. The voltage across a capacitor increases when the
   a. resistance increases
   b. charge decreases
   c. distance between plates decreases
   d. stored charge increases

13. The total capacitance of two series capacitances is
   a. greater than either capacitance
   b. the sum of the capacitances
   c. the product over the sum of the capacitances
   d. none of the above

14. When a capacitor is discharging, *dv/dt* is
   a. decreasing
   b. not changing
   c. increasing
   d. negative

15. What type of capacitor has the greatest capacitance?
   a. trimmer
   b. paper
   c. polystyrene
   d. electrolytic

16. Which of the following capacitors can be be adjusted?
   a. trimmer
   b. paper
   c. polystyrene
   d. electrolytic

PROBLEMS

Sec. 5-2: Increasing the Stored Charge

5-1. A capacitor has a voltage rating of 25 V. If the distance between the capacitor plates is 0.0003 m, what is the dielectric strength?
5-2. If the dielectric strength is 1 MV/m and the distance between plates is 0.00005 m, what is the voltage rating?

Sec. 5-3: Capacitor Formulas

5-3. What is the capacitance if the stored charge is 12 μC and the voltage is 5 V?
5-4. What is the stored charge if $C = 470$ μF and $V = 15$ V?
5-5. What is the stored charge in Fig. 5-20a?
5-6. What is the voltage if $Q = 500$ μC and $C = 10$ μF?
5-7. What is the voltage in Fig. 5-20b?

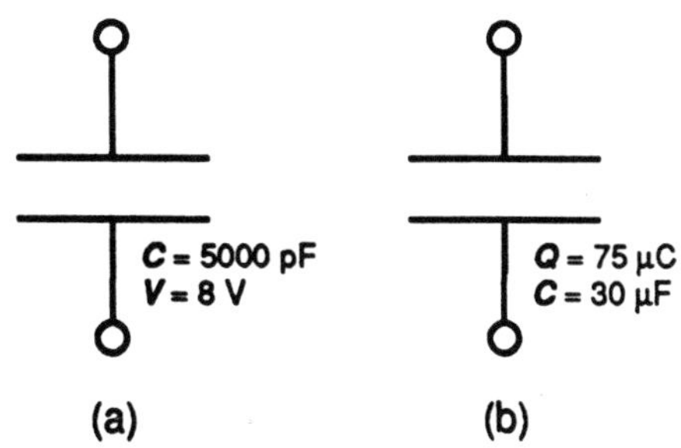

Fig. 5-20.

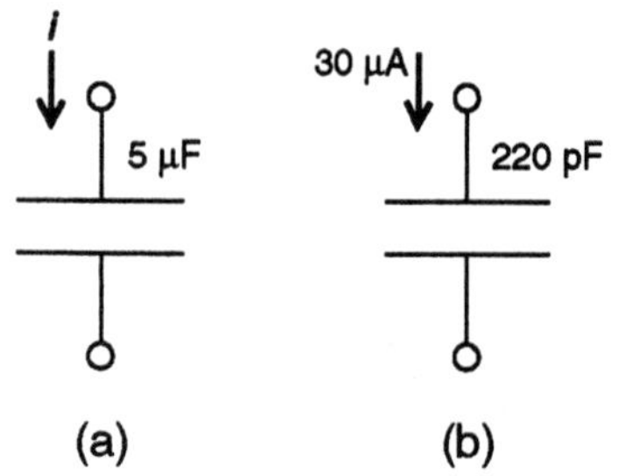

Fig. 5-21.

Sec. 5-4: Another Formula for Capacitance

5-8. A capacitor has a plate area of 0.75 m^2 and a distance between plates of 0.0005 m. If the dielectric constant is 1000, what is the capacitance?
5-9. A capacitor has a distance between plates of 0.001 m. If the dielectric constant is 10 and the capacitance is 1000 pF, what is the area of the plates?
5-10. A capacitor has a distance between plates of 0.0001 m. The dielectric is air. If the plate area can vary from 0.005 m^2 to 0.05 m^2, what is the range of capacitance?

Sec. 5-5: Current and Capacitance

5-11. A capacitance of 20 μF has dv/dt = 500 V/s. What is the charging current?
5-12. A sinusoidal voltage is across a capacitance of 470 pF. If the instantaneous rate of voltage change is 50 V/s at $t = 0$, what is the charging current at the beginning of the sine wave?
5-13. If dv/dt = –12 V/s across a capacitance of 220 μF, what is the charging current?
5-14. If dv/dt = 100 V/s in Fig. 5-21a, what is the current?

Sec. 5-6: Capacitor Opposes Change in Voltage

5-15. What is the capacitance if i = 5 μA and dv/dt = 12 V/s?
5-16. If i = 75 mA and dv/dt = 60 V/s, what is the capacitance?
5-17. What does dv/dt equal if i = 50 μA and C = 1000 pF?
5-18. If i = 50 μA and C = 1000 pF, what does dv/dt equal ?
5-19. What is the dv/dt in Fig. 5-21b?

Sec. 5-7. Series and Parallel Capacitors

5-20. A capacitance of 10 pF is in parallel with another capacitance of 15 pF. What is the total capacitance?
5-21. Three capacitances of 30 pF, 50 pF, and 65 pF are in parallel. What is the total capacitance?
5-22. A capacitance of 35 pF is in series with another capacitance of 25 pF. What is the total capacitance?
5-23. Three capacitances of 60 pF, 40 pF, and 35 pF are in series. What is the total capacitance?
5-24. What is the total capacitance in Fig. 5-22a?
5-25. What is the total capacitance in Fig. 5-22b?
5-26. If the capacitances are doubled in Fig. 5-22a, what is the total capacitance?

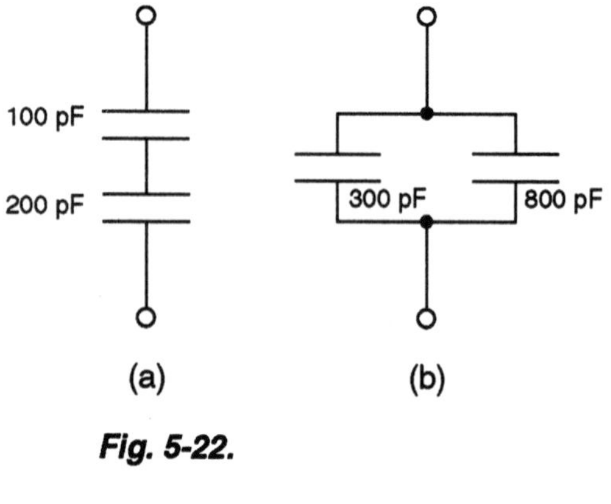

Fig. 5-22.

CHAPTER 6
CAPACITIVE REACTANCE

This chapter is about *capacitive reactance,* the opposition that a capacitor has to alternating current. Capacitive reactance is different from resistance because it depends on the charging and discharging of a capacitor rather than atomic collisions within the material. Since it takes time to charge and discharge a capacitor, capacitive reactance varies with frequency.

6-1. BASIC IDEAS

In Fig. 6-1, the sinusoidal current through the capacitor produces a voltage across the capacitor. Because of this voltage, the capacitor opposes the current. We call this opposition the capacitive reactance, symbolized by X_C.

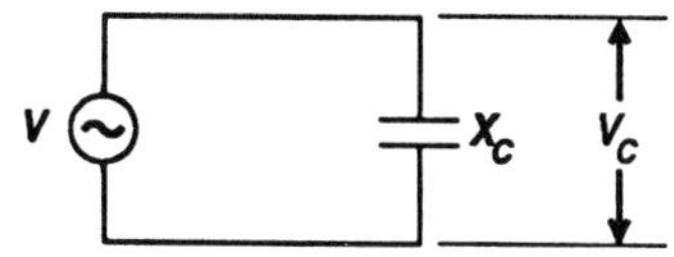

Fig. 6-1.

Definition

In Fig. 6-1, the rms value of the sinusoidal voltage across the capacitor is symbolized by V_C, and the rms value of the sinusoidal current is symbolized by I_C. The defining equation for capacitive reactance is

$$X_C = \frac{V_C}{I_C} \tag{6-1}$$

Since peak values are directly proportional to rms values, Eq. (6-1) may be used with peak values as well as rms values.

For instance, in Fig. 6-2 the rms voltage across the capacitor is 10 V. Assume the rms current is 1 mA. Divide the voltage by the current to get the capacitive reactance:

$$X_C = \frac{10\text{ V}}{1\text{ mA}} = 10\text{ k}\Omega$$

In Fig. 6-2, the peak values are 14.1 V and 1.41 mA. Therefore, another way to calculate the capacitive reactance is with peak values:

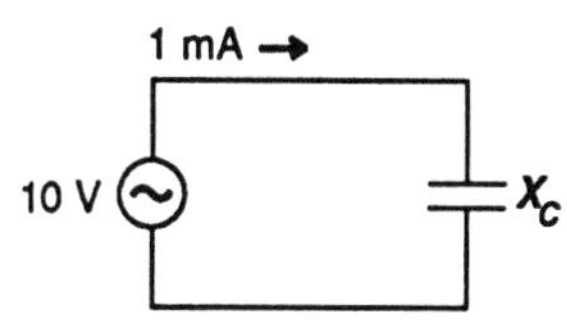

Fig. 6-2.

$$X_C = \frac{14.1\text{ V}}{1.41\text{ mA}} = 10\text{ k}\Omega$$

Ohm's Law

Equation (6-1) is a form of Ohm's law applied to an capacitor. We can rearrange the equation to get

IMPORTANT

$$I_C = \frac{V_C}{X_C}$$

$$I_C = \frac{V_C}{X_C} \qquad (6\text{-}2)$$

This form of Ohm's law for an capacitor is useful because sometimes we know the values of voltage and capacitive reactance. In this case, we can calculate the rms current through the capacitor.

By cross-multiplying Eq. (6-2), we get this alternative form of Ohm's law:

$$V_C = I_C X_C \qquad (6\text{-}3)$$

This equation is useful when we have the rms current and the capacitive reactance. Then, we can calculate the rms voltage across the capacitor.

EXAMPLE 6-1

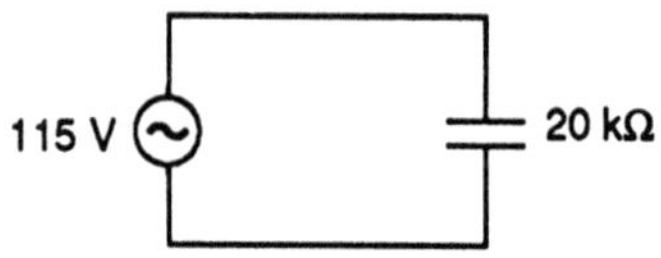

Fig. 6-3.

What is the rms current in Fig. 6-3?

SOLUTION

Divide the rms voltage by the capacitive reactance to get

$$I = \frac{115\text{ V}}{20\text{ k}\Omega} = 5.75\text{ mA}$$

EXAMPLE 6-2

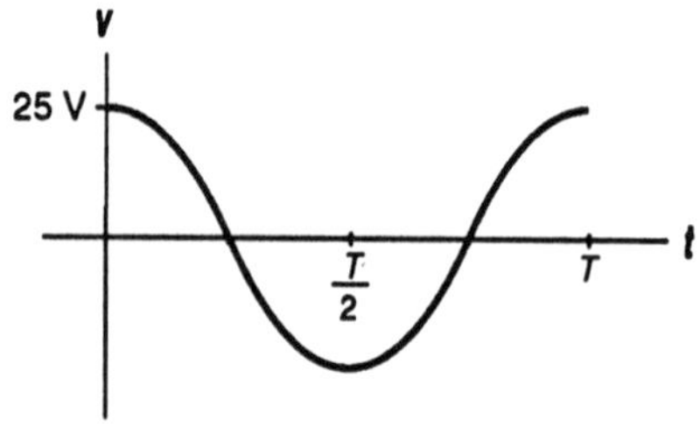

Fig. 6-4.

Suppose the sinusoidal voltage of Fig. 6-4 is across an capacitive reactance of 8 kΩ. What is the peak current? The rms current?

SOLUTION

Since the peak voltage is 25 V, the peak current is

$$I_p = \frac{25\text{ V}}{8\text{ k}\Omega} = 3.13\text{ mA}$$

The rms current equals 0.707 times the peak current:

$$I_{rms} = 0.707(3.13\text{ mA}) = 2.21\text{ mA}$$

EXAMPLE 6-3

If the current is 2 mA, what is the rms voltage in Fig. 6-5? The peak voltage?

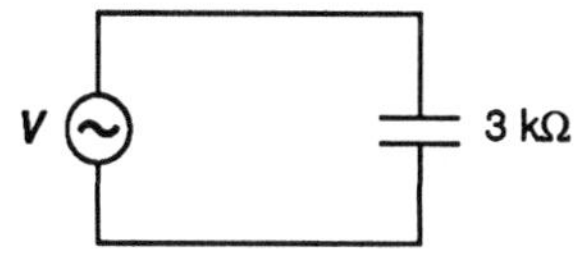

Fig. 6-5.

SOLUTION

Multiply the rms current by the capacitive reactance to get the rms voltage across the capacitor:

$V_{rms} = (2\text{ mA})(3\text{ k}\Omega) = 6\text{ V}$

The peak voltage is

$V_p = 1.414(6\text{ V}) = 8.48\text{ V}$

EXAMPLE 6-4

A capacitive reactance of 10 kΩ has a sinusoidal current with a peak value of 2.25 mA. What is the rms voltage across the capacitor?

SOLUTION

Do this in two steps. First, find the peak voltage:

$V_p = (2.25\text{ mA})(10\text{ k}\Omega)) = 22.5\text{ V}$

Second, find the rms voltage:

$V_{rms} = 0.707(22.5\text{ V}) = 15.9\text{ V}$

6-2. THE EFFECT OF *F* AND *C*

Two factors that affect the capacitive reactance are the *frequency* and the *capacitance*. First, increasing the frequency means there is less time each cycle to charge and discharge the capacitor. Second, increasing the capacitance means a given current will produce less voltage.

Formula for Capacitive Reactance

The following formula can be derived for the capacitive reactance:

$$X_C = \frac{1}{2\pi fC} \qquad (6\text{-}4)$$

MEMORIZE

$$X_C = \frac{1}{2\pi fC}$$

Memorize this formula because it is used constantly in the analysis of ac circuits. This formula is useful when you know the values of frequency and capacitance. For instance, if f = 60 Hz and C = 5 μF, the capacitive reactance is

$$X_C = \frac{1}{2\pi(60\ \text{Hz})(5\ \mu\text{F})} = 531\ \Omega$$

Calculating C and f

Equation (6-4) contains three variables: X_C, f, and C. Given any two of these quantities, you can solve for the third. For instance, if you know the values of f and C, you can substitute into the equation and calculate X_C.

On the other hand, if you have the values of X_C and f, you can rearrange the equation to get

$$C = \frac{1}{2\pi f X_C} \tag{6-5}$$

After substituting the values of X_C and f, you can calculate the value of C.

A third possibility is knowing the values of X_C and C. In this case, the equation can be rearranged as

$$f = \frac{1}{2\pi C X_C} \tag{6-6}$$

After substituting for X_C and C, you can solve for f.

Derivation

Where does Eq. (6-4) come from? The standard proof uses calculus. But we can prove the equation using only trigonometry and the method of small changes. If you are not interested in seeing the proof, skip to Example 6-5.

To begin with, one of identities discussed in trigonometry is this relation:

$$\sin x = x - \frac{x^3}{3!} + \frac{x^5}{5!} - \frac{x^7}{7!} + \ldots$$

where x is an angle in radians. When x is small, the higher-order terms drop out, leaving

$$\sin x = x$$

In terms of the rotation angle, this may be written as

$$\sin 2\pi ft = 2\pi ft$$

The equation for any sinusoidal voltage is

$$v = V_p \sin 2\pi ft$$

When the angle of rotation is small, the foregoing simplifies to

$$v = V_p(2\pi ft) = 2\pi fV_p t$$

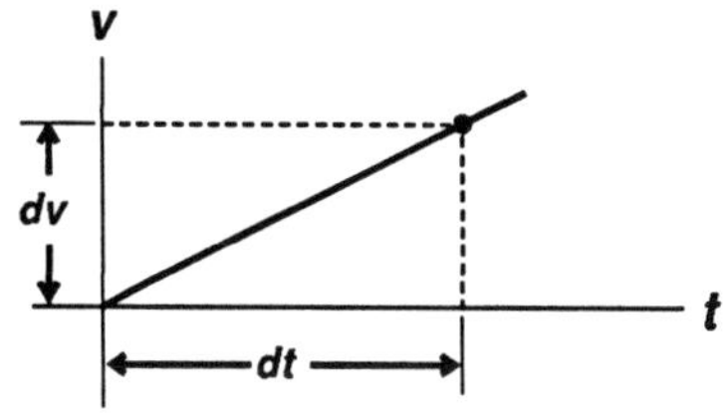

Fig. 6-6.

Recall the discussion of small changes. Figure 6-6 shows the earliest part of a sine wave. When dt is less than $0.01T$, the earliest part of a sine wave becomes linear. Because of this, variables v and t may be replaced by the changes dv and dt. In other words, the foregoing equation may be written as

$$dv = 2\pi fV_p dt$$

or

$$\frac{dv}{dt} = 2\pi fV_p \qquad (6\text{-}7)$$

Recall the physical meaning of this. It is slope of the line in Fig. 6-6. It tells you how the fast the sine wave is changing at the beginning of the cycle.

The relation between current and capacitance is given by

$$i = C\frac{dv}{dt}$$

When we substitute Eq. (6-7) into this, we get

$$I_p = C(2\pi fV_p)$$

which rearranges into

$$\frac{V_p}{I_p} = \frac{1}{2\pi fC}$$

or

$$X_C = \frac{1}{2\pi fC}$$

This final result gives us a way to calculate capacitance reactance using frequency and capacitance.

EXAMPLE 6-5

What is the rms current in Fig. 6-7?

SOLUTION

First, get the capacitive reactance:

$$X_C = \frac{1}{2\pi(60\text{ Hz})(10\ \mu\text{F})} = 265\ \Omega$$

Second, use Ohm's law to get the rms current:

$$I = \frac{115\text{ V}}{265\ \Omega} = 0.434\text{ A}$$

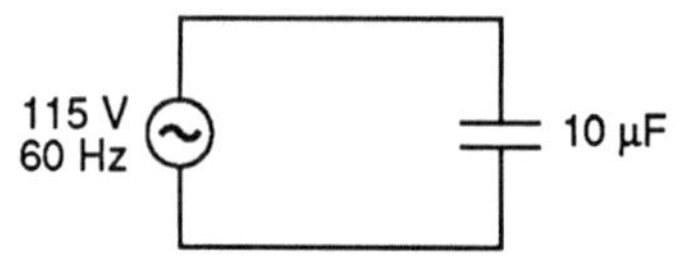

Fig. 6-7.

EXAMPLE 6-6

What is the rms current in Fig. 6-7 if the frequency is changed to 600 Hz?

SOLUTION

You can use the same approach as in the preceding example. Or you can use a shortcut. When the frequency increases by a factor of 10, the capacitive reactance decreases by a factor of 10 and the current increases by a factor of 10. This means we can multiply the current of the preceding example by 10 to get

$$I = 10(0.434\text{ A}) = 4.34\text{ A}$$

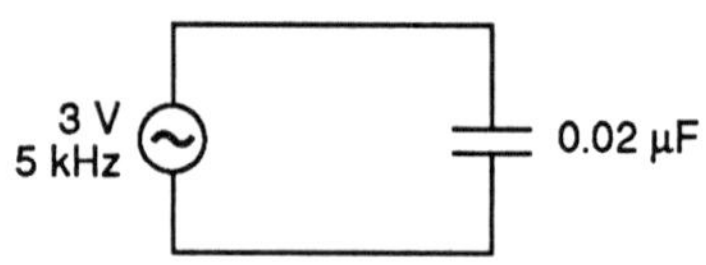

Fig. 6-8.

EXAMPLE 6-7

What is the rms current in Fig. 6-8?

SOLUTION

First, get the capacitive reactance:

$$X_C = \frac{1}{2\pi(5\text{ kHz})(0.02\ \mu\text{F})} = 1.59\text{ k}\Omega$$

Second, use Ohm's law to get the rms current:

$$I = \frac{3\text{ V}}{1.59\text{ k}\Omega} = 1.89\text{ mA}$$

6-3. SERIES AND PARALLEL CAPACITORS

Since capacitive reactance is measured in ohms, it makes sense to say that capacitive reactances in series can be added to get the total capacitive reactance. Likewise, capacitive reactances in parallel may be combined by the product-over-sum rule to get the total capacitive reactance.

SERIES

$$X_C = X_{C1} + X_{C2}$$

X_C for Series Capacitors

The total capacitive reactance of two series capacitors is

$$X_C = X_{C1} + X_{C2} \tag{6-8}$$

Figure 6-9a shows two capacitors in series. The total capacitive reactance is the sum of the two capacitive reactances as shown in Fig. 6-9b. For instance, if $X_{C1} = 50\ \Omega$ and $X_{C1} = 75\ \Omega$, the total capacitive reactance is

$$X_C = 50\ \Omega + 75\ \Omega = 125\ \Omega$$

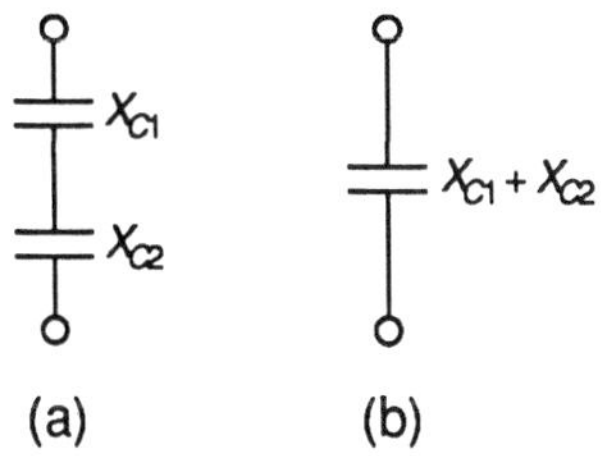

Fig. 6-9.

When two capacitors are in series, they combine by the product-over-the-sum rule:

$$C = \frac{C_1 C_2}{C_1 + C_2} \tag{6-9}$$

This equation is easy to prove. Equation (6-8) may be written as

$$\frac{1}{2\pi fC} = \frac{1}{2\pi fC_1} + \frac{1}{2\pi fC_2}$$

By combining the terms on the right side, we get

$$\frac{1}{2\pi fC} = \frac{C_1 + C_2}{2\pi fC_1 C_2}$$

Now, this can be written as

$$\frac{1}{2\pi fC} = \frac{1}{2\pi fC_1 C_2/(C_1 + C_2)}$$

This shows that the total capacitance equals the product over the sum.

X_C for Parallel Capacitors

The total capacitive reactance of two parallel capacitors is

$$X_C = \frac{X_{C1} X_{C2}}{X_{C1} + X_{C2}} \tag{6-10}$$

PARALLEL

$$X_C = \frac{X_{C1} X_{C2}}{X_{C1} + X_{C2}}$$

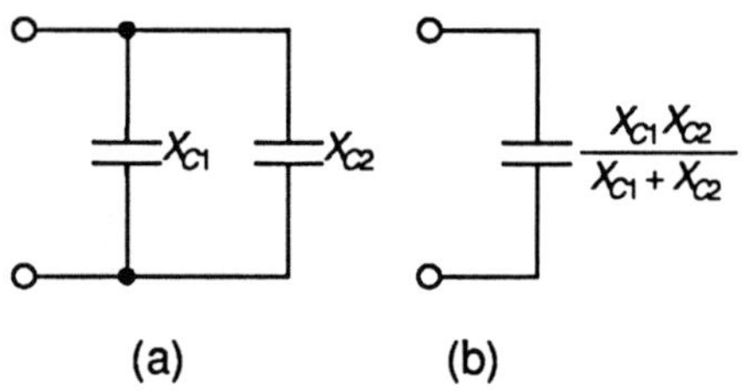

Fig. 6-10.

For instance, given two capacitive reactances as shown in Fig. 6-10a, the total capacitive reactance is given by the product over the sum as shown in Fig. 6-10b.

When two capacitors are in parallel, they act the same as a single capacitor with a value of

$$C = C_1 + C_2 \tag{6-11}$$

The proof of this is similar to that given for series capacitors. Express Eq. (6-10) in terms of $2\pi f$. Then, simplify the equation to get Eq. (6-11).

EXAMPLE 6-8

What is the rms current in Fig. 6-11?

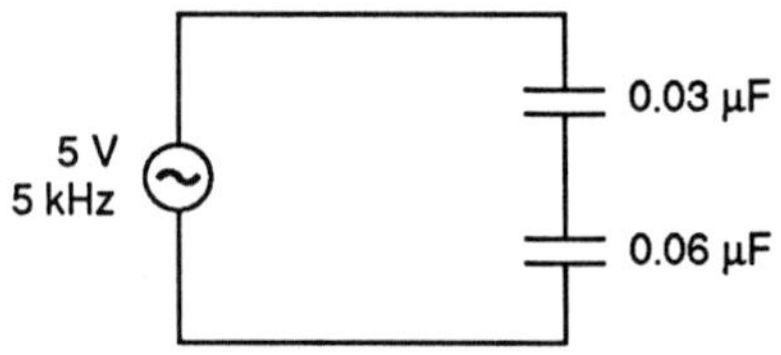

Fig. 6-11.

SOLUTION

The capacitors are in series. The total capacitance is

$$C = \frac{(0.03\ \mu\text{F})(0.06\ \mu\text{F})}{0.03\ \mu\text{F} + 0.06\ \mu\text{F}} = 0.02\ \mu\text{F}$$

The capacitive reactance is

$$X_C = \frac{1}{2\pi(5\ \text{kHz})(0.02\ \mu\text{F})} = 1.59\ \text{k}\Omega$$

Next, use Ohm's law to get the rms current:

$$I = \frac{5\ \text{V}}{1.59\ \text{k}\Omega} = 3.14\ \text{mA}$$

EXAMPLE 6-9

What is the rms current in Fig. 6-12?

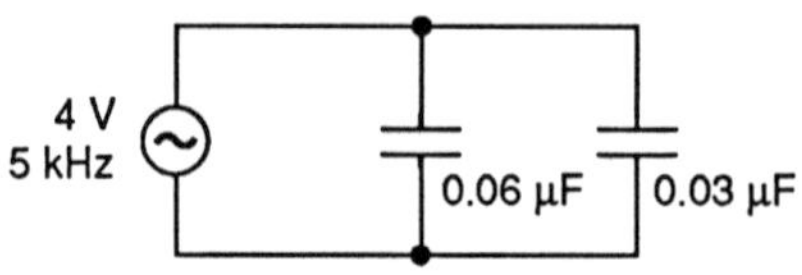

Fig. 6-12.

SOLUTION

The capacitors are in parallel. The total capacitance is

$$C = 0.06\ \mu\text{F} + 0.03\ \mu\text{F} = 0.09\ \mu\text{F}$$

The capacitive reactance is

$$X_C = \frac{1}{2\pi(5\ \text{kHz})(0.09\ \mu\text{F})} = 354\ \Omega$$

Now, use Ohm's law to get the rms current:

$$I = \frac{4\ \text{V}}{354\ \Omega} = 11.3\ \text{mA}$$

6-4. INSTANTANEOUS ANALYSIS

Typical ac voltmeters will indicate the voltages as rms values. But when you use an oscilloscope, you will see sinusoidal voltages on the screen. For this reason, you have to know more about the instantaneous operation of capacitive circuits. Many of the concepts in this section were discussed earlier in Chap. 3. What follows is a review of the key concepts as they apply to capacitive circuits.

Additive Property of Sine Waves

The additive property of sine waves is valid for capacitors as well as inductors. Recall the additive property means:

The sum of two or more in-phase sine waves is a sine wave.

Figure 6-13 illustrates the idea graphically. If we add the instantaneous values of v_1 and v_2, we get the new sine wave labeled v. At an angle of 90°, the peak value of the new sine wave is

$$V_p = V_{p1} + V_{p2}$$

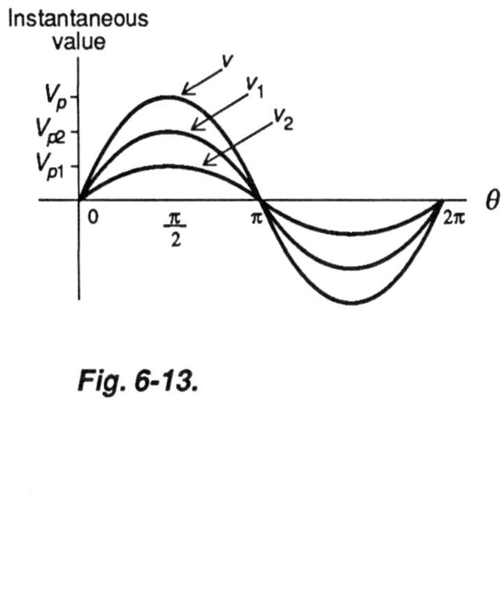

Fig. 6-13.

For instance, if $V_{p1} = 3$ V and $V_{p2} = 4$ V, the new sine wave has a peak value of

$$V_p = 3\ \text{V} + 4\ \text{V} = 7\ \text{V}$$

Capacitors in Series

Here are two properties that apply to capacitive circuits as well as inductive circuits:

The instantaneous current is the same in all parts of a series ac circuit.

and

The total instantaneous voltage in a series ac circuit is the sum of the individual instantaneous voltages.

For instance, Fig. 6-14 shows an ac voltage source across two capacitors. Because of the series connection, the instantaneous current is the same in all parts of the circuit. The values of this current can be calculated at any instant in time with this equation:

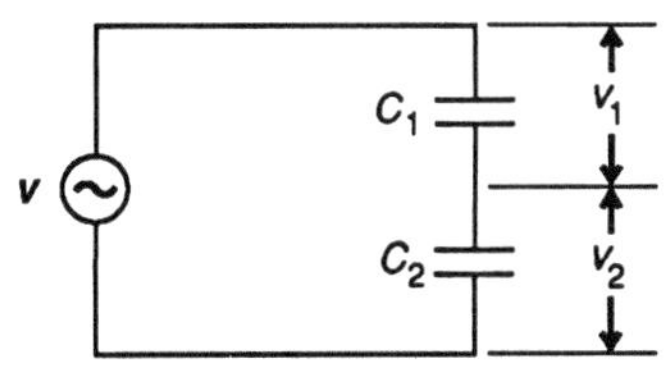

Fig. 6-14.

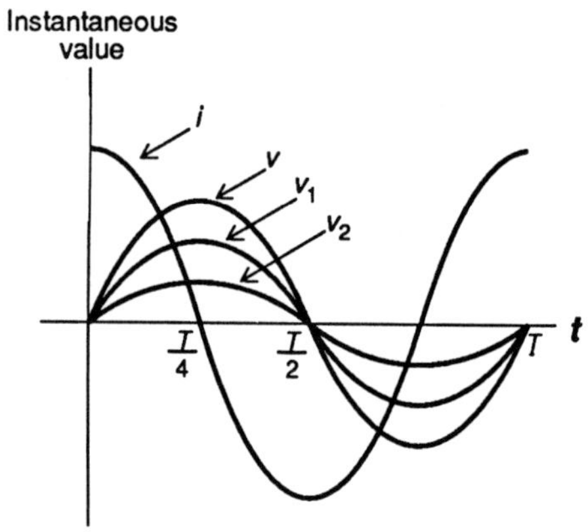

Fig. 6-15.

$i = I_p \cos 2\pi ft$

Figure 6-15 shows a sketch of this sinusoidal current.

The instantaneous current is the same in both capacitors. For instance, when the instantaneous current is zero, both capacitors have zero current. When the instantaneous current reaches its peak value, both capacitors have a maximum current.

Because the instantaneous current is the same in both capacitors, each capacitor voltage lags the current by 90°. Therefore, the capacitor voltages are given by

$$v_1 = V_{p1} \sin 2\pi ft$$
$$v_2 = V_{p2} \sin 2\pi ft$$

We can add these two voltages to get the total voltage:

$$v = v_1 + v_2 = V_{p1} \sin 2\pi ft + V_{p2} \sin 2\pi ft$$

or

$$v = (V_{p1} + V_{p2}) \sin 2\pi ft$$

This says the total voltage is a sine wave shifted by a quarter cycle. The peak value is the sum of the two peak values.

Figure 6-15 shows the capacitor voltages and the total voltage. As you can see, v_1 and v_2 lag i by a quarter cycle. Notice that the total voltage v is the sum of the individual voltages v_1 and v_2. Because of this, the total voltage v also lags the current i by a quarter cycle. Make sure you understand the waveforms in this figure. The key ideas are

1. The same current is in all series components
2. The individual voltages lag this current by a quarter cycle.
3. The total voltage is the sum of the individual voltages.

Capacitors in Parallel

Here are two more properties that apply to capacitive circuits as well as inductive circuits:

The instantaneous voltage is the same across all parts of a parallel ac circuit.

and

The total instantaneous current in a parallel ac circuit is the sum of the individual instantaneous currents.

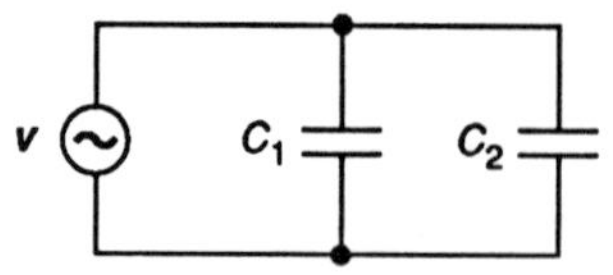

Fig. 6-16.

For instance, Fig. 6-16 shows an ac voltage source across two capacitors. Because of the parallel connection, the instantaneous voltage is the same across all parts of the circuit. This voltage can be calculated at any instant in time with this equation:

$$v = V_p \sin 2\pi ft$$

Figure 6-17 shows a sketch of this sinusoidal voltage.

Because the instantaneous voltage is the same across both capacitors, each capacitor current *leads* its voltage by 90°. Therefore, the capacitor currents are given by

$$i_1 = I_{p1} \cos 2\pi ft$$
$$i_2 = I_{p2} \cos 2\pi ft$$

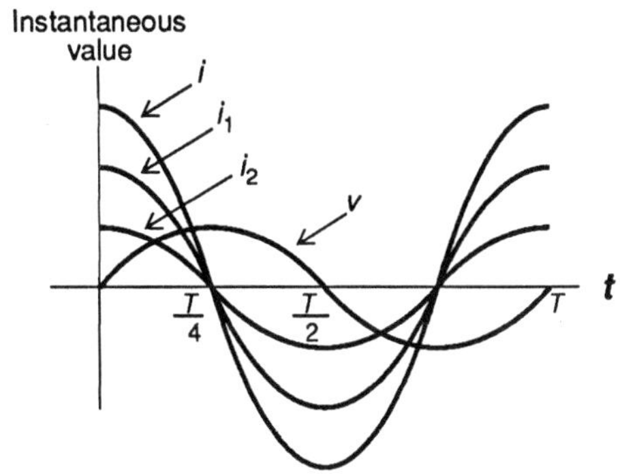

Fig. 6-17.

We can add these two currents to get the total current:

$$i = i_1 + i_2 = I_{p1} \cos 2\pi ft + I_{p2} \cos 2\pi ft$$

or

$$i = (I_{p1} + I_{p2}) \cos 2\pi ft$$

This says the total current is a sine wave with a peak value equal to the sum of the two peak values.

Figure 6-17 shows the capacitor currents and the total current. As you can see, i_1 and i_2 lead v by a quarter cycle. Notice that the total current i is the sum of the individual currents i_1 and i_2. Because of this, the total current i also leads the voltage v by a quarter cycle. Make sure you understand the waveforms in this figure. The key ideas are

1. *The same voltage is across all components.*
2. *The individual currents lead the voltage by a quarter cycle.*
3. *The total current is the sum of the individual currents.*

6-5. PHASORS

You use phasors in almost the same way with capacitors as with inductors. As before, we use phasors to represent sinusoidal quantities. To get a unique code, we draw each phasor as it appears at the beginning of the cycle. For instance, Fig. 6-18a shows the phasor for a sinusoidal voltage across a capacitor, and Fig. 6-18b shows the phasor for the sinusoidal current.

As you see, the phasor is drawn at whatever angle is necessary to produce the initial voltage. For instance, the sinusoidal voltage of Fig. 6-18a has an initial value of zero; therefore, the phasor is drawn at an angle of 0°. The sinusoidal current of Fig. 6-18b has an initial value of I_p; therefore, the phasor is drawn at an angle of 90°.

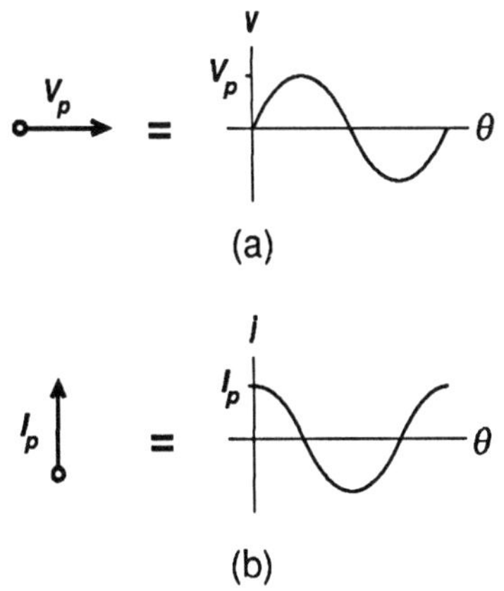

Fig. 6-18.

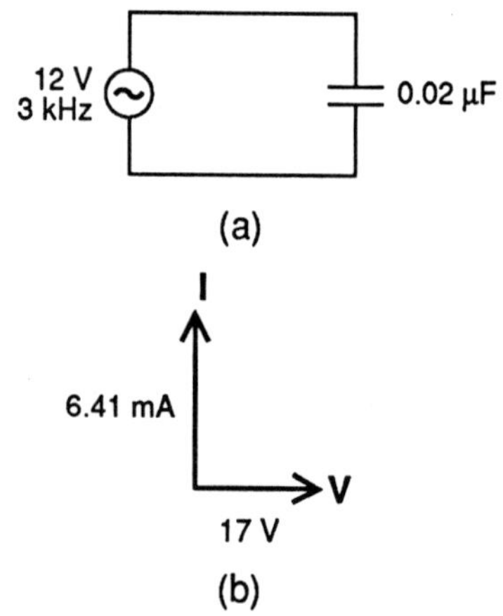

Fig. 6-19.

EXAMPLE 6-10

Show the phasors for peak current and peak voltage in Fig. 6-19a.

SOLUTION

The capacitive reactance is

$$X_C = \frac{1}{2\pi(3\text{ kHz})(0.02\ \mu\text{F})} = 2.65\text{ k}\Omega$$

The rms current is

$$I = \frac{12\text{ V}}{2.65\text{ k}\Omega} = 4.53\text{ mA}$$

The peak current is

$$I_p = 1.414(4.53\text{ mA}) = 6.41\text{ mA}$$

and the peak voltage is

$$V_p = 1.414(12\text{ V}) = 17\text{ V}$$

Figure 6-19b shows the phasors. We can see that the sinusoidal voltage has a peak value of 17 V, while the sinusoidal current has a peak value of 6.41 mA and a phase angle of 90°.

EXAMPLE 6-13

In Fig. 6-20a, an rms current of 32 mA flows through a capacitance of 0.005 μF. If the frequency is 50 kHz, draw the phasor diagram for the sinusoidal current and voltage.

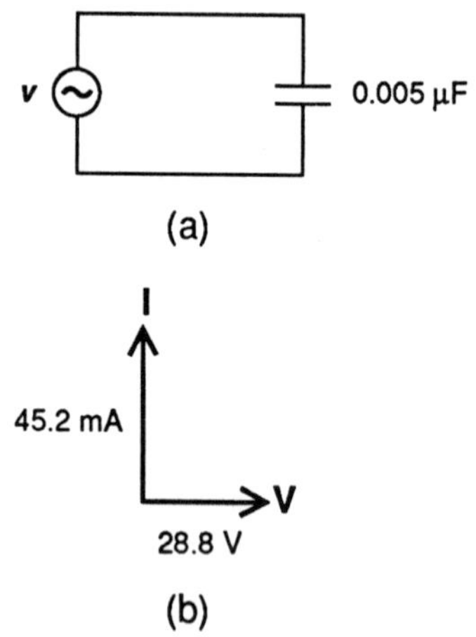

Fig. 6-20.

SOLUTION

The capacitive reactance is

$$X_C = \frac{1}{2\pi(50\text{ kHz})(0.005\ \mu\text{F})} = 637\ \Omega$$

The rms voltage across the capacitor is

$$V = (32\text{ mA})(637\ \Omega) = 20.4\text{ V}$$

Now, calculate the peak current and peak voltage:

$$I_p = 1.414(32\text{ mA}) = 45.2\text{ mA}$$

$$V_p = 1.414(20.4\text{ V}) = 28.8\text{ V}$$

Figure 6-20b shows the phasor diagram.

MULTIPLE-CHOICE TEST

The following questions may have more than one right answer. Select the best answer. This is the one that is always true, or covers more situations, or agrees with the material of this chapter, etc.

1. Capacitive reactance equals the rms voltage divided by the
   a. instantaneous current
   b. rms current
   c. peak current
   d. average current

2. When frequency increases, capacitive reactance
   a. decreases
   b. is unaffected
   c. increases
   d. none of the above

3. When capacitance increases, capacitive reactance
   a. decreases
   b. is unaffected
   c. increases
   d. none of the above

4. When two capacitors are in series, the total capacitive reactance is
   a. less than the smaller
   b. less than the larger
   c. equal to the sum of the two capacitive reactances
   d. equal to the product over the sum of the capacitive reactances

5. When two capacitors are in parallel, the total capacitive reactance is
   a. less than the smaller
   b. less than the larger
   c. equal to the sum of the two capacitive reactances
   d. equal to the product over the sum of the capacitive reactances

6. The sum of two in-phase sine waves is a
   a. sine wave
   b. cosine wave
   c. square wave
   d. none of the above

7. The instantaneous current is the same in all parts of a
   a. series ac circuit
   b. parallel ac circuit
   c. parallel dc circuit
   d. none of the above

8. The length of a phasor equals the
   a. voltage
   b. peak value
   c. rotation angle
   d. frequency

9. When the voltage across a capacitor has a phase angle of 0°, the current has a phase angle of
   a. 0°
   b. 90°
   c. 180°
   d. 270°

10. If you divide the peak voltage across an capacitor by the peak current, you get the
    a. resistance
    b. frequency
    c. period
    d. capacitive reactance

11. When the current into a capacitor has a phase angle of 0°, the voltage has a phase angle of
    a. 0°
    b. –90°
    c. –180°
    d. –270°

12. The rms voltage across an capacitor divided by the capacitive reactance gives the
    a. peak voltage
    b. rms voltage
    c. peak current
    d. rms current

13. The rms current through an capacitor times the capacitive reactance gives the
    a. peak voltage
    b. rms voltage
    c. peak current
    d. rms current

14. When the frequency doubles, the capacitive reactance
    a. drops in half
    b. stays the same
    c. doubles
    d. quadruples

15. If you divide 1 by $2\pi fX_C$, you get the
    a. peak voltage
    b. peak current
    c. rms current
    d. capacitance

16. Two series capacitors have reactances of 30 Ω and 60 Ω. The total capacitive reactance is
   a. 20 Ω
   b. 30 Ω
   c. 60 Ω
   d. 90 Ω

17. Two in-phase sinusoidal voltages have peak values of 10 V and 15 V. If these sinusoidal voltages are added, the total peak voltage is
   a. 0 V
   b. 10 V
   c. 15 V
   d. 25 V

18. A sinusoidal voltage has a phase angle of 90°. Its phasor points
   a. to the left
   b. straight up
   c. to the right
   d. straight down

PROBLEMS

Sec. 6-1: Basic Ideas

6-1. What is the rms current in Fig. 6-21 ?
6-2. If the applied voltage doubles in Fig. 6-21 , what is the rms current?
6-3. If the capacitive reactance doubles in Fig. 6-21 , what is the rms current?
6-4. If the capacitance decreases by a factor of 10 in Fig. 6-21, what is the current?
6-5. 50 μA flows through an capacitive reactance of 24 kΩ. What is the rms voltage across the capacitor?
6-6. A sinusoidal voltage with a peak value of 15 V is across an capacitive reactance of 2 kΩ. What is the rms current?

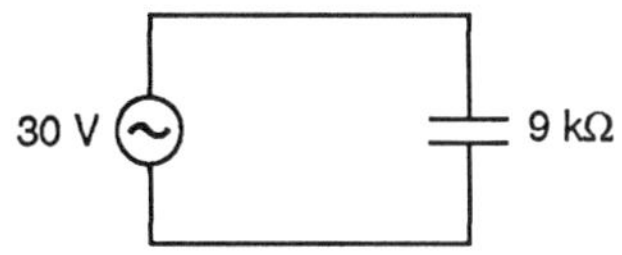

Fig. 6-21.

Sec. 6-2: The Effect of F and C

6-7. What is the rms current in Fig. 6-22 ?
6-8. If the frequency changes to 500 Hz in Fig. 6-22 , what is the rms current?
6-9. If the frequency changes to 5 kHz in Fig. 6-22 , what is the rms current?
6-10. 15 V rms is across an capacitance of 0.68 μF. What is the rms current if the frequency is 10 kHz?
6-11. 250 μA flows through an capacitance of 800 pF. What is the rms voltage across the capacitor if the frequency is 1 MHz?
6-12. A sinusoidal voltage with a peak value of 15 V is across an capacitance of 0.22 μF. If the frequency is 20 kHz, what is the rms current?

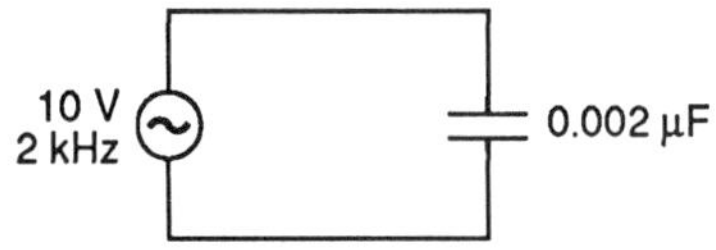

Fig. 6-22.

Sec. 6-3: Series and Parallel Capacitors

6-13. What is the rms current in Fig. 6-23 ?
6-14. If both capacitances are doubled in Fig. 6-23 , what is the rms current?

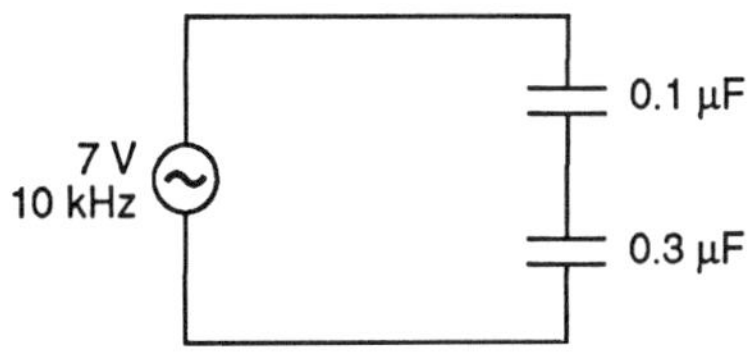

Fig. 6-23.

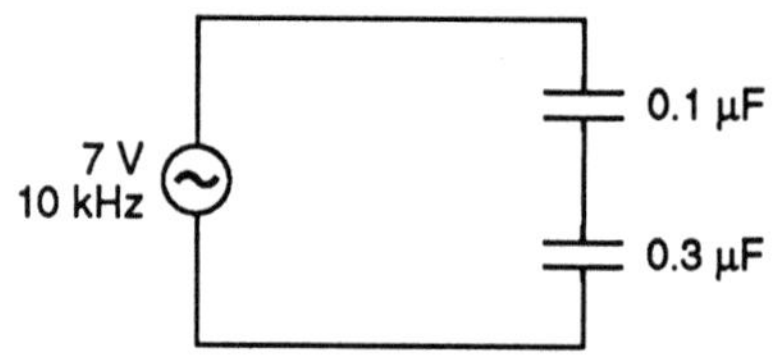

Fig. 6-23.

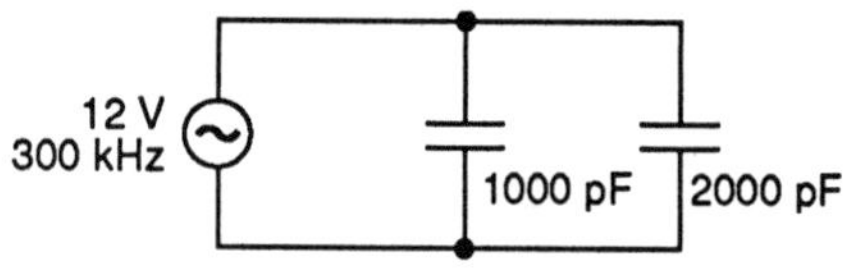

Fig. 6-24.

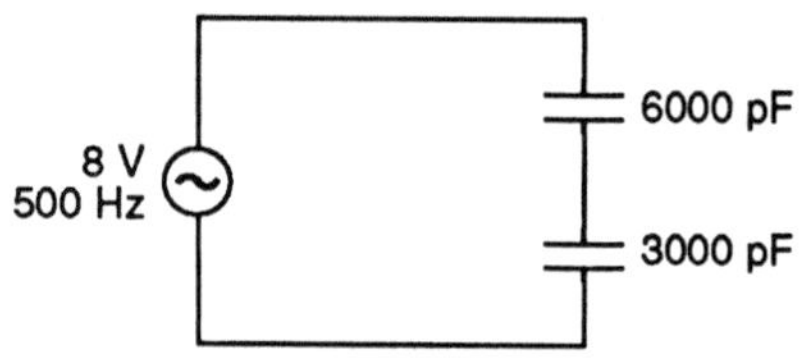

Fig. 6-25.

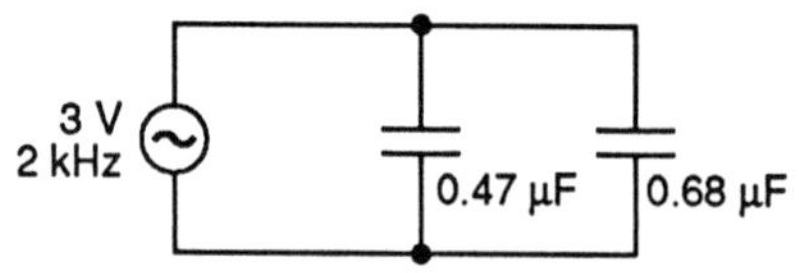

Fig. 6-26.

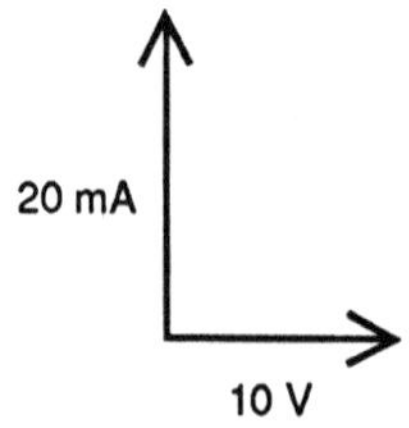

Fig. 6-27.

6-15. If the frequency decreases to 1 kHz, what is the rms current in Fig. 6-24 ?

6-16. What is the total rms current in Fig. 6-24 ?

6-17. In Fig. 6-24 , what is the rms current through the 1000 pF? Through the 2000 pF?

6-18. If the frequency doubles in Fig. 6-24 , what is the total rms current? The rms current through the left capacitor? Through the right capacitor?

Sec. 6-4. Instantaneous Analysis

6-19. What are the capacitive reactances for the capacitors of Fig. 6-25 ? The total capacitive reactance? The peak current? The peak voltages across the capacitors?

6-20. Sketch the current and voltage waveforms for Fig. 6-25 .

6-21. If the applied voltage doubles in Fig. 6-25 , what is the peak current? The peak voltages across the capacitors? Sketch the current and voltage waveforms.

6-22. If the frequency doubles in Fig. 6-25 , what is the peak current? The peak voltages across the capacitors? Sketch the current and voltage waveforms.

6-23. What are the capacitive reactances for the capacitors of Fig. 6-26 ? The total capacitive reactance? The peak current? The peak voltages across the capacitors?

6-24. Sketch the current and voltage waveforms for Fig. 6-26 .

6-25. If the applied voltage doubles in Fig. 6-26 , what is the peak current? The peak voltages across the capacitors? Sketch the current and voltage waveforms.

6-26. If the frequency doubles in Fig. 6-26 , what is the peak current? The peak voltages across the capacitors? Sketch the current and voltage waveforms.

Sec. 6-5. Phasors

6-27. Figure 6-27 shows the phasor diagram for a capacitor . What is the peak voltage across the capacitor? The rms voltage? The peak current? The rms current?

6-28. An capacitor has the phasor diagram of Fig. 6-27 . If the voltage remains at 10 V but the frequency doubles, what happens to the current phasor? The voltage phasor?

6-29. Draw the phasor diagram for Fig. 6-21 . Your drawing should look like Fig. 6-27 but the numbers should be different.

6-30. Draw the phasor diagram for Fig. 6-22 .

CHAPTER 7
CAPACITIVE CIRCUITS

This chapter shows you how to analyze ac circuits that have resistors and capacitors. The analysis of capacitive circuits is similar to that of inductive circuits. The main difference is the current phasors of capacitive circuits lead the voltage phasors.

7-1. PHASORS FOR A CAPACITOR

When a component is purely capacitive, the instantaneous voltage across the capacitance is given by

$$i = C\frac{dv}{dt} \quad (7\text{-}1)$$

If the voltage is sinusoidal, it is given by

$$v = V_p \sin \theta$$

As shown in Chap. 5, the current is sinusoidal and given by

$$i = I_p \cos \theta$$

where $I_p = V_p/X_C$. This means a sinusoidal voltage across a capacitor produces a sinusoidal current with a phase angle of 90°. In terms of phasors, it means the current leads the voltage by 90°, as shown in Fig. 7-1. An equivalent statement is the voltage lags the current by 90°.

Fig. 7-1.

EXAMPLE 7-1

Figure 7-2a shows a capacitive circuit. The first capacitor has a capacitive reactance of 1 kΩ, and the second has a capacitive reactance of 2 kΩ. If the source voltage has a peak value of 60 V, what is the current? The voltage across each resistor?

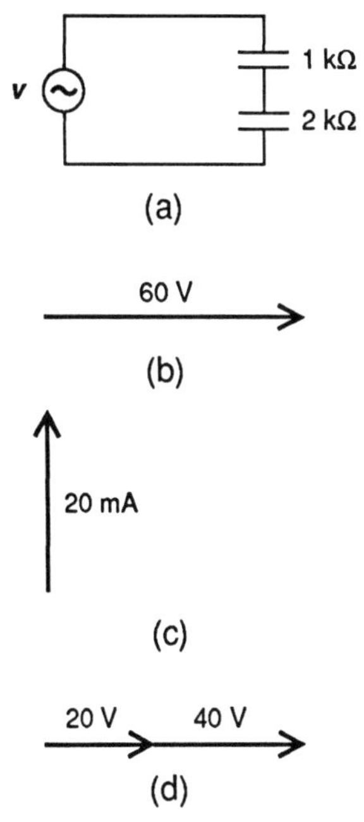

Fig. 7-2.

SOLUTION

Visualize the phasor of the source voltage as shown in Fig. 7-2b. This voltage produces a current with a peak value of

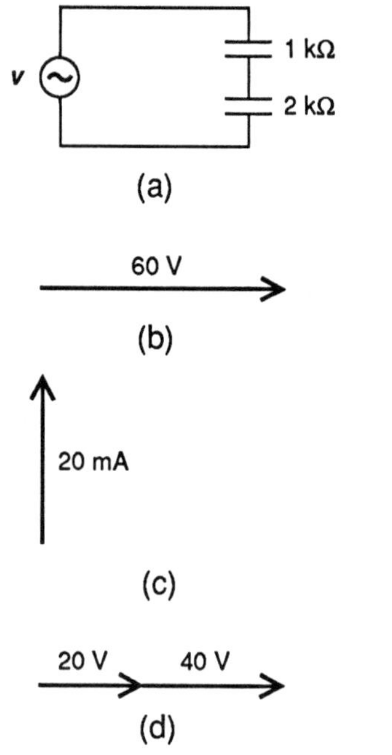

Fig. 7-2.

$$I_p = \frac{60\text{ V}}{3\text{ k}\Omega} = 20\text{ mA}$$

Figure 7-2c shows the phasor for this current. Notice that it leads the applied voltage by 90°. This is equivalent to saying the applied voltage lags the current by 90°.

Since the peak current is 20 mA, the first capacitor has a peak voltage of

$$V_{p1} = I_pX_{C1} = (20\text{ mA})(1\text{ k}\Omega) = 20\text{ V}$$

The second inductor has a peak voltage of

$$V_{p1} = I_pX_{C1} = (20\text{ mA})(2\text{ k}\Omega) = 40\text{ V}$$

Figure 7-2d shows the phasors for these two voltages. Notice that these voltages are in phase and add to a total of 60 V, the magnitude of the applied voltage.

7-2. SERIES CIRCUITS

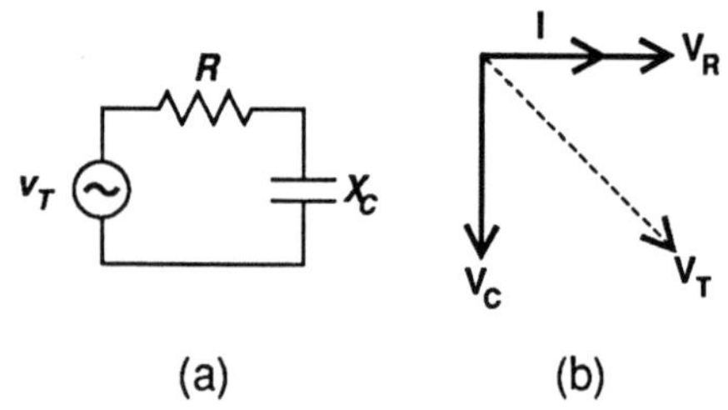

Fig. 7-3.

Figure 7-3a shows a series *RC* circuit where an ac source is connected to a resistor and capacitor in series. Because it is a series circuit, the same instantaneous current exists in the resistor and the capacitor. This means a sinusoidal source voltage produces a sinusoidal current in all parts of the circuit. At any instant in time, the current has the same value in the resistor and the capacitor. Stated another way, the current is in phase in both the resistor and capacitor.

To simplify the analysis, we will assume the instantaneous current has a phase angle of 0°. Therefore, it is given by

$$i = I_p \sin \theta$$

Figure 7-3b shows the phasor current. As you see, it is a horizontal vector pointing to the right. This current is the same for the resistor and the capacitor.

From earlier discussions, we know the voltage across the resistance is always in phase with the current through the resistance. This is why the phasor voltage across the resistor is drawn with a phase angle of 0° in Fig. 7-3b. On the other hand, the voltage across a capacitance lags the current by 90°. This is why the phasor voltage across the capacitance is shown with a phase angle of –90°.

Most of the time, we don't need to know the instantaneous voltages in a circuit. All we usually need to know are the peak or rms voltages. In this case, it is lot easier and faster to use phasor voltages. Kirchhoff's voltage law is valid for phasor voltages, provided we use vector addition. When phasor voltages are used, the sum of the phasor voltages is

$$\mathbf{V_T} = \mathbf{V_R} + \mathbf{V_C} \qquad (7\text{-}2)$$

This says the total phasor voltage of the circuit equals the sum of the individual phasor voltages. In this equation, the plus sign stands for vector addition because we are dealing with vector quantities.

When a resistance is in series with a capacitance, the phasor voltages are 90° apart as shown in Fig. 7-4. Because of the right triangle, we can use the Pythagorean theorem to calculate the total voltage:

$$V_T = \sqrt{V_R^2 + V_C^2} \tag{7-3}$$

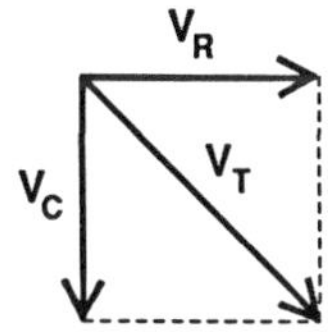

Fig. 7-4.

Again, the total voltage is not the ordinary sum of the two voltages. Instead, it is the quadrature sum.

QUADRATURE SUM

$$V_T = \sqrt{V_R^2 + V_C^2}$$

EXAMPLE 7-2

In Fig. 7-5, you measure an rms voltage of 3.75 V across the resistor and an rms voltage of 2.86 V across the capacitor. What is the total rms voltage?

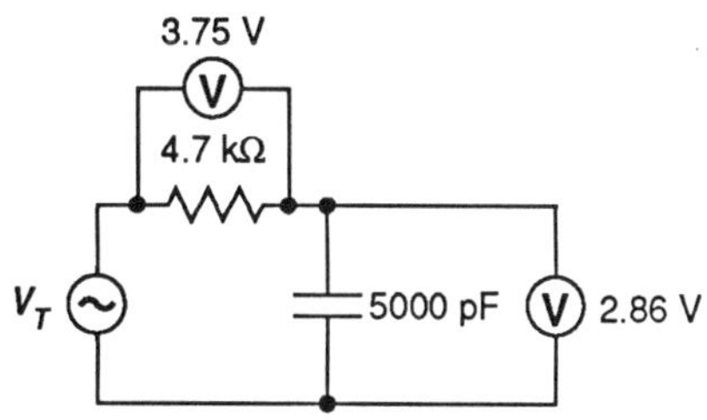

Fig. 7-5.

SOLUTION

The total voltage equals the quadrature sum, the square root of the sum of the squares:

$$V_T = \sqrt{(3.75\text{ V})^2 + (2.86\text{ V})^2} = 4.72\text{ V}$$

This is the total rms voltage. This is what you would measure if you connect an ac voltmeter across the voltage source. The reason this voltage is less than the sum is because the two voltages being measured are 90° out of phase.

EXAMPLE 7-3

In Fig. 7-6a, the rms current is 2 mA. Calculate the rms value of the applied voltage.

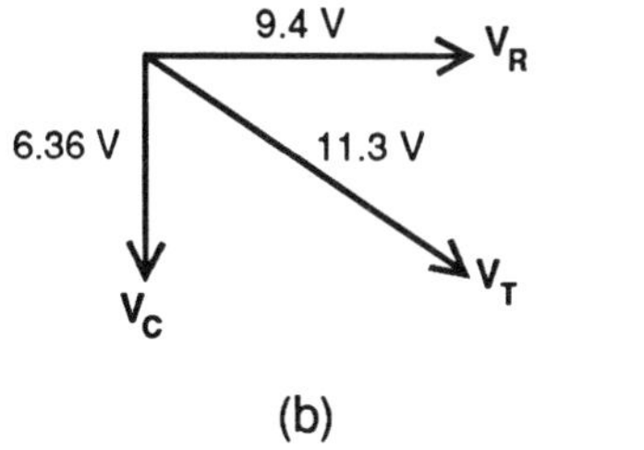

Fig. 7-6.

SOLUTION

The capacitive reactance is

$$X_C = \frac{1}{2\pi fC} = \frac{1}{2\pi(1\text{ kHz})(0.05\ \mu\text{F})} = 3.18\text{ k}\Omega$$

The current is the same in all parts of a series circuit. Therefore, an rms current of 2 mA flows through the resistor and the capacitor. The rms voltage across the resistance is

$$V_R = (2\text{ mA})(4.7\text{ k}\Omega) = 9.4\text{ V}$$

and the rms voltage across the capacitance is

$$V_C = (2\text{ mA})(3.18\text{ k}\Omega) = 6.36\text{ V}$$

With the Pythagorean theorem, the total voltage is

$$V_T = \sqrt{(9.4\ \text{V})^2 + (6.36\ \text{V})^2} = 11.3\ \text{V}$$

Figure 7-6b shows the phasor voltages.

EXAMPLE 7-4

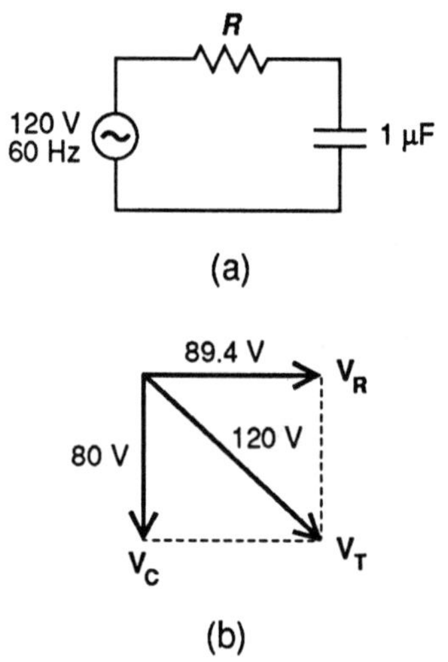

Fig. 7-7.

In Fig. 7-7a, the rms voltage across the capacitance is 80 V. What is the resistance?

SOLUTION

With the Pythagorean theorem,

$$120\ \text{V} = \sqrt{V_R^2 + (80\ \text{V})^2}$$

Solve for V_R as follows. Square both sides to get

$$14400 = V_R^2 + 6400$$

Rearrange to get

$$V_R^2 = 14400 - 6400 = 8000$$

Take the square root of both sides to get

$$V_R = 89.4\ \text{V}$$

Figure 7-7b shows the phasor voltages.
The capacitive reactance is

$$X_C = \frac{1}{2\pi(60\ \text{Hz})(1\ \mu\text{F})} = 2.65\ \text{k}\Omega$$

The rms current through the capacitance is

$$I = \frac{80\ \text{V}}{2.65\ \text{k}\Omega} = 30.2\ \text{mA}$$

Since we have the rms voltage across the resistance and the current through it, we can calculate the resistance:

$$R = \frac{89.4\ \text{V}}{30.2\ \text{mA}} = 2.96\ \text{k}\Omega$$

EXAMPLE 7-5

In Fig. 7-8a, the rms current is 4 mA. What is the capacitance?

SOLUTION

The rms voltage across the resistance is

$$V_R = (4 \text{ mA})(3 \text{ k}\Omega) = 12 \text{ V}$$

With the Pythagorean theorem,

$$15 \text{ V} = \sqrt{(12 \text{ V})^2 + V_C^2}$$

Solving for V_C, we get

$$V_C = 9 \text{ V}$$

Figure 7-8b shows the phasor voltages.

Since we have the rms voltage across the capacitance and the current through it, we can calculate the capacitive reactance:

$$X_C = \frac{9 \text{ V}}{4 \text{ mA}} = 2.25 \text{ k}\Omega$$

Now, solve for the capacitance as follows:

$$C = \frac{1}{2\pi f X_C} = \frac{1}{2\pi(2 \text{ kHz})(2.25 \text{ k}\Omega)} = 0.0354 \ \mu\text{F}$$

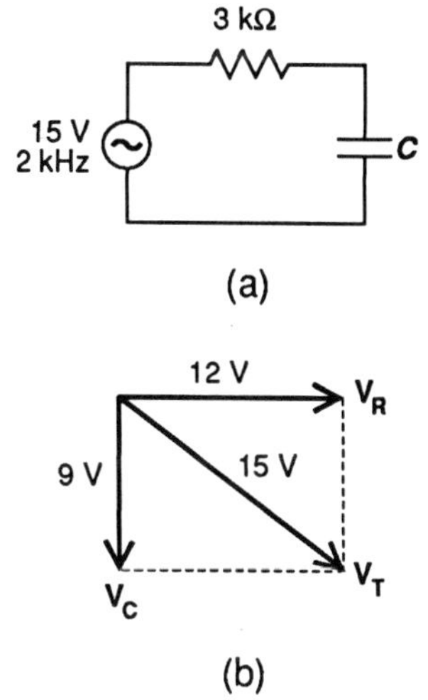

Fig. 7-8.

7-3. IMPEDANCE

Figure 7-9a shows a series *RC* circuit. The phasor voltages for the resistor and the capacitor are 90° out of phase as shown in Fig. 7-9b. As you see, the capacitive voltage lags the resistive voltage by 90°. The impedance of the circuit is defined as the total opposition to current. This opposition is a combined effect produced by the resistance and capacitance acting together. Mathematically, the impedance is defined as

$$Z = \frac{V_T}{I} \tag{7-4}$$

This says the impedance equals the total voltage divided by the current.

Since the total voltage is the quadrature sum of the individual voltages, we can substitute to get

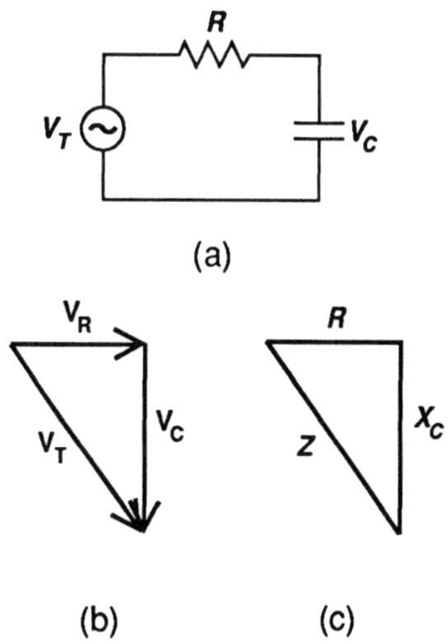

Fig. 7-9.

$$Z = \frac{\sqrt{V_R^2 + V_C^2}}{I} \tag{7-5}$$

Next, we can divide numerator and denominator by the current to get

$$Z = \sqrt{(V_R/I)^2 + (V_C/I)^2}$$

Since $R = V_R/I$ and $X_C = V_C/I$, the foregoing equation becomes

$$Z = \sqrt{R^2 + X_C^2} \tag{7-6}$$

MEMORIZE

$$Z = \sqrt{R^2 + X_C^2}$$

This says the impedance of the circuit equals the square root of the sum of the square of resistance and the square of capacitive reactance.

Equation (7-6) is often visualized by drawing the impedance triangle shown in Fig. 7-9c. The impedance represents the total opposition to current, the combined effect of the resistance and capacitive reactance. Because of this, we can calculate the current with this equation:

$$I = \frac{V_T}{Z} \tag{7-7}$$

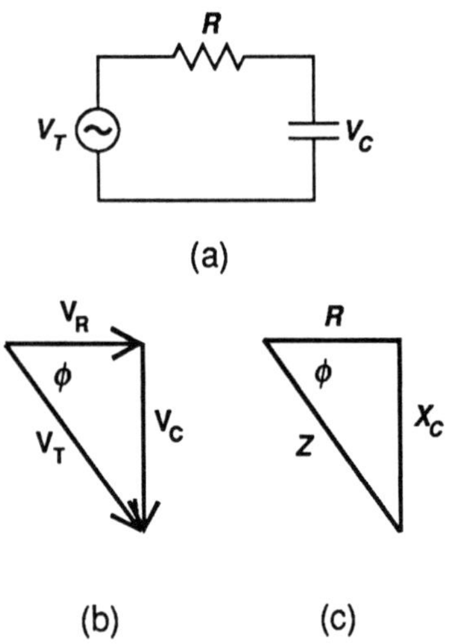

Fig. 7-9.

Given the total voltage, we can divide by the impedance to find the current. An alternative form of Eq. (7-7) is

$$Z = \frac{V_T}{I} \tag{7-8}$$

You can use this to calculate the impedance when you know the total voltage and current.

EXAMPLE 7-6

What is the current in Fig. 7-10a? Draw the phasor diagram for the voltages.

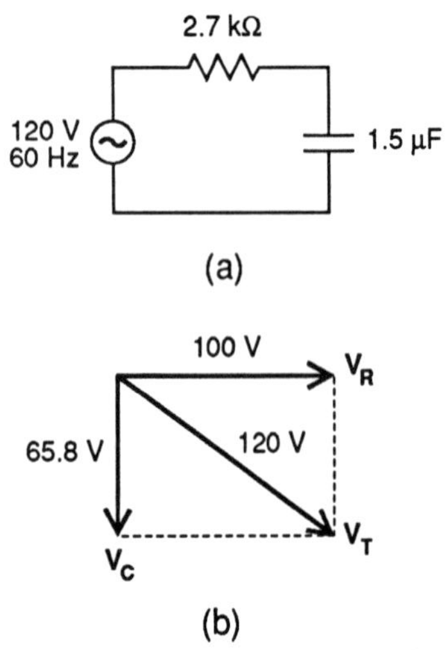

Fig. 7-10.

SOLUTION

Start by calculating the capacitive reactance:

$$X_C = \frac{1}{2\pi(60\text{ Hz})(1.5\ \mu\text{F})} = 1.77\text{ k}\Omega$$

Next, calculate the impedance:

$$Z = \sqrt{(2.7\text{ k}\Omega)^2 + (1.77\text{ k}\Omega)^2} = 3.23\text{ k}\Omega$$

Now, calculate the current:

$$I = \frac{120\ \text{V}}{3.23\ \text{k}\Omega} = 37.2\ \text{mA}$$

After you have the current, you can use Ohm's law to calculate the individual voltages as follows:

$$V_R = (37.2\ \text{mA})(2.7\ \text{k}\Omega) = 100\ \text{V}$$

$$V_C = (37.2\ \text{mA})(1.77\ \text{k}\Omega) = 65.8\ \text{V}$$

Figure 7-10b shows the phasor diagram.

7-4. ANGLE OF THE IMPEDANCE

The preceding section showed you how to calculate the impedance of a series *RC* circuit. It is also important to be able to calculate the angle between the total voltage and current.

Finding the Angle

Figure 7-11a shows the phasor voltages for a series *RC* circuit. Figure 7-11b shows the impedance triangle for the same circuit. The tangent of the angle equals the ratio of X_C to R:

$$\tan\phi = \frac{X_C}{R} \tag{7-9}$$

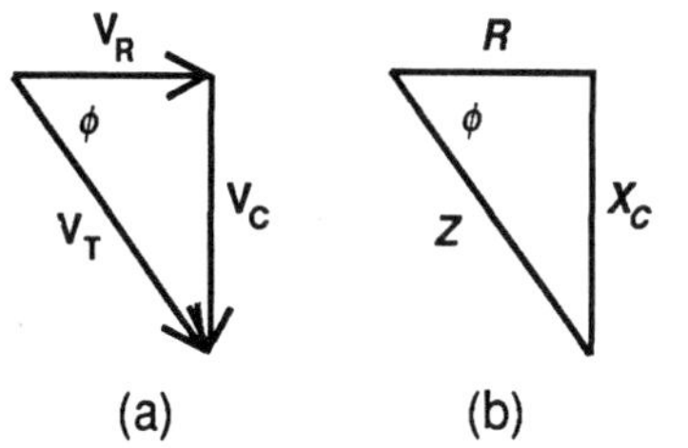

Fig. 7-11.

The angle ϕ may vary from 0° to 90°.

The tangent of the angle can also be calculated with

$$\tan\phi = \frac{V_C}{V_R} \tag{7-10}$$

Figure 7-11a shows that the angle of the impedance represents the angle between the total voltage and the current.

Critical Frequency

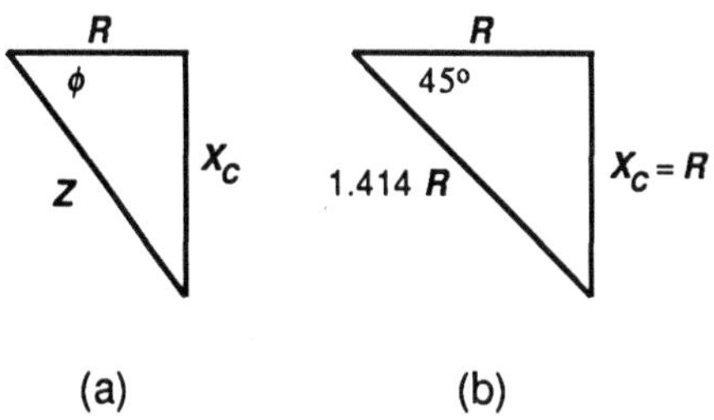

Fig. 7-12.

The impedance triangle of Fig. 7-12a will change when the frequency changes. When the frequency is very low, the angle is near 90°. On the other hand, when the frequency is very high, the angle approaches 0°.

There is an important frequency called the *critical* frequency, symbolized f_c. This is the frequency where X_C equals R. At this frequency,

MEMORIZE

$$f_c = \frac{1}{2\pi RC}$$

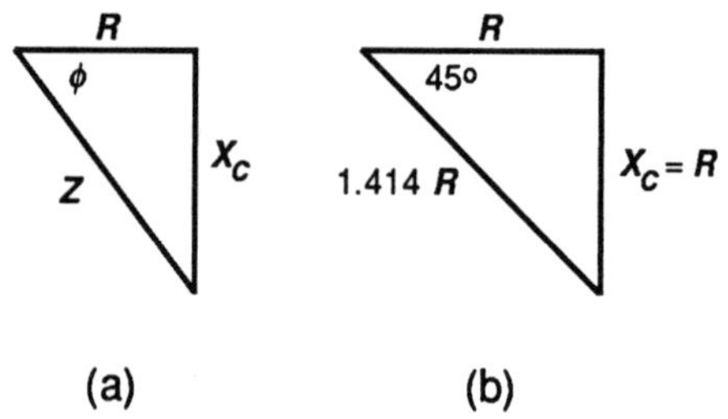

Fig. 7-12.

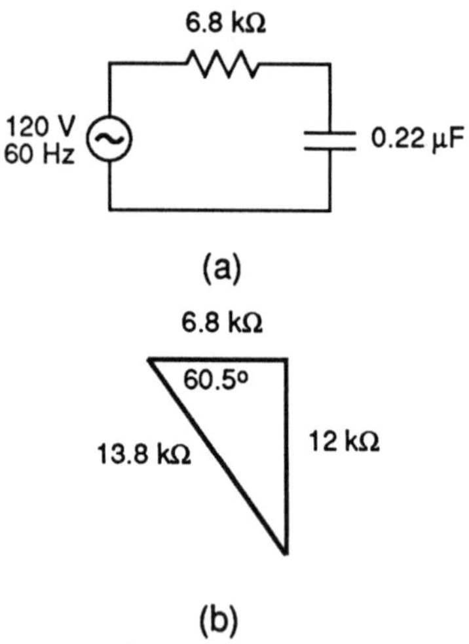

Fig. 7-13.

$$X_C = \frac{1}{2\pi fC} = R$$

Solving for the critical frequency gives

$$f_c = \frac{1}{2\pi RC} \tag{7-11}$$

Figure 7-12b shows the impedance triangle for the critical frequency. In this case, the impedance is

$$Z = \sqrt{R^2 + X_C^2} = \sqrt{R^2 + R^2} = \sqrt{2R^2} = 1.414R$$

Also, the angle of impedance is

$$\tan\phi = \frac{X_C}{R} = \frac{R}{R} = 1$$

or

$$\phi = 45^\circ$$

EXAMPLE 7-7

What is the angle of the impedance in Fig. 7-13a? Draw the impedance triangle.

SOLUTION

Start by calculating the capacitive reactance:

$$X_C = \frac{1}{2\pi(60\text{ Hz})(0.22\ \mu\text{F})} = 12\text{ k}\Omega$$

Next, get the impedance:

$$Z = \sqrt{(6.8\text{ k}\Omega)^2 + (12\text{ k}\Omega)^2} = 13.8\text{ k}\Omega$$

The tangent of the angle is

$$\tan\phi = \frac{12\text{ k}\Omega}{6.8\text{ k}\Omega}$$

Solving for ϕ, we get

$$\phi = 60.5^\circ$$

Figure 7-13b shows the impedance triangle.

EXAMPLE 7-8

Draw the phasor diagram for Fig. 7-13a.

SOLUTION

In the preceding example, we calculated the impedance and the angle:

$Z = 13.8\ \text{k}\Omega$

$\phi = 60.5^\circ$

Now, calculate the current in Fig. 7-13a:

$$I = \frac{120\ \text{V}}{13.8\ \text{k}\Omega} = 8.7\ \text{mA}$$

This current flows through the resistor and capacitor to produce voltages of

$V_R = (8.7\ \text{mA})(6.8\ \text{k}\Omega) = 59.2\ \text{V}$

$V_C = (8.7\ \text{mA})(12\ \text{k}\Omega) = 104\ \text{V}$

Figure 7-14a shows the phasor diagram of the voltages. Since the current is in phase with the resistor voltage, Fig. 7-14b shows the phasor diagram of total voltage and current. As you can see, the total voltage lags the current by 60.5°.

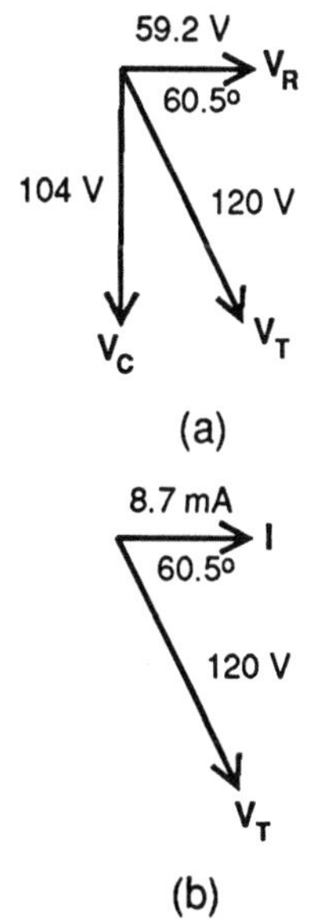

Fig. 7-14.

EXAMPLE 7-9

What is the critical frequency in Fig. 7-15?

SOLUTION

At the critical frequency, the capacitive reactance equals the resistance. With Eq. (7-11),

$$f_C = \frac{1}{2\pi RC} = \frac{1}{2\pi(10\ \text{k}\Omega)(150\ \text{pF})} = 106\ \text{kHz}$$

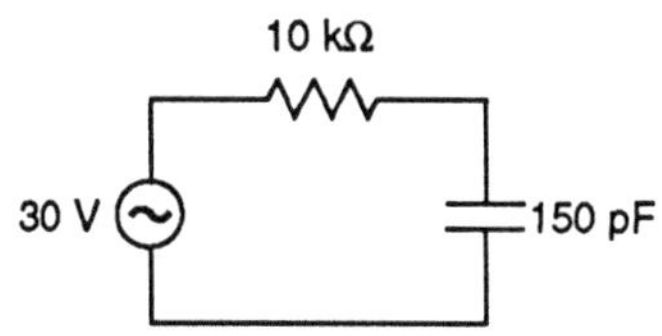

Fig. 7-15.

7-5. PARALLEL CIRCUITS

With parallel circuits, the applied voltage has the same value across each branch. But the branch currents are usually different, and their phasor sum equals the total phasor current. Again, phasors are used because they are easier to work with than sine waves.

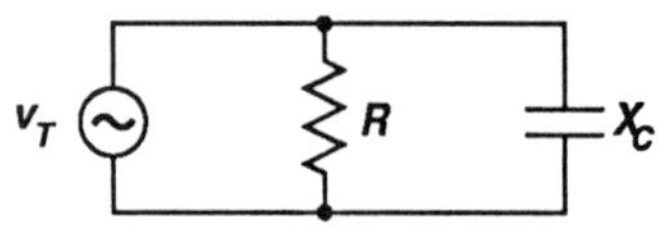

(a)

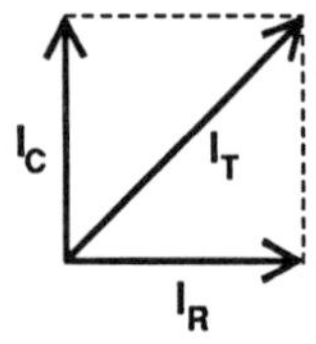

(b)

Fig. 7-16.

Figure 7-16a shows an ac source across a resistor and capacitor in parallel. Because of the parallel connection, the same instantaneous voltage appears across the resistor and the capacitor simultaneously. Visualize the applied voltage as a sine wave with a phase angle of 0°. Then, the corresponding phasor voltage V_T has an angle of 0°.

Since the voltage and current are in phase in a resistor, the phasor current I_R in the resistor has an angle of 0° as shown in Fig. 7-16b. Since the current leads the voltage by 90° in an ideal capacitor, the phasor current I_C has an angle of 90° as shown. The total phasor current is the phasor sum of the individual phasor currents. As before, the Pythagorean theorem gives the length of the hypotenuse, equivalent to the magnitude of the total phasor current.

When analyzing parallel *RC* circuits, the magnitude of each phasor is either an rms value or a peak value, whichever is convenient in the particular circuit you are analyzing. If you use rms values for the phasors, then all phasors represent rms values. If you use peak values, then all are peak values.

Here are the equations used in analyzing parallel circuits. The current in the resistive branch is

$$I_R = \frac{V_T}{R} \tag{7-12}$$

PARALLEL CIRCUIT

$$I_T = \sqrt{I_R^2 + I_C^2}$$

The current in the capacitive branch is

$$I_C = \frac{V_T}{X_C} \tag{7-13}$$

The formula for the total current is

$$I_T = \sqrt{I_R^2 + I_C^2} \tag{7-14}$$

Again, the quadrature sum refers to taking the square root of the sum of the squares of individual quantities to find the total quantity.

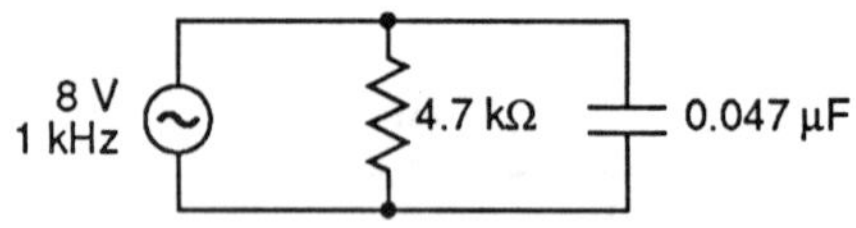

(a)

EXAMPLE 7-10

In Fig. 7-17a, the rms voltage is 8 V. What is the total rms current?

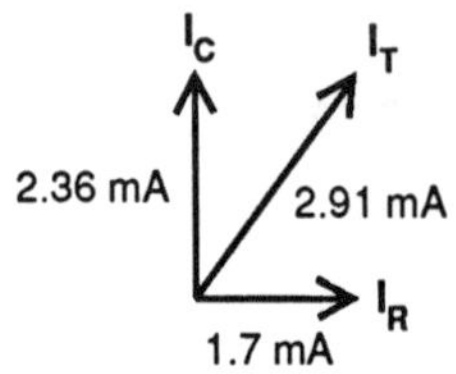

(b)

Fig. 7-17.

SOLUTION

The capacitive reactance is

$$X_C = \frac{1}{2\pi(1\text{ kHz})(0.047\ \mu\text{F})} = 3.39\text{ k}\Omega$$

The current through the resistive branch is

$$I_R = \frac{8\ \text{V}}{4.7\ \text{k}\Omega} = 1.7\ \text{mA}$$

The current through the capacitive branch is

$$I_C = \frac{8\ \text{V}}{3.39\ \text{k}\Omega} = 2.36\ \text{mA}$$

With the Pythagorean theorem, the total current is

$$I_T = \sqrt{(1.7\ \text{mA})^2 + (2.36\ \text{mA})^2} = 2.91\ \text{mA}$$

Figure 7-17b shows the phasor currents.

7-6. PARALLEL IMPEDANCE AND ANGLE

The impedance of a parallel circuit equals the total voltage divided by the total current. The formula for impedance is

$$Z = \frac{V_T}{I_T} \tag{7-15}$$

where the voltage and current are expressed in either rms or peak values. The easiest way to find the impedance of parallel RC circuit is to calculate each branch current. Then, apply the Pythagorean theorem to find the total current (Fig. 7-18). After you find the total current, you can calculate the impedance of the circuit with Eq. (7-15).

In Fig. 7-18, the total current leads the resistive current by an angle of ϕ. The tangent of this angle equals the opposite side over the adjacent side:

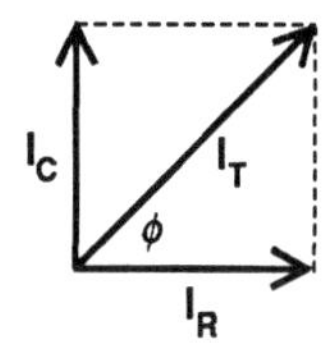

Fig. 7-18.

$$\tan \phi = \frac{I_C}{I_R} \tag{7-16}$$

Because $I_C = V/X_C$ and $I_R = V/R$, the foregoing equation can be rewritten as

$$\tan \phi = \frac{R}{X_C} \tag{7-17}$$

When you know the values of I_C and I_R, or the values of R and X_C, you can calculate the tangent of the angle. Then, you can use trig tables or a calculator to find the angle.

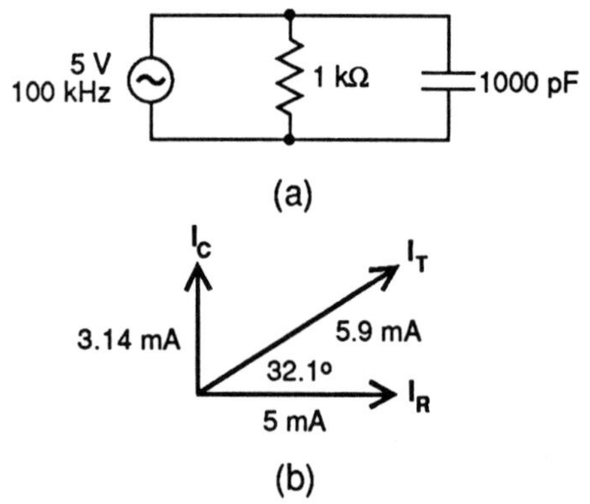

Fig. 7-19.

EXAMPLE 7-11

What is the impedance and angle in Fig. 7-19a?

SOLUTION

The approach is to calculate each branch current, the total current, the impedance, and the angle. Here is how it looks. The current through the resistor equals the total voltage divided by the resistance:

$$I_R = \frac{5\text{ V}}{1\text{ k}\Omega} = 5\text{ mA}$$

The capacitive reactance is

$$X_C = \frac{1}{2\pi(100\text{ kHz})(1000\text{ pF})} = 1.59\text{ k}\Omega$$

The current through the capacitive reactance is the total voltage divided by the capacitive reactance:

$$I_C = \frac{5\text{ V}}{1.59\text{ k}\Omega} = 3.14\text{ mA}$$

With the Pythagorean theorem, the quadrature sum is

$$I_T = \sqrt{(5\text{ mA})^2 + (3.14\text{ mA})^2} = 5.9\text{ mA}$$

Figure 7-19b shows the phasor currents.

Now, divide the total voltage by the total current to get the impedance:

$$Z = \frac{5\text{ V}}{5.9\text{ mA}} = 847\ \Omega$$

The tangent of the angle equals the capacitive current divided by the resistive current:

$$\tan\phi = \frac{3.14\text{ mA}}{5\text{ mA}} = 0.628$$

Solving for the angle, we get

$$\phi = 32.1^\circ$$

We can get the same angle as follows. The tangent of the angle equals the resistance divided by the capacitive reactance:

$$\tan \phi = \frac{1\ \text{k}\Omega}{1.59\ \text{k}\Omega} = 0.628$$

which gives same angle, 32.1°. Therefore, we can find the angle by either method.

In summary, we have calculated an impedance of 847 Ω and an angle of 32.1°. In Fig. 7-19b, notice that the total current leads the resistive current by 32.1°. An equivalent statement is that the resistive current lags the total current by 32.1°.

7-7. MORE ABOUT PARALLEL IMPEDANCE

In the preceding section, you learned one way to find the total impedance of a parallel *RC* circuit. There is another way to calculate the impedance of a parallel *RC* circuit.

The Formula Approach

The total impedance is given by

$$Z = \frac{V_T}{I_T}$$

We can substitute the quadrature sum for the total current to get

$$Z = \frac{V_T}{\sqrt{I_R^2 + I_C^2}}$$

Now, divide the numerator and the denominator by V_T to get

$$Z = \frac{1}{\sqrt{(I_R/V_T)^2 + (I_C/V_T)^2}}$$

Since $R = V_T/I_R$ and $X_C = V_T/I_C$, we can rewrite the foregoing equation as

$$Z = \frac{1}{\sqrt{(1/R)^2 + (1/X_C)^2}} \qquad (7\text{-}18)$$

This equation allows us to calculate the impedance of a parallel *RC* circuit.

PRODUCT OVER QUADRATURE SUM

$$Z = \frac{RX_C}{\sqrt{(R^2 + X_C^2}}$$

By applying some algebra to Eq. (7-18), we can expand the denominator and rearrange the equation to get this equivalent form:

$$Z = \frac{RX_C}{\sqrt{R^2 + X_C^2}} \qquad (7\text{-}19)$$

This is similar to the product-over-sum rule you learned in dc circuits. The equation says the impedance equals the product over the quadrature sum. You can use this equation instead of Eq. (7-18) if you prefer.

Calculating the Branch Currents

After you calculate the impedance with Eq. (7-18) or (7-19), you can calculate the total current with

$$I_T = \frac{V_T}{Z} \qquad (7\text{-}20)$$

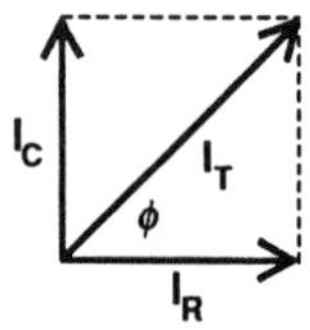

Fig. 7-20.

One way to get the branch currents is like this. The resistive current equals the total voltage divided by the resistance:

$$I_R = \frac{V_T}{R}$$

The capacitive current equals the total voltage divided by the capacitive reactance:

$$I_C = \frac{V_T}{X_C}$$

But there is another way to find the branch currents. In Fig. 7-20, we can use trigonometry to calculate the resistive and capacitive currents. Since the cosine of the angle is given by the adjacent side over the hypotenuse, we can calculate the resistive current with

$$I_R = I_T \cos \phi \qquad (7\text{-}21)$$

Since the sine of the angle is given by the opposite side over the hypotenuse, we can calculate the capacitive current with

$$I_C = I_T \sin \phi \qquad (7\text{-}22)$$

EXAMPLE 7-12

What is the total current in Fig. 7-21? The branch currents?

SOLUTION

First, get the capacitive reactance:

$$X_C = \frac{1}{2\pi(60\text{ Hz})(0.68\ \mu\text{F})} = 3.9\text{ k}\Omega$$

Next, calculate the impedance with Eq. (7-18):

$$Z = \frac{1}{\sqrt{(1/4.7\text{ k}\Omega)^2 + (1/3.9\text{ k}\Omega)^2}} = 3\text{ k}\Omega$$

The total current equals the total voltage divided by the impedance:

$$I_T = \frac{120\text{ V}}{3\text{ k}\Omega} = 40\text{ mA}$$

The tangent of the angle equals the resistance divided by the capacitive reactance:

$$\tan\phi = \frac{4.7\text{ k}\Omega}{3.9\text{ k}\Omega} = 1.21$$

which gives an angle of

$$\phi = 50.3^\circ$$

Figure 7-21b shows the phasor currents. Here you see that the total current leads the resistive current by 50.3°.

Now, we can use trigonometry to calculate the branch currents. With Fig. 7-21b,

$$I_R = (40\text{ mA})\cos 50.3^\circ = 25.6\text{ mA}$$

and

$$I_C = (40\text{ mA})\sin 50.3^\circ = 30.8\text{ mA}$$

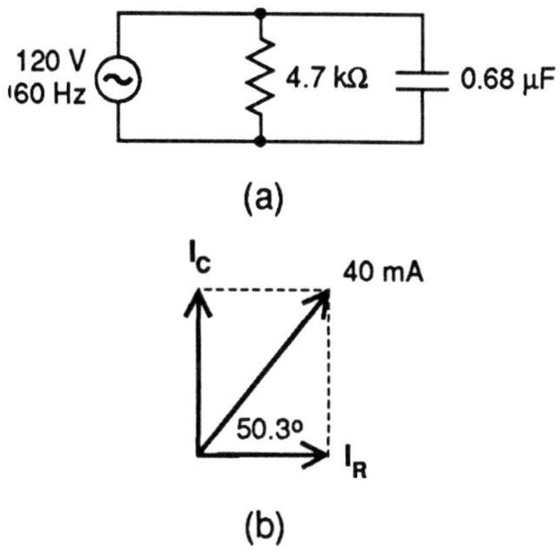

Fig. 7-21.

7-8. THE Q OF CAPACITIVE CIRCUITS

The figure of merit, symbolized by Q, tells you how capacitive an RC circuit is. When Q is much greater than 1, the circuit is highly capacitive. When Q is much less than 1, the circuit is highly resistive. Let's find out what this means.

Q of Series RC Circuits

The Q of a series RC circuit is defined as the ratio of capacitive voltage to resistive voltage:

$$Q = \frac{V_C}{V_R} \tag{7-23}$$

When the capacitive voltage equals the resistive voltage, Q equals 1. When the capacitive voltage is larger than the resistive voltage, Q is greater than 1. When the capacitive voltage is less than the resistive voltage, Q is less than 1. Because of this, Q indicates whether a circuit is primarily capacitive or resistive. When Q is greater than 1, the circuit is primarily capacitive because more voltage is dropped across the capacitor. When Q is less than 1, the circuit is primarily resistive because more voltage is dropped across the resistor.

Since the current is the same throughout a series circuit, Eq. (7-23) can be rewritten as

SERIES

$$Q = \frac{X_C}{R}$$

$$Q = \frac{V_C}{V_R} = \frac{IX_C}{IR}$$

or

$$Q = \frac{X_C}{R} \tag{7-24}$$

This says the Q of a series RC circuit equals the capacitive reactance divided by the resistance.

In fact, the Q of a series RC circuit is equal to the tangent of the angle. You can see this by looking at Fig. 7-22. In symbols,

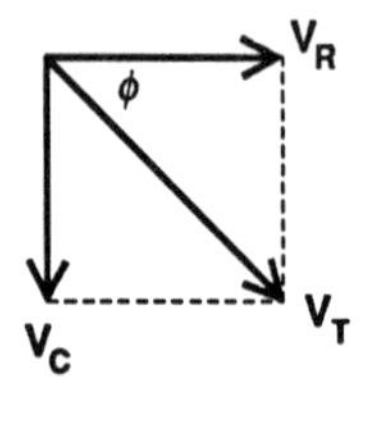

Fig. 7-22.

$$Q = \tan \phi \tag{7-25}$$

When the angle is 45°, the Q of the circuit is 1. When the angle is greater than 45°, the Q of the circuit is greater than 1 and the circuit is more capacitive than resistive. When the angle is less than 45°, the Q of the circuit is less than 1 and the circuit is more resistive than capacitive.

Q of Parallel RC Circuits

The Q of a parallel RC circuit is defined as the ratio of capacitive current to resistive current:

$$Q = \frac{I_C}{I_R} \tag{7-26}$$

When the capacitive current equals the resistive current, Q equals 1. When the capacitive current is larger than the resistive current, Q is greater than 1. When the capacitive current is less than the resistive current, Q is less than 1. When Q is greater than 1, the circuit is primarily capacitive because more current flows through the capacitor. When Q is less than 1, the circuit is primarily resistive because more current flows through the resistor.

Since the voltage is the same across the branches of a parallel circuit, Eq. (7-26) can be rewritten as

$$Q = \frac{I_C}{I_R} = \frac{V_T/X_C}{V_T/R}$$

or

$$Q = \frac{R}{X_C} \qquad (7\text{-}27)$$

PARALLEL

$$Q = \frac{R}{X_C}$$

This says the Q of a parallel RC circuit equals the resistance divided by capacitive reactance. This is exactly the opposite to the Q of a series RC circuit. In a parallel RC circuit, the Q increases when the resistance increases or when the capacitive reactance decreases.

Again, the Q of a series RC circuit is equal to the tangent of the angle. You can see this by looking at Fig. 7-23. In symbols,

$$Q = \tan \phi \qquad (7\text{-}28)$$

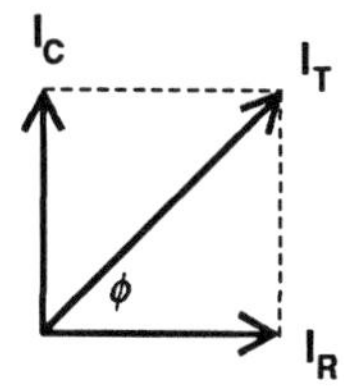

Fig. 7-23.

When the angle is 45°, the Q of the circuit is 1. When the angle is greater than 45°, the Q of the circuit is greater than 1 and the circuit is more capacitive than resistive. When the angle is less than 45°, the Q of the circuit is less than 1 and the circuit is more resistive than capacitive.

EXAMPLE 7-13

A series RC circuit has R = 6.8 kΩ and X_C = 108 kΩ. What is the Q of the circuit?

SOLUTION

Since this is a series circuit, Q equals the capacitive reactance divided by the resistance:

$$Q = \frac{108\ \text{k}\Omega}{6.8\ \text{k}\Omega} = 15.9$$

EXAMPLE 7-14

A parallel RC circuit has R = 25 kΩ and X_C = 400 Ω. What is the Q of the circuit?

SOLUTION

Since this is a parallel circuit, Q equals the resistance divided by the capacitive reactance:

$$Q = \frac{25\ \text{k}\Omega}{400\ \Omega} = 62.5$$

EXAMPLE 7-15

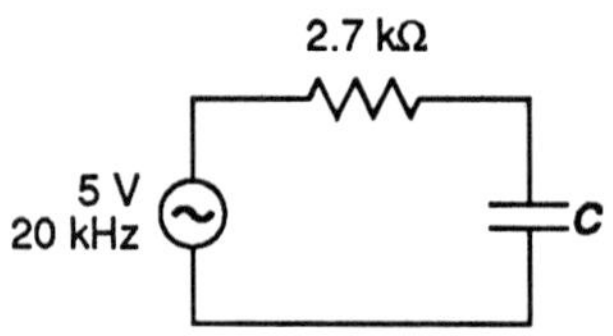

Fig. 7-24.

The circuit of Fig. 7-24 has a Q of 15. What is the capacitance?

SOLUTION

This is a series circuit. Substitute into the Q equation to get

$$15 = \frac{X_C}{2.7\ \text{k}\Omega}$$

Solve for the capacitive reactance to get

$$X_C = 15(2.7\ \text{k}\Omega) = 40.5\ \text{k}\Omega$$

Now, solve for the capacitance:

$$C = \frac{1}{2\pi f X_C} = \frac{1}{2\pi(20\ \text{kHz})(40.5\ \text{k}\Omega)} = 196\ \text{pF}$$

EXAMPLE 7-16

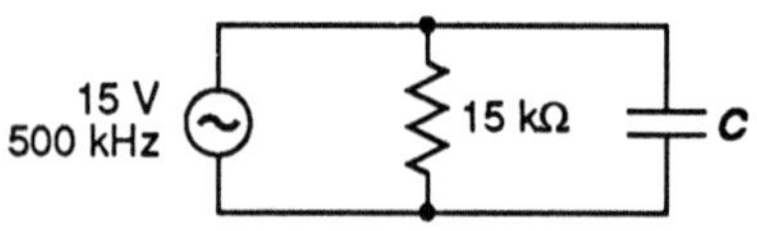

Fig. 7-25.

The circuit of Fig. 7-25 has a Q of 75. What is the capacitance?

SOLUTION

This is a parallel circuit. Substitute into the Q equation to get

$$75 = \frac{15\ \text{k}\Omega}{X_C}$$

Solve for the capacitive reactance to get

$$X_C = \frac{15\ \text{k}\Omega}{75} = 200\ \Omega$$

Now, solve the capacitance:

$$C = \frac{1}{2\pi f X_C} = \frac{1}{2\pi(500\ \text{kHz})(200\ \Omega)} = 0.00159\ \mu\text{F}$$

MULTIPLE-CHOICE TEST

The following questions may have more than one right answer. Select the best answer. This is the one that is always true, or covers more situations, or agrees with the material of this chapter, etc.

1. In a capacitive circuit, the current
   a. leads the voltage
   b. lags the voltage
   c. is in phase with the voltage
   d. none of the above

2. A phasor is equivalent to
   a. a scalar
   b. a sinusoidal quantity
   c. a phase angle
   d. a quadrature sum

3. When two vectors are separated by 90°, the resultant vector has a length that is given by the
   a. algebraic sum
   b. tangent of the angle
   c. quadrature sum
   d. critical frequency

4. The impedance of a circuit is the
   a. tangent of the angle between current and voltage
   b. angle between the current and the voltage
   c. quadrature sum of the current and voltage
   d. total opposition to current

5. In a series *RC* circuit, the impedance equals the quadrature sum of
   a. current and voltage
   b. resistance and capacitance
   c. frequency and capacitive reactance
   d. capacitive reactance and resistance

6. When the capacitive voltage increases in a series *RC* circuit, the angle
   a. decreases
   b. stays the same
   c. increases
   d. none of the above

7. When the frequency decreases in a series *RC* circuit, the angle
   a. decreases
   b. stays the same
   c. increases
   d. none of the above

8. When the frequency increases in a series *RC* circuit, the current
   a. decreases
   b. stays the same
   c. increases
   d. none of the above

9. At the critical frequency, how is the capacitive reactance related to the resistance?
   a. less than
   b. equal
   c. greater than
   d. none of the above

10. In a parallel *RC* circuit, the total current equals the quadrature sum of
    a. branch currents
    b. branch voltages
    c. resistance and capacitance
    d. capacitive reactance and resistance

11. When the frequency doubles, the total current in a parallel *RC* circuit
    a. drops in half
    b. stays the same
    c. doubles
    d. increases

12. The impedance of an ac circuit is defined as the
    a. total voltage divided by the total current
    b. quadrature sum of the resistance and capacitive reactance
    c. product over the quadrature sum
    d. total power divided by the total voltage

13. The Q of an ac circuit equals the
    a. capacitive reactance divided by the resistance
    b. resistance divided by the capacitive reactance
    c. tangent of the angle
    d. angle

14. The Q of a parallel *RC* circuit equals the
    a. capacitive reactance divided by the resistance
    b. resistance divided by the capacitive reactance
    c. impedance times the sine of the angle
    d. impedance divided by the angle

15. The Q of a series *RC* circuit equals the
    a. capacitive reactance divided by the resistance
    b. resistance divided by the capacitive reactance
    c. impedance divided by the angle
    d. impedance times the sine of the angle

16. In a parallel *RC* circuit the total current times the sine of the angle is the
    a. resistive current
    b. capacitive current
    c. resistive voltage
    d. capacitive voltage

17. In a series *RC* circuit the total voltage times the cosine of the angle is the
    a. resistive current
    b. capacitive current
    c. resistive voltage
    d. capacitive voltage

PROBLEMS

Sec. 7-2: Series Circuits

7-1. In Fig. 7-26, you measure an rms voltage of 3.75 V across the resistor and an rms voltage of 5 V across the capacitor. What is the total rms voltage?

7-2. In Fig. 7-27, the rms current is 2 mA. What is the rms value of the applied voltage? The rms voltage across the resistor? The rms voltage across the capacitor?

7-3. The capacitance of Fig. 7-27 is doubled and the applied voltage is changed to get an rms current of 5 mA. What is the applied voltage?

7-4. The frequency is doubled in Fig. 7-27 and the applied voltage is increased to get an rms current of 4 mA. What is the applied voltage?

7-5. In Fig. 7-28, the rms voltage across the capacitance is 30 V. What is the voltage across the resistance?

7-6. In Fig. 7-28, the rms voltage across the capacitance is 35.4 V. What is the resistance?

7-7. In Fig. 7-29, the rms current is 3 mA. What is the voltage across the capacitance?

7-8. In Fig. 7-29, the applied voltage is increased to get a current of 5 mA. What is the capacitance?

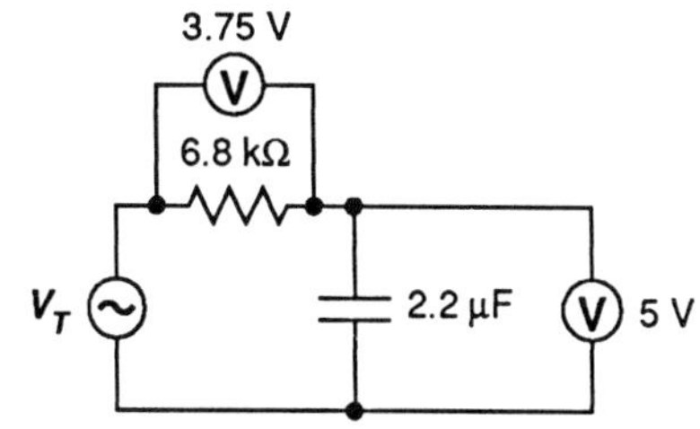

Fig. 7-26.

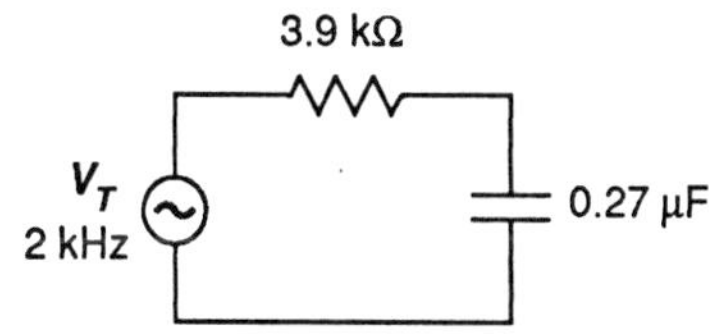

Fig. 7-27.

Sec. 7-3. Impedance

7-9. The total voltage is 15 V rms in Fig. 7-27. What is the current? The voltage across the resistor? The voltage across the capacitor?

7-10. The frequency is doubled in Fig. 7-27. What is the current if the total voltage is 5 V rms? The voltage across the resistor? The voltage across the capacitor?

7-11. The capacitance is reduced to 400 pF in Fig. 7-27. If the applied voltage is 20 V, what is the current? The voltage across the resistor? The voltage across the capacitor?

7-12. What is the current in Fig. 7-28 if the resistance is 15 kΩ?

7-13. What is the voltage across the resistor of Fig. 7-28 if the resistance is 8 kΩ?

7-14. If the resistance is 8 kΩ in Fig. 7-28, what is the voltage across the capacitor?

7-15. What is the current in Fig. 7-30? The voltage across the resistor? The voltage across the capacitor?

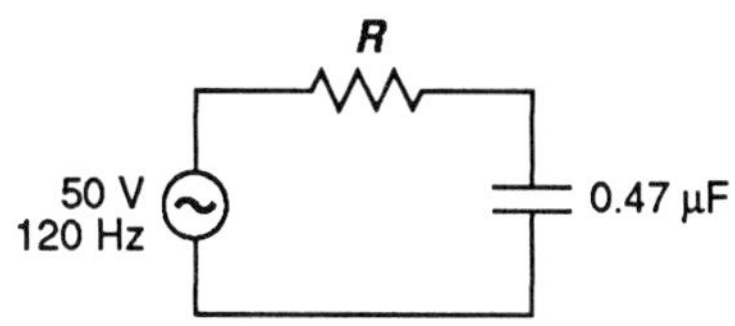

Fig. 7-28.

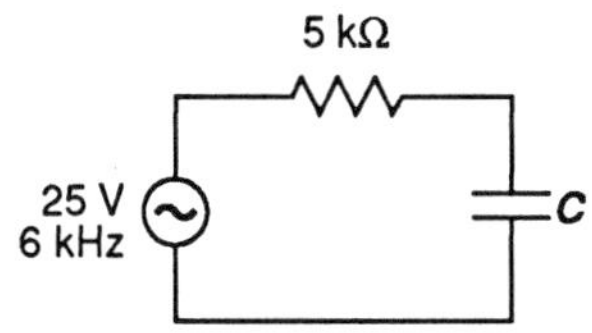

Fig. 7-29.

Sec. 7-4. Angle of Impedance

7-16. What is the angle of the impedance in Fig. 7-27?

7-17. What is the angle of impedance in Fig. 7-28 if the resistance is half the capacitive reactance?

7-18. What is the angle of the impedance in Fig. 7-29 if the voltage across the capacitor is 18 V?

7-19. What is the critical frequency in Fig. 7-27?

7-20. What is the critical frequency in Fig. 7-30?

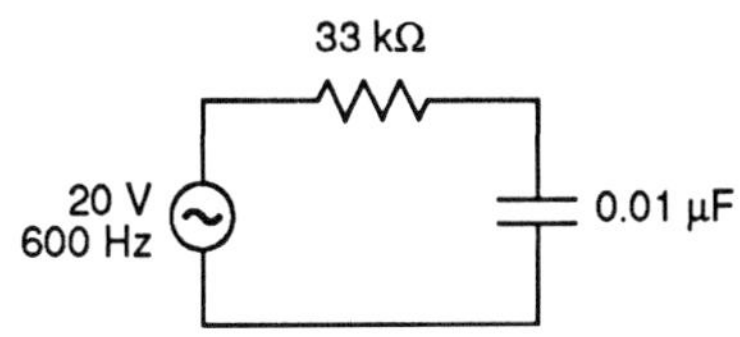

Fig. 7-30.

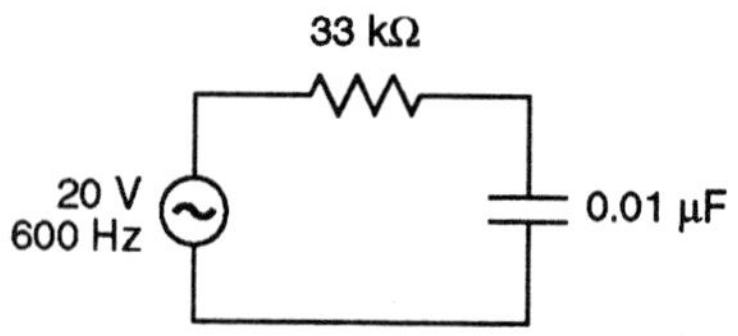

Fig. 7-30.

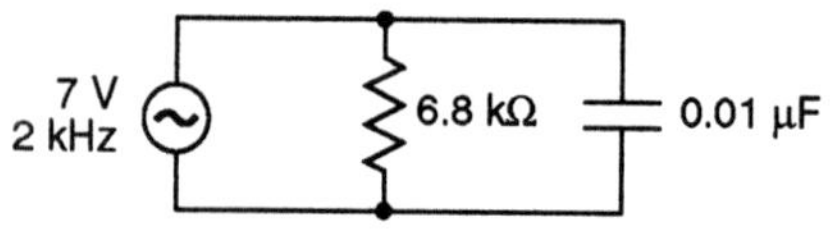

Fig. 7-31.

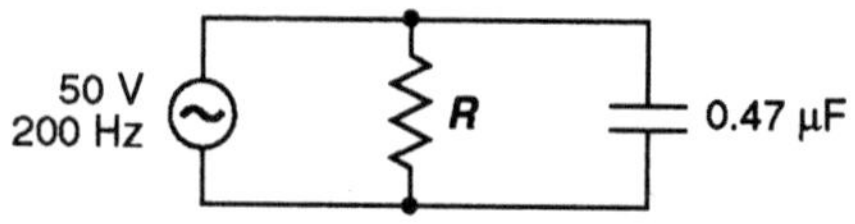

Fig. 7-32.

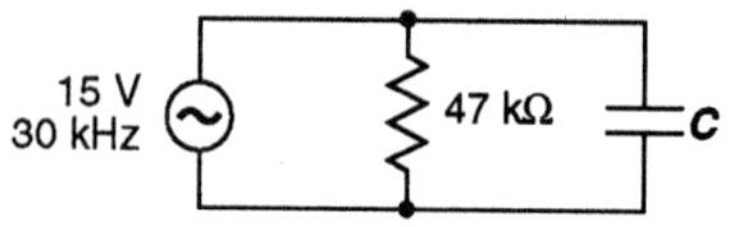

Fig. 7-33.

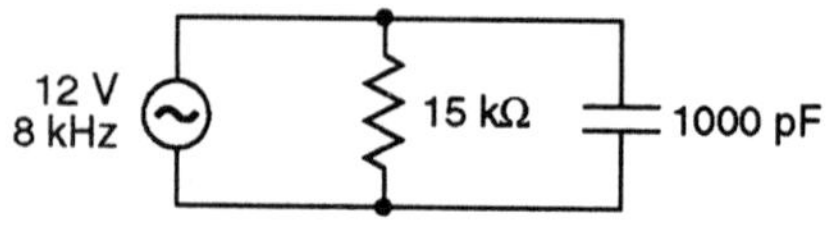

Fig. 7-34.

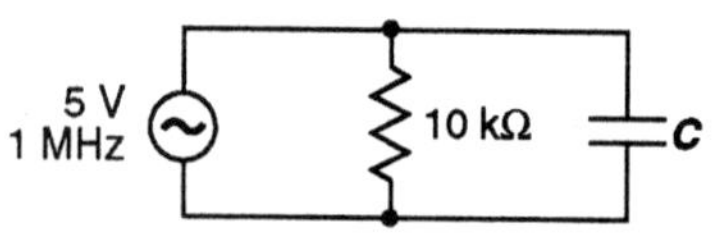

Fig. 7-35.

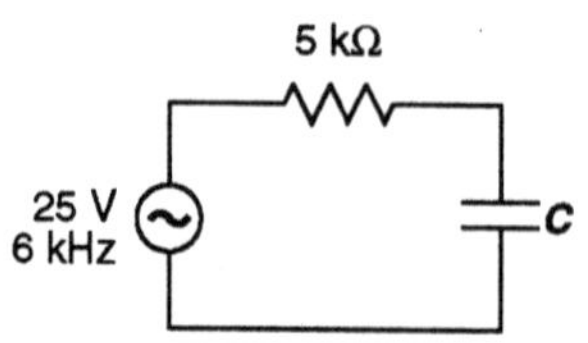

Fig. 7-36.

Sec. 7-5. Parallel Circuits

7-21. In Fig. 7-31, the rms voltage is 7 V. What is the total rms current?
7-22. If the rms voltage of Fig. 7-31 is increased to 14 V, what is the total rms current?
7-23. If the frequency is doubled in Fig. 7-31, what is the total rms current?
7-24. If the capacitance is tripled in Fig. 7-31, what is the total rms current?
7-25. In Fig. 7-32, the total rms current is 37.3 mA. What is the resistance?
7-26. In Fig. 7-33, the total rms current is 35 mA. What is the capacitance?

Sec. 7-6. Parallel Impedance and Angle

7-27. What is the impedance and angle in Fig. 7-31?
7-28. If the frequency doubles in Fig. 7-31, what is the impedance and the angle?
7-29. If the capacitance triples in Fig. 7-31, what is the impedance and the angle?

Sec. 7-7. More About Parallel Impedance

7-30. Calculate the impedance of Fig. 7-34 using Eq. (7-18).
7-31. Calculate the impedance of Fig. 7-34 using Eq. (7-19).
7-32. Use Eq. (7-18) to calculate the impedance of Fig. 7-35 for a capacitance of 10 pF. Then, calculate the total current and the branch currents.

Sec. 7-8. The *Q* of Capacitive Circuits

7-33. A series *RC* circuit has $R = 2.7\ \text{k}\Omega$ and $X_C = 18\ \text{k}\Omega$. What is the Q of the circuit?
7-34. A parallel *RC* circuit has $R = 33\ \text{k}\Omega$ and $X_C = 425\ \Omega$. What is the Q of the circuit?
7-35. The circuit of Fig. 7-36 has a Q of 20. What is the capacitance?
7-36. The circuit of Fig. 7-35 has a Q of 120. What is the capacitance?

CHAPTER 8
AC ANALYSIS

This chapter discusses a variety of ac circuits. The earlier topics are straightforward discussions of many circuits seen before. But later, you will see new circuits that combine resistance, inductance, and capacitance.

8-1. RESISTIVE CIRCUITS

Let us start with purely resistive circuits. This will allow us to review and consolidate what we know about circuits that contain only resistances. In a series circuit, the current is the same through all components. When the circuit is purely resistive, the individual voltages are the product of this current and the individual resistances. The sum of these individual voltages equals the applied voltage. Furthermore, the individual voltages are in phase with the current.

In a parallel circuit, the voltage is the same across all components. When the circuit is purely resistive, each branch current equals the applied voltage divided by the branch resistance. The sum of the branch currents equals the total current. Furthermore, the branch currents are in phase with the applied voltage.

EXAMPLE 8-1

Show the phasors for Fig. 8-1a.

SOLUTION

The resistances are in series. The total resistance is 9 kΩ, the sum of the individual resistances. Therefore, the total current is 18 V divided by 9 kΩ, which equals 2 mA. When this 2 mA flows through the 3 kΩ, it produces a voltage drop of 6 V. Similarly, 2 mA flowing through 6 kΩ produces 12 V.

Figure 8-1b shows the phasors. A phase angle of 0° is normally used for the current in a series circuit. Since there is no phase shift between voltage and current in a purely resistive circuit, the two voltages have a phase angle of 0°. The vector sum of the two voltages equals the applied voltage.

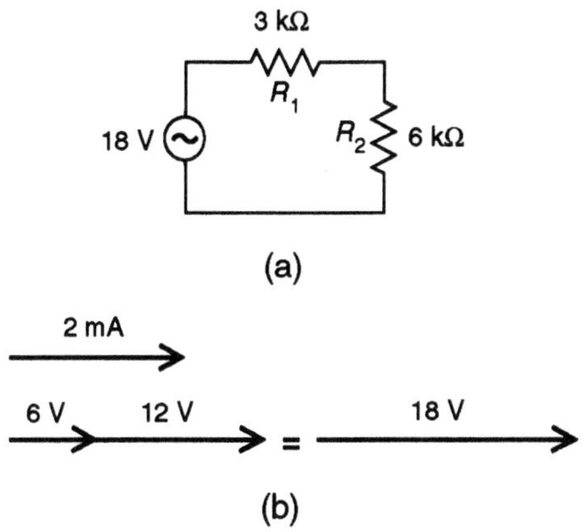

Fig. 8-1.

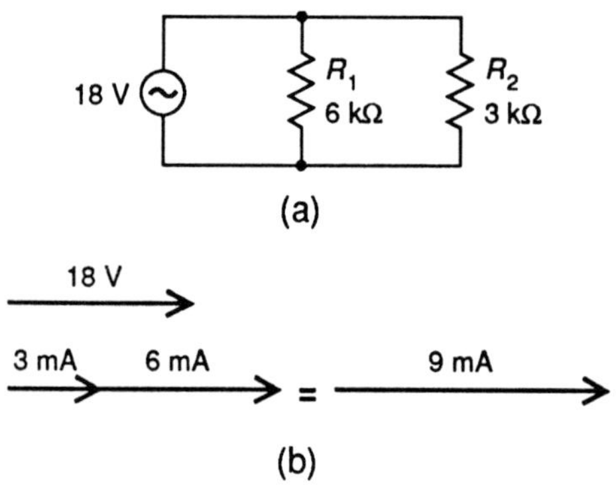

Fig. 8-2.

EXAMPLE 8-2

Show the phasors for Fig. 8-2a.

SOLUTION

The resistances are in parallel. The current through R_1 is 3 mA, and the current through R_2 is 6 mA. The total current is 3 mA plus 6 mA, or 9 mA. The total resistance equals 18 V divided by 9 mA, or 2 kΩ.

Figure 8-2b shows the phasors. A phase angle of 0° is normally used for the applied voltage in a parallel circuit. Since there is no phase shift between current and voltage in a purely resistive circuit, the two branch currents have a phase angle of 0°. The vector sum of the two branch currents equals the total current.

8-2. INDUCTIVE CIRCUITS

In purely inductive circuits, all voltages lead the currents by 90°. Furthermore, Ohm's law is valid for rms, peak, and peak-to-peak values. Throughout this chapter, you may assume voltages and currents are expressed as rms values, unless otherwise indicated.

In a series circuit with ideal inductors, the voltages across the inductors are the product of the current and the inductive reactances. The sum of these individual voltages equals the applied voltage. Furthermore, all voltages lead the current by 90°.

In a parallel circuit with ideal inductors, each branch current equals the applied voltage divided by the reactance of the branch. The sum of the branch currents equals the total current. Furthermore, all currents lag the applied voltage by 90°.

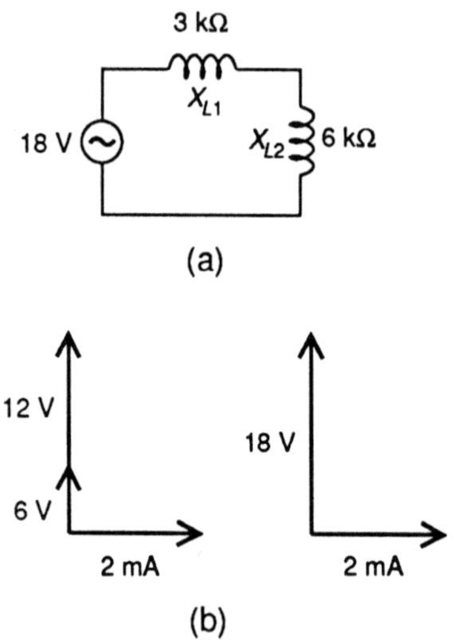

Fig. 8-3.

EXAMPLE 8-3

Show the phasors for Fig. 8-3a.

SOLUTION

The inductors are in series. The total inductive reactance is 9 kΩ, the sum of the individual reactances. The total current is 18 V divided by 9 kΩ, or 2 mA. The product of this 2 mA and 3 kΩ gives a voltage of 6 V across X_{L1}. Likewise, 2 mA through 6 kΩ gives a voltage of 12 V across X_{L2}.

Figure 8-3b shows the phasors. The current has a phase angle of 0°. All voltages lead this current by 90° because we are assuming ideal inductors. The vector sum of the two voltages equals the applied voltage. In other words, the applied voltage has a magnitude of 18 V and an angle of 90°.

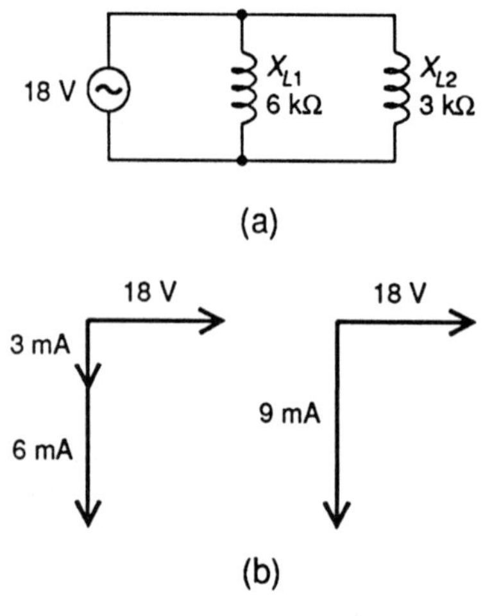

Fig. 8-4.

EXAMPLE 8-4

Show the phasors for Fig. 8-4a.

SOLUTION

The inductive reactances are in parallel. The current through the first inductor is 18 V divided by 6 kΩ, or 3 mA. The current through the second inductor is 18 V divided by 3 kΩ, or 6 mA. The total current is 3 mA plus 6 mA, or 9 mA. The total or equivalent inductive reactance equals 18 V divided by 9 mA, or 2 kΩ.

Figure 8-4b shows the phasors. The applied voltage has a phase angle of 0°. All currents lag this voltage by 90° because we are assuming ideal inductors. The vector sum of the two currents equals the total current. This means the total current is 9 mA at an angle of -90°.

8-3. CAPACITIVE CIRCUITS

In purely capacitive circuits, all currents lead the voltages by 90°. Furthermore, Ohm's law is valid for rms, peak, and peak-to-peak values. In a series circuit with ideal capacitors, the voltages across the capacitors are the product of the current and the capacitive reactances. The sum of these individual voltages equals the applied voltage. Furthermore, all voltages lag the current by 90°.

In a parallel circuit with ideal capacitors, the currents into each capacitor equal the applied voltage divided by the capacitive reactance. The sum of these branch currents equals the total current. Furthermore, all currents lead the applied voltage by 90°.

EXAMPLE 8-5

Show the phasors for Fig. 8-5a.

SOLUTION

The capacitors are in series. The total capacitive reactance is 9 kΩ, the sum of the individual reactances. The total current is 18 V divided by 9 kΩ, or 2 mA. The product of this 2 mA and 3 kΩ gives a voltage of 6 V across X_{C1}. Likewise, 2 mA through 6 kΩ gives a voltage of 12 V across X_{C2}.

Figure 8-5b shows the phasors. The current has a phase angle of 0°. All voltages lag this current by 90° because we are assuming ideal capacitors. The vector sum of the two voltages equals the applied voltage, which means the applied voltage is 18 V at an angle of -90°.

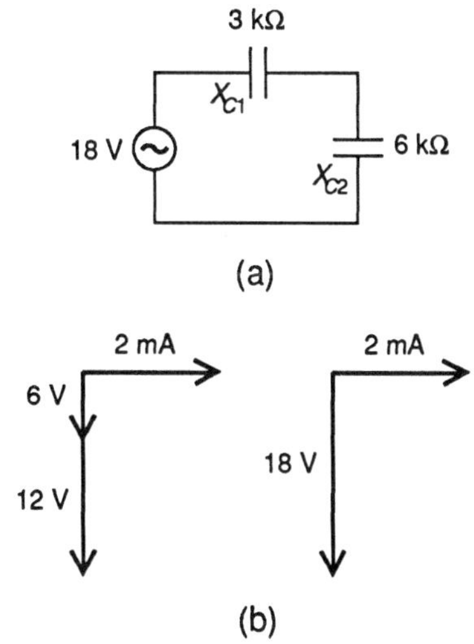

Fig. 8-5.

EXAMPLE 8-6

Show the phasors for Fig. 8-6a.

SOLUTION

The capacitors are in parallel. The current into the first capacitor is 3 mA, and the current into the second capacitor is 6 mA. The total current is 3 mA + 6 mA, or 9 mA. The total capacitive reactance equals 18 V/9 mA, or 2 kΩ.

Figure 8-6b shows the phasors. The applied voltage has a phase angle of 0°. All currents lead this voltage by 90° because we are assuming ideal capacitors. Also, the vector sum of the two currents equals the total current.

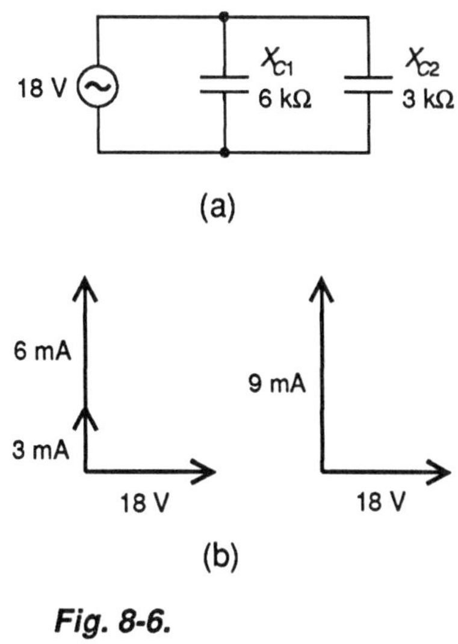

Fig. 8-6.

8-4. SERIES *LC* CIRCUITS

LC circuits combine ideal inductors and ideal capacitors. In these circuits, the currents may lead or lag the voltages by 90°, depending on the values of X_L and X_C. As usual, Ohm's law is valid for rms, peak, and peak-to-peak values.

In a series circuit with an ideal inductor and an ideal capacitor, the voltage across each component is the product of the current and the reactance. When X_L is greater than X_C, the inductive voltage is greater than the capacitive voltage. Furthermore, the inductive voltage leads the current by 90°, while the capacitive voltage lags the current by 90°. Because of this, the inductive voltage leads the capacitive voltage by 180°. As a result, the combined effect of the two reactances is a total impedance given by

$$Z = X_L - X_C \tag{8-1}$$

On the other hand, when X_C is greater than X_L, the capacitive voltage is larger than the inductive voltage. Again, the two voltage phasors are 180° apart. In this case, the combined effect of the two reactances produces a total impedance given by

$$Z = X_C - X_L \tag{8-2}$$

In either case, the applied voltage equals the phasor sum of the individual voltages. You can calculate this applied voltage by adding the individual voltages vectorially. The following examples shows you how it is done.

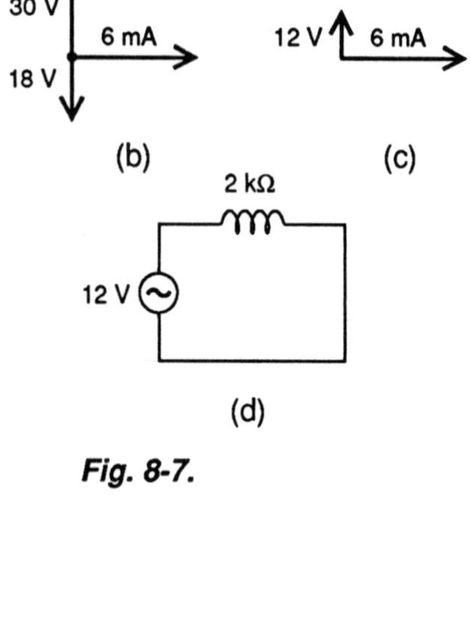

Fig. 8-7.

EXAMPLE 8-7

Show the phasors for Fig. 8-7a.

SOLUTION

An inductor and capacitor are in series. In Fig. 8-7a, X_L = 5 kΩ and X_C = 3 kΩ. Subtract the smaller from the larger to get

$$\text{Net } X_L = X_L - X_C = 5 \text{ k}\Omega - 3 \text{ k}\Omega = 2 \text{ k}\Omega$$

The total current equals the total voltage divided by the total or net reactance. The applied voltage is 12 V and the current is

$$I = \frac{V}{Z} = \frac{12 \text{ V}}{2 \text{ k}\Omega} = 6 \text{ mA}$$

This implies individual voltages of

$$V_L = IX_L = (6 \text{ mA})(5 \text{ k}\Omega) = 30 \text{ V}$$

and

$V_C = IX_C = (6 \text{ mA})(3 \text{ k}\Omega) = 18 \text{ V}$

Figure 8-7b shows the phasor diagram. The inductive voltage leads the current by 90°, while the capacitive voltage lags the current by 90°. Since these phasors are 180° out of phase, their magnitudes may be subtracted to get a net upward phasor of 12 V as shown in Fig. 8-7c. In this final phasor diagram, an applied voltage of 12 V leads a current of 6 mA by 90°. Because of this, the circuit acts inductive rather than capacitive. In other words, the circuit is equivalent to Fig. 8-7d.

EXAMPLE 8-8

Show the phasors for Fig. 8-8a.

SOLUTION

This time, X_C is greater than X_L. Subtract the smaller from the larger to get

$Z = X_C - X_L = 3 \text{ k}\Omega - 2 \text{ k}\Omega = 1 \text{ k}\Omega$

The reason the reactances can be subtracted is because the inductive voltage and capacitive voltages are 180° out of phase, which means the phasors point in opposite directions, equivalent to subtraction of magnitudes.

In Fig. 8-8a, the total current equals the applied voltage divided by the total reactance. Therefore, the current is

$$I = \frac{V}{Z} = \frac{10 \text{ V}}{1 \text{ k}\Omega} = 10 \text{ mA}$$

This implies individual voltages of

$V_L = IX_L = (10 \text{ mA})(2 \text{ k}\Omega) = 20 \text{ V}$

and

$V_C = IX_C = (10 \text{ mA})(3 \text{ k}\Omega) = 30 \text{ V}$

Figure 8-8b shows the phasor diagram. The inductive voltage leads the current by 90°, while the capacitive voltage lags the current by 90°. Since these phasors are 180° out of phase, the magnitudes may be subtracted to get a net downward phasor of 10 V as shown in Fig. 8-8c. In this final phasor diagram, an applied voltage of 10 V lags a current of 10 mA by 90°. Because of this, the circuit acts capacitive rather than inductive. In other words, the circuit is equivalent to Fig. 8-8d.

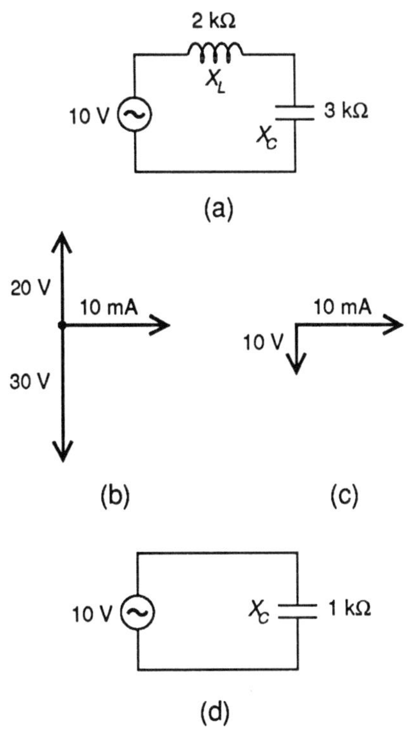

Fig. 8-8.

8-5. PARALLEL *LC* CIRCUITS

In a parallel circuit with an inductor and capacitor, the current into each component equals the applied voltage divided by the reactance of the component. When the inductive current is larger than the capacitive current, the circuit is inductive. When the capacitive current is larger than the inductive current, the circuit is capacitive.

Inductive Current Larger

When X_L is smaller than X_C, the inductive current is larger than the capacitive current. In this case, the equivalent circuit is inductive. Furthermore, since the inductive current lags the applied voltage by 90° and the capacitive current leads the applied voltage by 90°, the total current is the difference of the two:

$$I_T = I_L - I_C \qquad (8\text{-}3)$$

This can be written as

$$\frac{V_T}{Z} = \frac{V_T}{X_L} - \frac{V_T}{X_C}$$

The V_T term can be divided out to get

$$\frac{1}{Z} = \frac{1}{X_L} - \frac{1}{X_C}$$

By rearranging the right side,

$$\frac{1}{Z} = \frac{X_C - X_L}{X_L X_C}$$

Taking the reciprocal gives a useful equation for the impedance:

$$Z = \frac{X_L X_C}{X_C - X_L} \qquad (8\text{-}4)$$

Capacitive Current Larger

When X_C is smaller than X_L, the capacitive current is larger than the inductive current. In this case, the equivalent circuit is capacitive. Furthermore, since the inductive current lags the applied voltage by 90° and the capacitive current leads the applied voltage by 90°, the total current is the difference of the two:

$$I_T = I_C - I_L \qquad (8\text{-}5)$$

This can be written as

$$\frac{V_T}{Z} = \frac{V_T}{X_C} - \frac{V_T}{X_L}$$

As before, this can be rearranged to get

$$Z = \frac{X_L X_C}{X_L - X_C} \qquad (8\text{-}6)$$

Notice that this is almost identical to Eq. (8-4), except for subtracting the smaller reactance from the larger.

EXAMPLE 8-9

Show the phasors for Fig. 8-9a.

SOLUTION

The currents are

$$I_L = \frac{V}{X_L} = \frac{12\text{ V}}{3\text{ k}\Omega} = 4\text{ mA}$$

and

$$I_C = \frac{V}{X_C} = \frac{12\text{ V}}{4\text{ k}\Omega} = 3\text{ mA}$$

Figure 8-9b shows the phasor diagram. The inductive current lags the applied voltage by 90°, while the capacitive current leads the applied voltage by 90°. Since these phasors are 180° out of phase, the magnitudes may be subtracted to get a net downward phasor of 1 mA as shown in Fig. 8-9c. In this final phasor diagram, a total current of 1 mA lags an applied voltage of 12 V by angle of 90°.

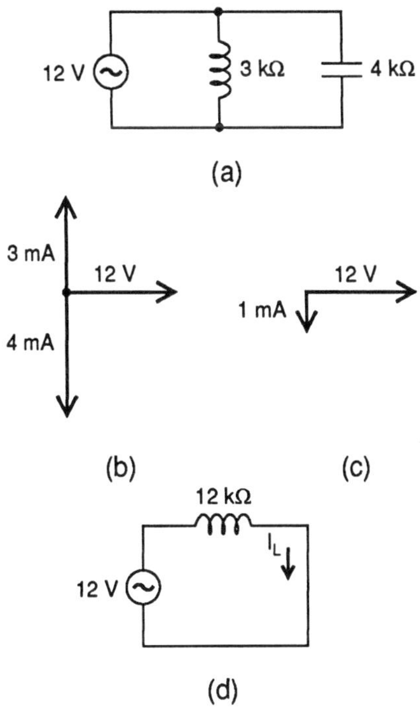

Fig. 8-9.

EXAMPLE 8-10

What is the impedance in Fig. 8-9a?

SOLUTION

There are two ways to find it. When you have the branch currents, you can subtract the smaller from the larger to get the total current. In the preceding example, we found $I_L = 4$ mA and $I_C = 3$ mA. Therefore, the total current is

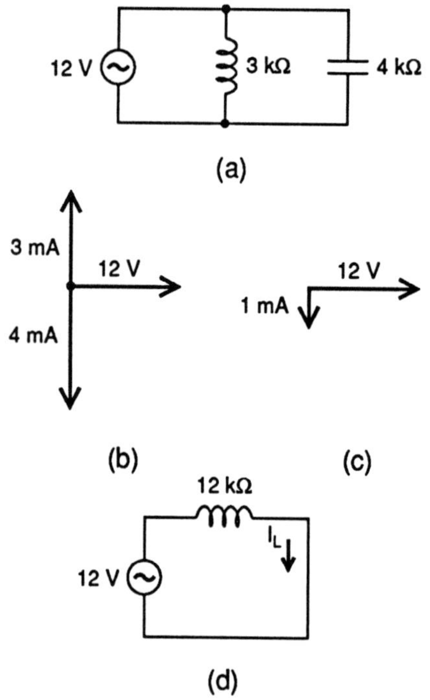

Fig. 8-9.

$$I_T = I_L - I_C = 4 \text{ mA} - 3 \text{ mA} = 1 \text{ mA}$$

The impedance equals the total voltage divided by the total current:

$$Z = \frac{12 \text{ V}}{1 \text{ mA}} = 12 \text{ k}\Omega$$

If you don't have the branch currents worked out, you can get the impedance directly from the reactances by using Eq. (8-4):

$$Z = \frac{X_L X_C}{X_C - X_L} = \frac{(3 \text{ k}\Omega)(4 \text{ k}\Omega)}{4 \text{ k}\Omega - 3 \text{ k}\Omega} = 12 \text{ k}\Omega$$

Figure 8-9d shows the equivalent circuit with this impedance. Now, you can calculate the total current as follows:

$$I_T = \frac{12 \text{ V}}{12 \text{ k}\Omega} = 1 \text{ mA}$$

EXAMPLE 8-11

Show the phasors for Fig. 8-10a.

SOLUTION

X_C is smaller than X_L. In a case like this, the capacitive current is larger than the inductive current. The currents are

$$I_L = \frac{V}{X_L} = \frac{10 \text{ V}}{5 \text{ k}\Omega} = 2 \text{ mA}$$

and

$$I_C = \frac{V}{X_C} = \frac{10 \text{ V}}{3 \text{ k}\Omega} = 3.33 \text{ mA}$$

Figure 8-10b shows the phasor diagram. The capacitive current leads the applied voltage by 90°, while the inductive current lags the voltage by 90°. Since these phasors are 180° out of phase, the magnitudes may be subtracted to get a net upward phasor of 1.33 mA as shown in Fig. 8-10c. In this final phasor diagram, a total current of 1.33 mA leads an applied voltage of 10 V by 90°.

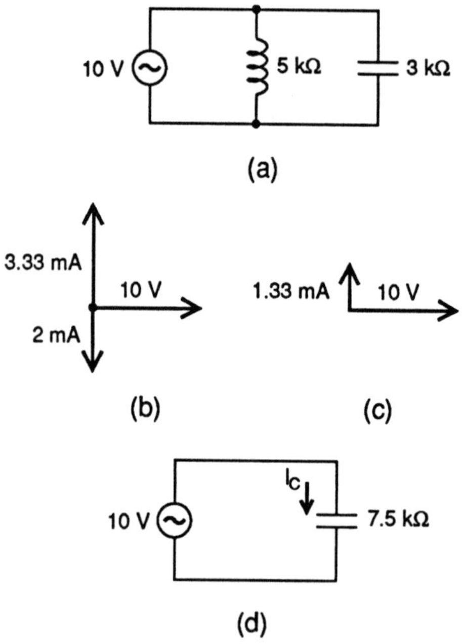

Fig. 8-10.

EXAMPLE 8-12

What is the impedance in Fig. 8-10a?

SOLUTION

In the preceding example, we found $I_L = 2$ mA and $I_C = 3.33$ mA. Therefore, the total current is

$$I_T = I_C - I_L = 3.33 \text{ mA} - 2 \text{ mA} = 1.33 \text{ mA}$$

The impedance equals the total voltage divided by the total current:

$$Z = \frac{10 \text{ V}}{1.33 \text{ mA}} = 7.5 \text{ k}\Omega$$

You can get the impedance directly from the reactances by using Eq. (8-6):

$$Z = \frac{X_L X_C}{X_L - X_C} = \frac{(5 \text{ k}\Omega)(3 \text{ k}\Omega)}{5 \text{ k}\Omega - 3 \text{ k}\Omega} = 7.5 \text{ k}\Omega$$

Figure 8-10d shows the equivalent circuit with this impedance. Now, you can calculate the total current as follows:

$$I_T = \frac{10 \text{ V}}{7.5 \text{ k}\Omega} = 1.33 \text{ mA}$$

8-6. SERIES *RLC* CIRCUITS

RLC circuits combine resistors, inductors and capacitors. In these circuits, the currents may lead or lag the voltages by 90°. As usual, Ohm's law is valid for rms, peak, and peak-to-peak values of individual components. You also need to use the quadrature sums when analyzing these circuits.

When a resistor, inductor, and capacitor are in series, the voltage across each component is the product of the current and the impedance of the component. When X_L is greater than X_C, the equivalent circuit is the series *RL* circuit discussed in Chap. 4. On the other hand, when X_C is greater than X_L, the equivalent circuit is the series *RC* circuit discussed in Chap. 7. In either case, the applied voltage equals the phasor sum of the individual voltages.

The impedance of a series *RLC* circuit is given by

$$Z = \sqrt{R^2 + (X_L - X_C)^2} \qquad (8\text{-}7)$$

This makes sense. Since the inductive and capacitive phasor voltages are 180° apart, we must subtract the reactances when working out the impedance. After we have the impedance, we can calculate the total current with

$$I = \frac{V_T}{Z} \qquad (8\text{-}8)$$

Next, we can calculate the individual voltages with

$$V_R = IR \qquad (8\text{-}9)$$

$$V_L = IX_L \qquad (8\text{-}10)$$

$$V_C = IX_C \qquad (8\text{-}11)$$

When the inductive reactance is larger than the capacitive reactance, the equivalent circuit is inductive. In other words, the equivalent circuit appears to have a net inductive reactance of

$$\text{Net } X_L = X_L - X_C \qquad (8\text{-}12)$$

When the capacitive reactance is larger than the inductive reactance, the equivalent circuit is capacitive. In other words, the equivalent circuit appears to have a net capacitive reactance of

$$\text{Net } X_C = X_C - X_L \qquad (8\text{-}13)$$

Read the following examples to see how to use all of the foregoing ideas and equations.

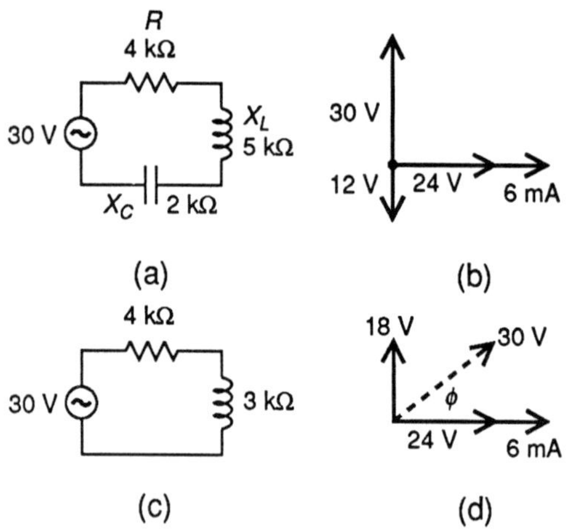

Fig. 8-11.

EXAMPLE 8-13

Show the phasors for Fig. 8-11a.

SOLUTION

A resistor, inductor, and capacitor are in series. The impedance is

$$Z = \sqrt{(4\text{ k}\Omega)^2 + (5\text{ k}\Omega - 2\text{ k}\Omega)^2} = \sqrt{(4\text{ k}\Omega)^2 + (3\text{ k}\Omega)^2} = 5\text{ k}\Omega$$

Therefore, the current is

$$I = \frac{30\text{ V}}{5\text{ k}\Omega} = 6\text{ mA}$$

This current times the impedance of each component gives the voltage across that component as follows:

$$V_R = (6\text{ mA})(4\text{ k}\Omega) = 24\text{ V}$$

$$V_L = (6\text{ mA})(5\text{ k}\Omega) = 30\text{ V}$$

$$V_C = (6\text{ mA})(2\text{ k}\Omega) = 12\text{ V}$$

Figure 8-11b shows the phasor diagram. Since the voltage and current are in phase in a resistor, the resistive voltage has a phase angle of 0° as shown. The inductive voltage leads the resistive voltage by 90°, and the capacitive voltage lags the resistive voltage by 90°.

Because the inductive reactance is larger than the capacitive reactance, the equivalent circuit has an inductive reactance of

$$\text{Net } X_L = X_L - X_C = 5\text{ k}\Omega - 2\text{ k}\Omega = 3\text{ k}\Omega$$

Figure 8-11c shows the equivalent circuit. Since 6 mA flows through 4 kΩ and 3 kΩ, the resistive voltage is 24 V and the inductive voltage is 18 V. The corresponding phasor voltages are shown in Fig. 8-11d.

EXAMPLE 8-14

What is the angle in Fig. 8-11d?

SOLUTION

Since the equivalent circuit is a series *RL* circuit, the tangent of the angle is given by

$$\tan f = \frac{V_L}{V_R} = \frac{18\text{ V}}{24\text{ V}} = 0.75$$

You get the same tangent using the reactances:

$$\tan f = \frac{X_L}{R} = \frac{3\text{ k}\Omega}{4\text{ k}\Omega} = 0.75$$

With trig tables or a calculator, the angle is

$$f = 36.9^\circ$$

EXAMPLE 8-15

A series *RLC* circuit has these values: $R = 4\text{ k}\Omega$, $X_L = 2\text{ k}\Omega$, and $X_C = 5\text{ k}\Omega$. Describe the phasor diagram.

SOLUTION

The calculations are similar to Example 8-13 because the values of X_L and X_C have been reversed. The impedance is

$$Z = \sqrt{(4\text{ k}\Omega)^2 + (2\text{ k}\Omega - 5\text{ k}\Omega)^2} = \sqrt{(4\text{ k}\Omega)^2 + (-3\text{ k}\Omega)^2} = 5\text{ k}\Omega$$

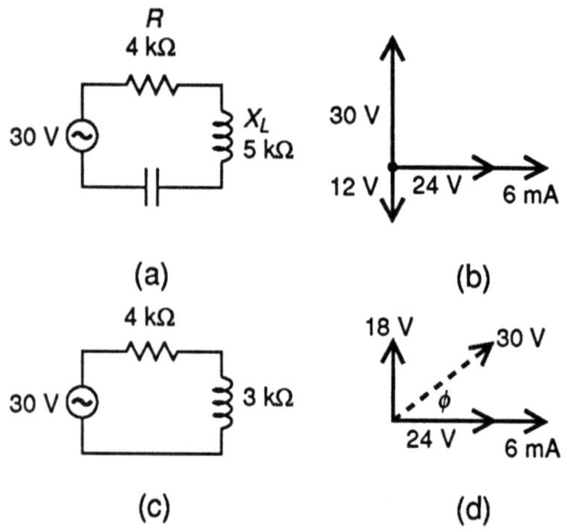

Fig. 8-11.

Therefore, the current is the same as before:

$$I = \frac{30\text{ V}}{5\text{ k}\Omega} = 6\text{ mA}$$

The voltages across the components are as follows:

$$V_R = (6\text{ mA})(4\text{ k}\Omega) = 24\text{ V}$$

$$V_L = (6\text{ mA})(2\text{ k}\Omega) = 12\text{ V}$$

$$V_C = (6\text{ mA})(5\text{ k}\Omega) = 30\text{ V}$$

The phasor diagram is similar to Fig. 8-11b, except that 12 V points up and 30 V points down. Also, the equivalent phasor diagram of Fig. 8-11d has the 18 V pointing down instead of up, and the applied voltage has a phase angle of -36.9° instead of +36.9°. Because the capacitive reactance is larger than the inductive reactance, the equivalent circuit has a capacitive reactance of

$$\text{Net } X_C = X_C - X_L = 5\text{ k}\Omega - 2\text{ k}\Omega = 3\text{ k}\Omega$$

The equivalent circuit is like Fig. 8-11c, except that a capacitor with a reactance of 3 kΩ is used instead of the inductor.

8-7. PARALLEL *RLC* CIRCUITS

When a resistor, inductor, and capacitor are in parallel, the current into each component equals the applied voltage divided by the impedance of the component. The easiest way to analyze a parallel *RLC* circuit is to calculate each branch current as follows:

$$I_R = \frac{V_T}{R} \tag{8-14}$$

$$I_L = \frac{V_T}{X_L} \tag{8-15}$$

$$I_C = \frac{V_T}{X_C} \tag{8-16}$$

When the inductive current is larger than the capacitive current, the equivalent circuit is the parallel *RL* circuit discussed in Chap. 4. In other words, the equivalent circuit appears to have a net inductive reactance of

$$\text{Net } X_L = \frac{V_T}{I_L - I_C} \qquad (8\text{-}17)$$

On the other hand, when the capacitive current is larger than the inductive current, the equivalent circuit is the parallel *RC* circuit discussed in Chap. 7. The net capacitive reactance is

$$\text{Net } X_C = \frac{V_T}{I_C - I_L} \qquad (8\text{-}18)$$

Read the following examples to see how to use the foregoing ideas.

EXAMPLE 8-16

What is the total current in Fig. 8-12a?

SOLUTION

A resistor, inductor, and capacitor are in parallel. The branch currents are

$$I_R = \frac{V}{R} = \frac{12\text{ V}}{3\text{ k}\Omega} = 4\text{ mA}$$

$$I_L = \frac{V}{X_L} = \frac{12\text{ V}}{2\text{ k}\Omega} = 6\text{ mA}$$

and

$$I_C = \frac{V}{X_C} = \frac{12\text{ V}}{4\text{ k}\Omega} = 3\text{ mA}$$

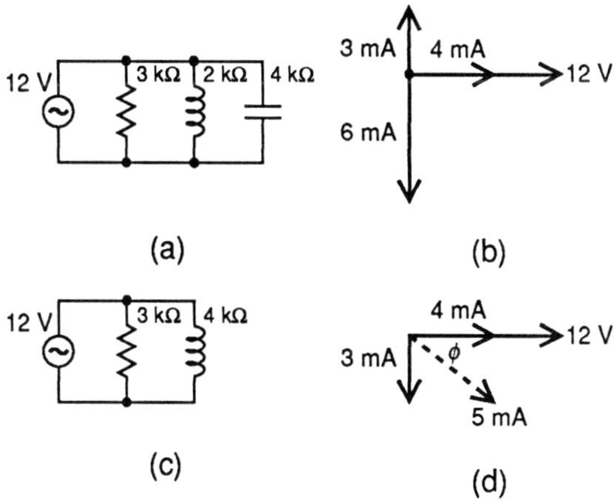

Fig. 8-12.

Figure 8-12b shows the phasor currents. The resistive current is in phase with the applied voltage, the inductive current lags the applied voltage by 90°, and the capacitive current leads the applied voltage by 90°.

Because the inductive current is larger than the capacitive current, the equivalent circuit has an inductive reactance of

$$\text{Net } X_L = \frac{V_T}{I_L - I_C} = \frac{12\text{ V}}{6\text{ mA} - 3\text{ mA}} = 4\text{ k}\Omega$$

Figure 8-12c shows the equivalent circuit. Since 12 V is across 3 kΩ and 4 kΩ, the resistive current is 4 mA and the inductive current is 3 mA. The corresponding phasor currents are shown in Fig. 8-12d. With the Pythagorean theorem, we can calculate a total current of

$$I = \sqrt{(4\text{ mA})^2 + (3\text{ mA})^2} = 5\text{ mA}$$

8-8. MANY SERIES COMPONENTS

The current is the same in all parts of a series circuit. Whenever many components are in series, this current will produce individual voltages equal to the current times the impedance of the component. All resistive voltages are in phase with the current, all inductive voltages lead the current by 90°, and all capacitive voltages lag the current by 90°.

To analyze the series circuit, you can add all the resistances to get the total resistance, add all the inductive reactances to get the total inductive reactance, and add all the capacitive reactances to get the total capacitive reactance. The equivalent circuit is then the series *RLC* circuit discussed in the preceding section.

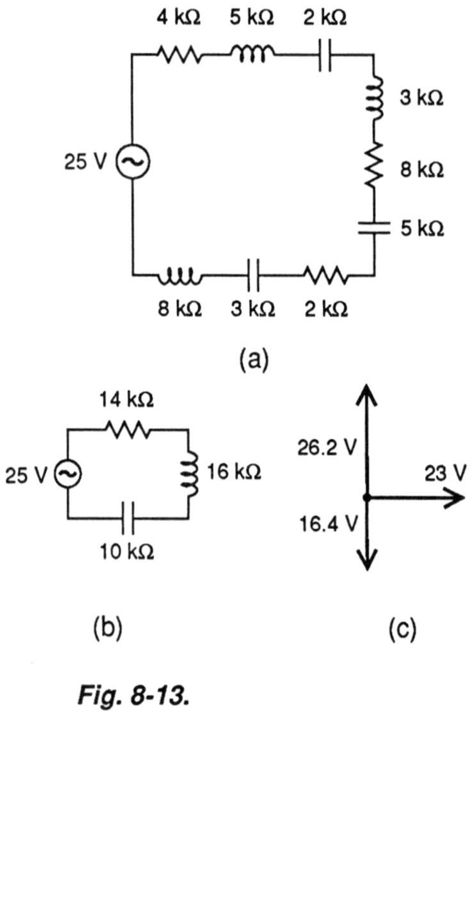

Fig. 8-13.

EXAMPLE 8-17

What is the current in Fig. 8-13a? The phasor voltages?

SOLUTION

Add all the resistances to get

$$R = 4\ \text{k}\Omega + 8\ \text{k}\Omega + 2\ \text{k}\Omega = 14\ \text{k}\Omega$$

Add all the inductive reactances to get

$$X_L = 5\ \text{k}\Omega + 3\ \text{k}\Omega + 8\ \text{k}\Omega = 16\ \text{k}\Omega$$

Add all the capacitive reactances to get

$$X_C = 2\ \text{k}\Omega + 5\ \text{k}\Omega + 3\ \text{k}\Omega = 10\ \text{k}\Omega$$

Figure 8-13b shows the equivalent circuit. At this point, the analysis reverts to the case of a series *RLC* circuit where the inductive reactance is larger than the capacitive reactance. The impedance is

$$Z = \sqrt{(14\ \text{k}\Omega)^2 + (16\ \text{k}\Omega - 10\ \text{k}\Omega)^2} = \sqrt{(14\ \text{k}\Omega)^2 + (6\ \text{k}\Omega)^2} = 15.2\ \text{k}\Omega$$

The current is

$$I_T = \frac{V_T}{Z} = \frac{25\ \text{V}}{15.2\ \text{k}\Omega} = 1.64\ \text{mA}$$

When this current flows through the equivalent circuit of Fig. 8-13b, it produces voltages of

$V_R = (1.64 \text{ mA})(14 \text{ k}\Omega) = 23 \text{ V}$

$V_L = (1.64 \text{ mA})(16 \text{ k}\Omega) = 26.2 \text{ V}$

$V_C = (1.64 \text{ mA})(10 \text{ k}\Omega) = 16.4 \text{ V}$

Figure 8-13c shows the phasor voltages.

8-9. MANY PARALLEL COMPONENTS

The voltage is the same across all branches of a parallel circuit. Whenever many components are in parallel, this voltage will produce branch currents equal to the voltage divided by the impedance of each branch. All resistive currents are in phase with the voltage, all inductive currents lag the voltage by 90°, and all capacitive currents lead the voltage by 90°.

To analyze the parallel circuit, you can add all the resistive currents to get the total resistive current, add all the inductive currents to get the total inductive current, and add all capacitive currents to get the total capacitive current. Dividing the applied voltage by each type of current leads to an equivalent parallel *RLC* circuit that you can analyze with the methods discussed earlier. The formulas for the equivalent branch impedances are

$$R = \frac{V_T}{\text{Total } I_R} \qquad (8\text{-}19)$$

$$X_L = \frac{V_T}{\text{Total } I_L} \qquad (8\text{-}20)$$

$$X_C = \frac{V_T}{\text{Total } I_C} \qquad (8\text{-}21)$$

An alternative way to find the equivalent resistance is with

$$R = \frac{1}{1/R_1 + 1/R_2 + \ldots + 1/R_n} \qquad (8\text{-}22)$$

An alternative way to find the equivalent reactances is with

$$X_L = \frac{1}{1/X_{L1} + 1/X_{L2} + \ldots + 1/X_{Ln}} \qquad (8\text{-}23)$$

and

$$X_C = \frac{1}{1/X_{C1} + 1/X_{C2} + \ldots + 1/X_{Cn}} \qquad (8\text{-}24)$$

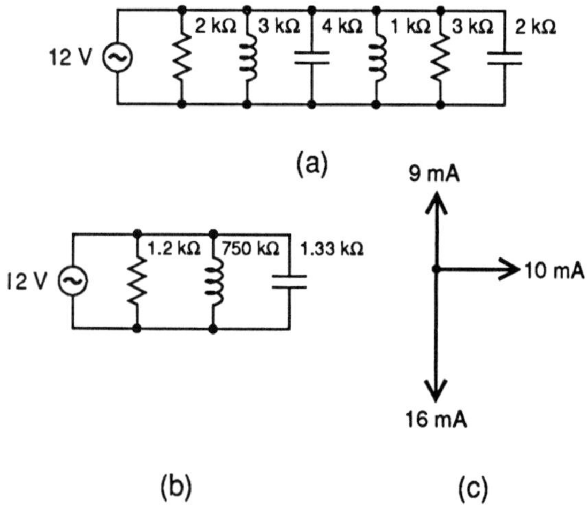

Fig. 8-14.

EXAMPLE 8-19

What is the equivalent circuit for Fig. 8-14a? The phasor currents?

SOLUTION

The total resistive current is

$$\text{Total } I_R = 6 \text{ mA} + 4 \text{ mA} = 10 \text{ mA}$$

Therefore, the equivalent resistance is

$$R = \frac{V}{\text{Total } I_R} = \frac{12 \text{ V}}{10 \text{ mA}} = 1.2 \text{ k}\Omega$$

By similar calculations, the total inductive current is 16 mA, and the equivalent inductive reactance is

$$X_L = \frac{V}{\text{Total } I_L} = \frac{12 \text{ V}}{16 \text{ mA}} = 750 \text{ }\Omega$$

The total capacitive current is 9 mA, and the equivalent capacitive reactance is

$$X_C = \frac{V}{\text{Total } I_C} = \frac{12 \text{ V}}{9 \text{ mA}} = 1.33 \text{ k}\Omega$$

Figure 8-14b shows the equivalent circuit. At this point, the analysis reverts to the case of a parallel *RLC* circuit where the inductive reactance is smaller than the capacitive reactance. Figure 8-14c shows the phasor currents.

8-10. POSITIVE AND NEGATIVE POWER

The power in ac circuits is more complicated than in dc circuits. To begin with, you have to use rms values to calculate average power over a complete cycle. Furthermore, only resistors *dissipate* power. This means they convert electrical energy to heat energy which escapes permanently to the outside air. On the other hand, ideal inductors and ideal capacitors do not dissipate power. Instead, they absorb electrical energy during one part of the cycle and return this energy during another part of the cycle.

Positive Power

The voltage across a resistor produces a current through the resistor. The product of the voltage and current is the power, or rate of heat loss by the resistor. For instance, if a resistor has a power dissipation of 2 W, the resistor is producing 2 joules of heat per second. This heat radiates into the surrounding air and is permanently lost by the resistor.

Resistive power is always *positive* because the product of the voltage and current is always a positive number. This is true even for alternating current. During the positive half cycle, the voltage and current are in one direction. During the negative half cycle, both the voltage and the current reverse; therefore, the product is still positive.

For instance, with an ideal resistor the sinusoidal voltage is in phase with the sinusoidal current (see Fig. 8-15a). The instantaneous power is the product of the instantaneous voltage and instantaneous current:

$$p = vi \qquad (8\text{-}25)$$

During the positive half cycle, the voltage and current are both positive values, so the product is positive. During the negative half cycle, the voltage and current are both negative, so the product is still positive.

You can see the foregoing concept mathematically as follows. Since the voltage and current are in-phase sine waves, Equation (8-25) may be written as

$$p = vi = (V_p \sin \phi)(I_p \sin \phi) = V_p I_p (\sin \phi)(\sin \phi) = V_p I_p (\sin \phi)^2$$

or

$$p = 0.5 V_p I_p (1 - \cos 2\phi)$$

Figure 8-15b shows the graph of this equation. It is a graph of instantaneous power versus angle ϕ. The power is positive throughout the cycle. This means the resistor absorbs electrical energy from the circuit at every instant during the cycle except at 0°, 180°, 360°, etc. The resistor converts this electrical energy into heat, which radiates into the surrounding air.

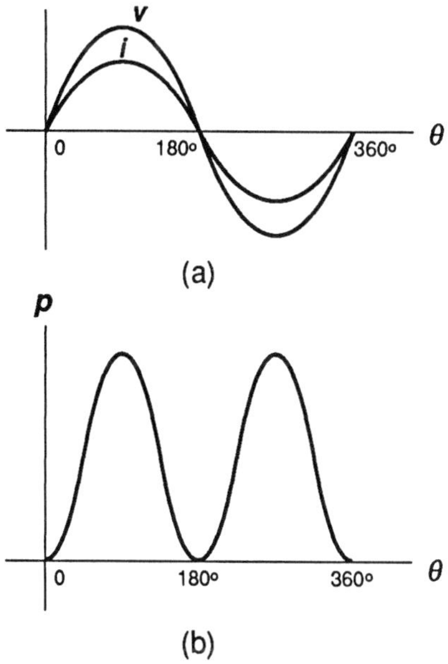

Fig. 8-15.

Negative Power

Power can be *negative*. Instead of absorbing energy, some devices return electrical energy to a circuit. An inductor is an example of a device that can produce negative power during the cycle. Figure 8-16a shows the voltage and current sine waves for an ideal inductor. Because of the 90° phase difference, the product of v and i is positive during the first quarter cycle, negative during the next quarter cycle, positive during the third quarter cycle, and negative during the final quarter cycle.

Here is the mathematical proof. Since the sinusoidal voltage and current are 90° out of phase, the instantaneous power in an inductor is given by

$$p = vi = (V_p \cos \phi)(I_p \sin \phi) = V_p I_p (\cos \phi)(\sin \phi)$$

or

$$p = 0.5 V_p I_p \sin 2\phi$$

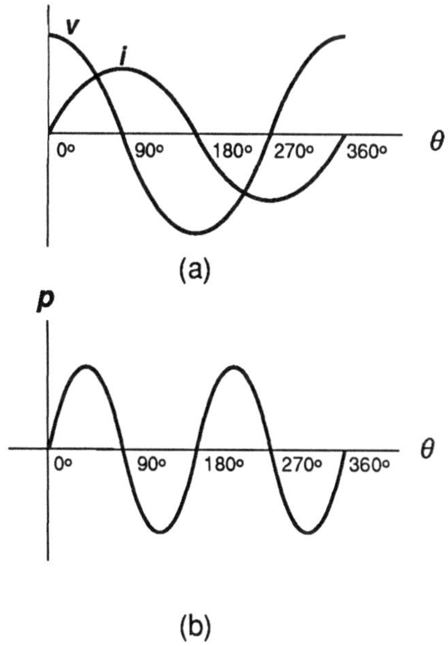

Fig. 8-16.

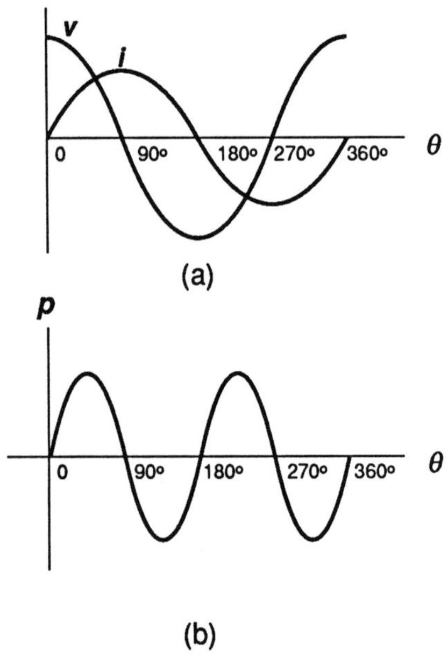

Fig. 8-16.

Figure 8-16b shows the graph of this equation. What does the graph mean? When the power is positive, the inductor absorbs energy from the circuit and stores it in the magnetic field. When the power is negative, the magnetic field decreases and the inductor returns energy to the circuit.

A similar statement applies to an ideal capacitor. Because the voltage and current sinusoids are 90° out of phase, the product of instantaneous voltage and current produce an instantaneous power like Fig. 8-16b. When the power is positive, energy is being stored in the electric field between the plates. When the power is negative, energy is being returned by the electric field to the circuit.

Look at Fig. 8-16b. The average value of this waveform is zero over the cycle. Because of this, we can state the following: an ideal inductor or ideal capacitor has zero power dissipation over a complete cycle. Ideal inductors and ideal capacitors don't waste any of the energy supplied by the source. Over a cycle, they return the same amount of energy that they absorb.

In summary, the power into a device is positive when the device absorbs energy from a circuit, and negative when the device returns energy to the circuit. A resistor has a positive power throughout the cycle because it absorbs energy continuously. Ideal inductors and ideal capacitors have positive power during the part of the cycle that they absorb energy, and negative power during the part of the cycle they return their stored energy to the circuit.

8-11. REAL, REACTIVE, AND APPARENT POWER

Resistors dissipate power, but inductors and capacitors do not. Because of this, we can have positive power and negative power. The differences can be also described by real and apparent power.

Real Power

Since a resistor produces heat that escapes to the surroundings, the average power in a resistor over an entire cycle is called *real power*. You can always calculate the real power by

$$\text{Real power} = V_R I_R \tag{8-26}$$

where V_R = rms voltage across resistor
I_R = rms current through resistor

Two other formulas that you can use in any circuit are

$$\text{Real power} = I_R{}^2 R \tag{8-27}$$

$$\text{Real power} = \frac{V_R{}^2}{R} \tag{8-28}$$

where I_R = rms current through the resistor
V_R = rms voltage across the resistor

Reactive Power

The product of the rms voltage across a reactive component and the rms current through it is called the *reactive power*. Since there are two types of reactance, the two formulas for reactive power are

Reactive power = $V_L I$ (8-29)

and

Reactive power = $V_C I$ (8-30)

The unit of measure for reactive power is *voltamperes reactive* (VAR). This helps to distinguish the two types of power. In the case of real power, the energy goes off in the form of heat (watts). In the case of reactive power, the energy is stored in the magnetic field (inductor) or electric field (capacitor) during one part of the cycle and returned to the circuit during another part of the cycle.

Apparent Power

Figure 8-17 shows the phasor diagram for a series *RL* circuit. The applied voltage V leads the current I by an angle of ϕ. The *apparent power* is defined as the product of total voltage and current:

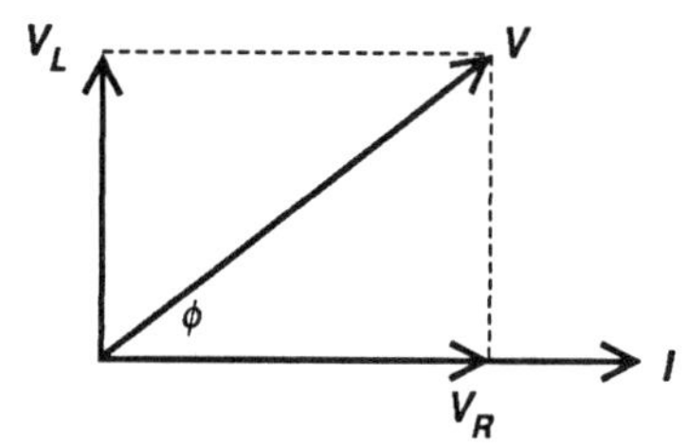

Fig. 8-17.

Apparent power = VI (8-31)

The unit of measure for apparent power is *voltamperes* (VA).

In Fig. 8-17, the resistive voltage is given by

$$V_R = V \cos \phi$$

Since $V_R = V \cos \phi$, the real power may be written as

Real power = $VI \cos \phi$ (8-32)

Similarly, since $V_L = V \sin \phi$, the reactive power may be written as

Reactive power = $VI \sin \phi$ (8-33)

The quantity $\cos \phi$ is called the *power factor*. To maximize the real power, the power factor should be unity, equivalent to have a phase angle of 0° between the total voltage and current. Equation (8-32) also applies to a series *RC* circuit, to series *RLC* circuits, parallel *RLC* circuits, and any other circuit where V is the applied voltage, I is the total current, and ϕ is the angle between the voltage and current.

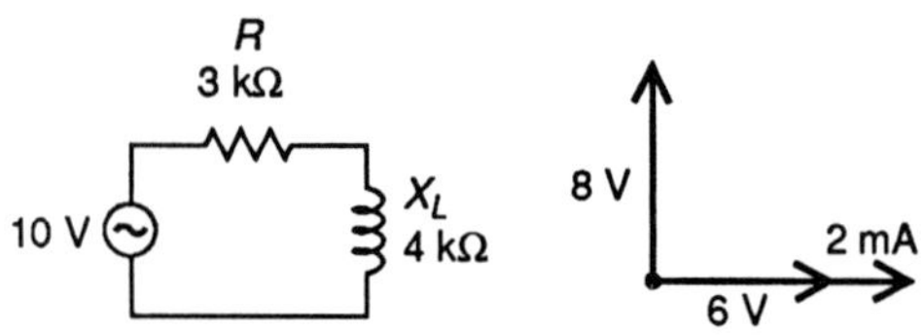

(a) (b)

Fig. 8-18.

EXAMPLE 8-19

The current is 2 mA in Fig. 8-18a. What is the real power? The reactive power? The apparent power?

SOLUTION

Since the current is 2 mA, the resistor has a voltage of 6 V and the inductor has a voltage of 8 V. The real power in the resistor is

Real power = (6 V)(2 mA) = 12 mW

The reactive power in the inductor is

Reactive power = (8 V)(2 mA) = 16 mVAR

The apparent power in the circuit is

Apparent power = (10 V)(2 mA) = 20 mVA

EXAMPLE 8-20

Calculate the real and reactive power of Fig. 8-18a using the power factor.

SOLUTION

Figure 8-18b shows the phasor diagram. The tangent of the angle is

$$\tan \phi = \frac{8\text{ V}}{6\text{ V}} = 1.33$$

and the angle is

$\phi = 53.1^\circ$

The power factor is

$\cos \phi = \cos (53.1^\circ) = 0.6$

With Eq. (8-32), the real power is

Real power = $VI \cos \phi$ = (10 V)(2 mA) cos (53.1°) = 12 mW

With Eq. (8-33), the reactive power is

Reactive power = $VI \sin \phi$ = (10 V)(2 mA) sin (53.1°) = 16 mVAR

MULTIPLE-CHOICE TEST

The following questions may have more than one right answer. Select the best answer. This is the one that is always true, or covers more situations, or agrees with the material of this chapter, etc.

1. In a series resistive circuit, the individual voltages are in phase with the
   a. current
   b. inductive reactance
   c. capacitive reactance
   d. none of the above

2. In a series circuit with ideal inductors, the voltages lead the current by
   a. 0°
   b. 90°
   c. -90°
   d. 180°

3. In a series circuit with ideal capacitors, the voltages lag the current by
   a. 0°
   b. 90°
   c. -90°
   d. 180°

4. In a series circuit with an ideal inductor and capacitor, the inductive and capacitive voltages are out of phase by
   a. 0°
   b. 90°
   c. -90°
   d. 180°

5. In a parallel circuit with an ideal inductor and capacitor, the inductive and capacitive currents are out of phase by
   a. 0°
   b. 90°
   c. -90°
   d. 180°

6. When X_C is greater than X_L, a series *RLC* circuit simplifies to a
   a. series *RL* circuit
   b. series *RC* circuit
   c. parallel *RL* circuit
   d. parallel *RC* circuit

7. When X_L is greater than X_C, a series *RLC* circuit simplifies to a
   a. series *RL* circuit
   b. series *RC* circuit
   c. parallel *RL* circuit
   d. parallel *RC* circuit

8. When X_L is greater than X_C, a parallel *RLC* circuit simplifies to a
   a. series *RL* circuit
   b. series *RC* circuit
   c. parallel *RL* circuit
   d. parallel *RC* circuit

9. When X_C is greater than X_L, a parallel *RLC* circuit simplifies to a
   a. series *RL* circuit
   b. series *RC* circuit
   c. parallel *RL* circuit
   d. parallel *RC* circuit

10. When many components are in series, you can add all the
    a. resistances to get the total resistance
    b. inductive reactances to get the total inductive reactance
    c. capacitive reactances to get the total capacitive reactance
    d. all of the above

11. When many components are in parallel, you can divide the applied voltage by the
    a. total resistive current to get the equivalent resistance
    b. total inductive current to get the equivalent inductive reactance
    c. total capacitive current to get the equivalent capacitive reactance
    d. all of the above

12. A device with positive power is
    a. absorbing energy from the circuit
    b. returning energy from the circuit
    c. resistive
    d. reactive

13. A device with negative power is
    a. absorbing energy from the circuit
    b. returning energy from the circuit
    c. resistive
    d. reactive

14. Real power refers to the power in
    a. a resistor
    b. an inductor
    c. a capacitor
    d. all of the above

15. Reactive power refers to the power in
    a. a coil
    b. an inductor
    c. a capacitor
    d. all of the above

PROBLEMS

Sec. 8-1: Resistive Circuits

8-1. The applied rms voltage of Fig. 8-19 is 12 V. What is the current? The voltage across each resistor? Show all phasors for this circuit.
8-2. The applied rms voltage of Fig. 8-20 is 24 V. What is each branch current? The total current? Show all phasors for this circuit.
8-3. If the applied rms voltage of Fig. 8-19 changes to 36 V, what is the current? The voltage across each resistor? Show all phasors for this circuit.
8-4. If the applied rms voltage of Fig. 8-20 changes to 8 V, what is each branch current? The total current? Show all phasors for this circuit.

Sec. 8-2: Inductive Circuits

8-5. The applied rms voltage of Fig. 8-21 is 5 V. What is the current? The voltage across each inductor? Show all phasors for this circuit.
8-6. The applied rms voltage of Fig. 8-22 is 14 V. What is each branch current? The total current? Show all phasors for this circuit.
8-7. If the applied rms voltage of Fig. 8-21 changes to 36 V, what is the current? The voltage across each inductor? Show all phasors for this circuit.
8-8. If the applied rms voltage of Fig. 8-22 changes to 8 V, what is each branch current? The total current? Show all phasors for this circuit.

Sec. 8-3: Capacitive Circuits

8-9. The applied rms voltage of Fig. 8-23 is 18 V. What is the current? The voltage across each capacitor? Show all phasors for this circuit.
8-10. The applied rms voltage of Fig. 8-24 is 24 V. What is each branch current? The total current? Show all phasors for this circuit.
8-11. If the applied rms voltage of Fig. 8-23 changes to 36 V, what is the current? The voltage across each capacitor? Show all phasors for this circuit.
8-12. If the applied rms voltage of Fig. 8-24 changes to 8 V, What is each branch current? The total current? Show all phasors for this circuit.

Sec. 8-4: Series *LC* Circuits

8-13. The applied rms voltage of Fig. 8-25 is 20 V. What is the current? The voltage across each component? Show all phasors for this circuit.
8-14. If the applied rms voltage of Fig. 8-25 changes to 36 V, what is the current? The voltage across each component? Show all phasors for this circuit.
8-15. The frequency of Fig. 8-25 doubles. What is the current? The voltage across each component? Show all phasors for this circuit.
8-16. The inductance of Fig. 8-25 doubles. What is the current? The voltage across each component? Show all phasors for this circuit.

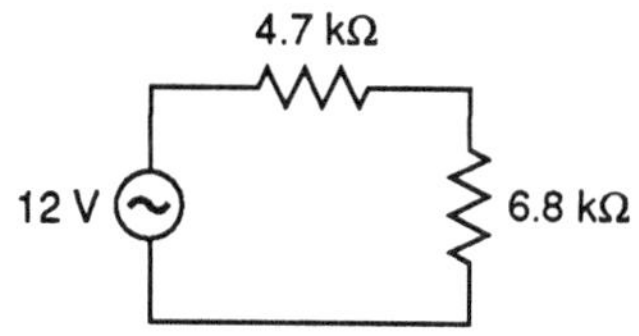

Fig. 8-19.

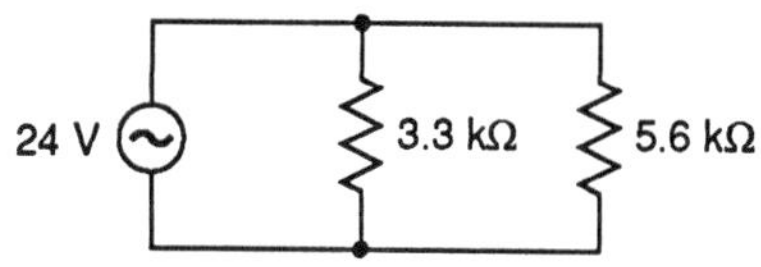

Fig. 8-20.

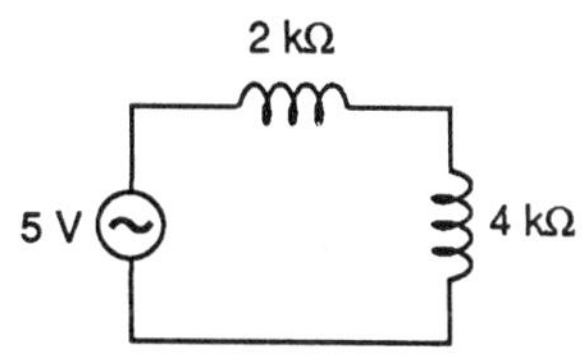

Fig. 8-21

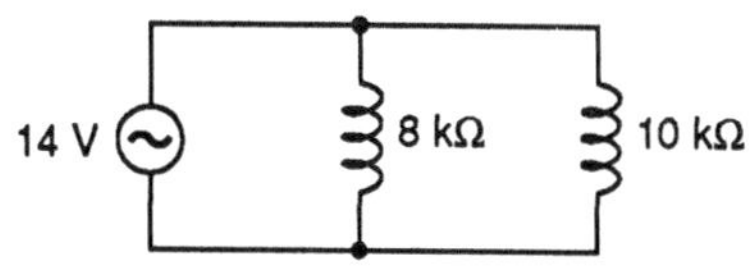

Fig. 8-22.

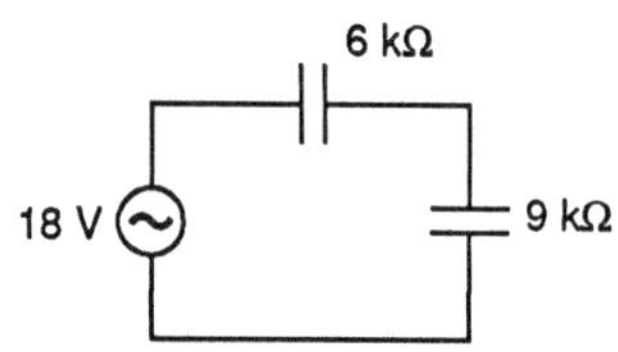

Fig. 8-23.

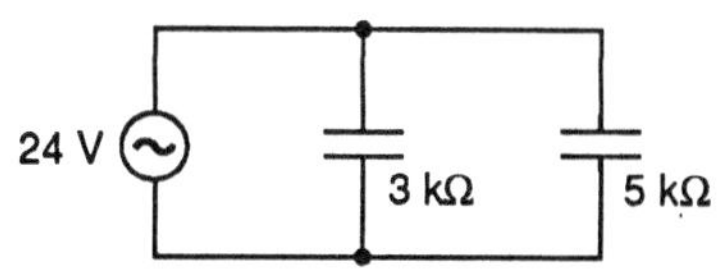

Fig. 8-24.

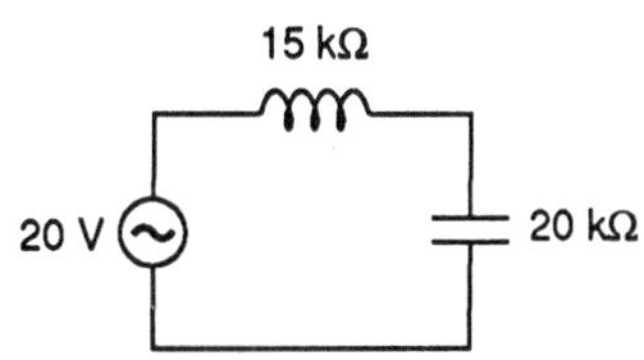

Fig. 8-25.

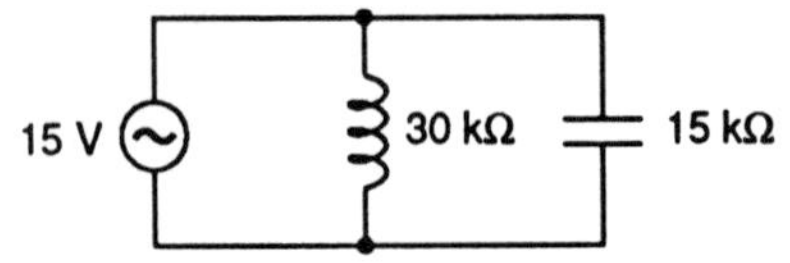

Fig. 8-26.

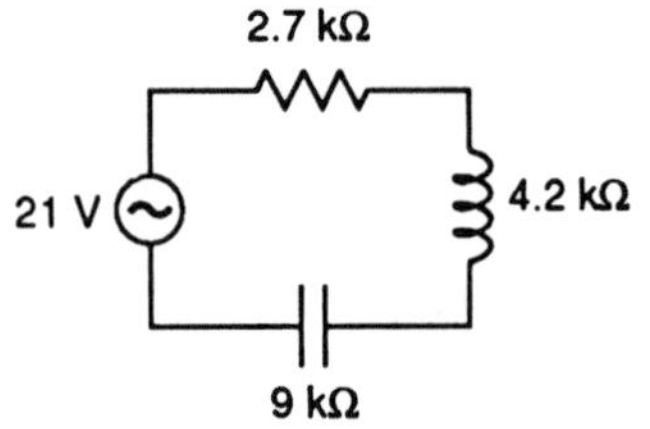

Fig. 8-27.

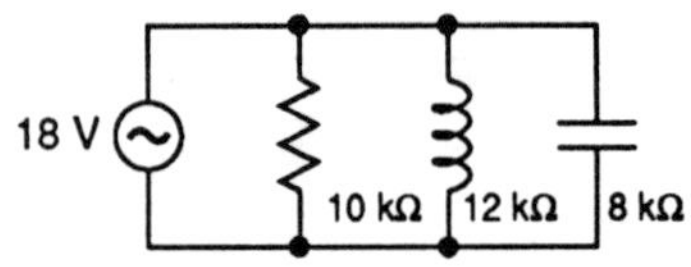

Fig. 8-28.

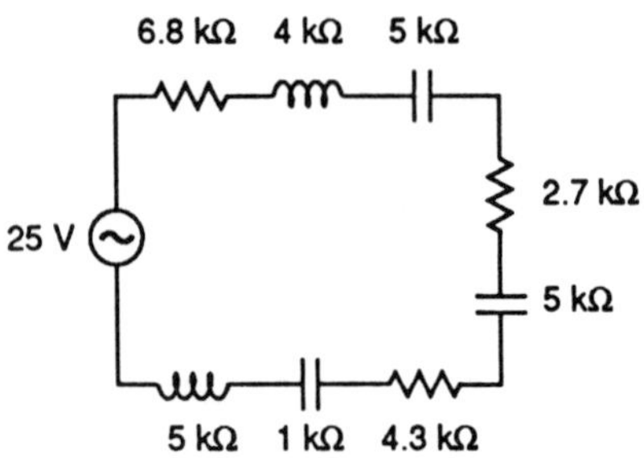

Fig. 8-29.

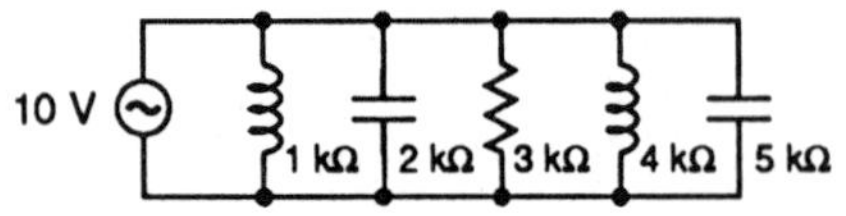

Fig. 8-30.

Sec. 8-5: Parallel *LC* Circuits

8-17. The applied rms voltage of Fig. 8-26 is 15 V. What is each branch current? The total current? Show all phasors for this circuit.

8-18. If the applied rms voltage of Fig. 8-26 doubles, what is each branch current? The total current? Show all phasors for this circuit.

8-19. The frequency of Fig. 8-26 doubles. What is each branch current? The total current? Show all phasors for this circuit.

8-20. The capacitance of Fig. 8-26 doubles. What is each branch current? The total current? Show all phasors for this circuit.

Sec. 8-6: Series *RLC* Circuits

8-21. The applied rms voltage of Fig. 8-27 is 21 V. What is the current? The voltage across each component? Show all phasors for this circuit.

8-22. If the applied rms voltage of Fig. 8-27 doubles, what is the current? The voltage across each component? Show all phasors for this circuit.

8-23. The frequency of Fig. 8-27 doubles. What is the current? The voltage across each component? Show all phasors for this circuit.

8-24. The inductance of Fig. 8-27 doubles. What is the current? The voltage across each component? Show all phasors for this circuit.

Sec. 8-7: Parallel *RLC* Circuits

8-25. The applied rms voltage of Fig. 8-28 is 18 V. What is each branch current? The total current? Show all phasors for this circuit.

8-26. If the applied rms voltage of Fig. 8-28 doubles, what is each branch current? The total current? Show all phasors for this circuit.

8-27. The frequency of Fig. 8-28 doubles. What is each branch current? The total current? Show all phasors for this circuit.

8-28. The capacitance of Fig. 8-28 doubles. What is each branch current? The total current? Show all phasors for this circuit.

Sec. 8-8: Many Series Components

8-29. What is the current in Fig. 8-29?

8-30. If the frequency doubles, what is the current in Fig. 8-29?

Sec. 8-9: Many Parallel Components

8-31. What is the total current in Fig. 8-30?

8-32. If the frequency doubles, what is the current in Fig. 8-30?

Sec. 8-11: Real, Reactive, and Apparent Power

8-33. What is the real power in Fig. 8-27? The apparent power?

8-34. If the frequency doubles, what is the real power in Fig. 8-27? The apparent power?

CHAPTER 9
COMPLEX NUMBERS

There are some ac circuits that cannot be analyzed with the methods of earlier chapters. These circuits have to be analyzed with *complex numbers*. It will be easier for you to understand complex numbers if you first learn about *codes*. This chapter begins with the concept of a code, followed by the definition of complex numbers, complex impedance, and Ohm's law using complex numbers.

9-1. CODES

A *set* is a collection of similar but distinct objects such as letters, numbers, symbols, etc. These objects are referred to as the *elements* of the set. The English alphabet is an example of a set because it is a collection of letters as shown in the left half of Fig. 9-1a. Other examples of sets include the Roman numerals, all the positive integers, real numbers, etc.

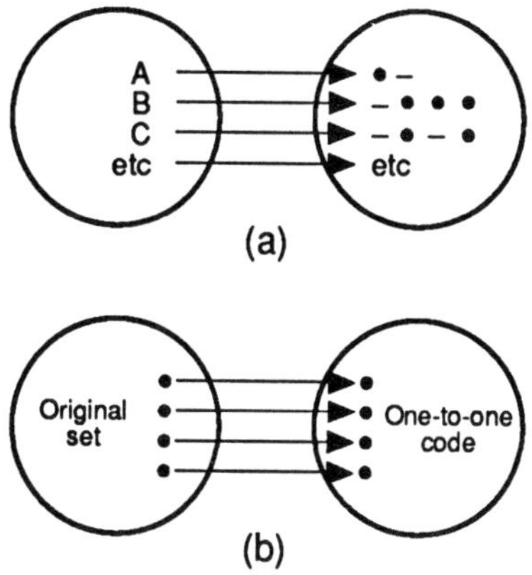

Fig. 9-1.

Definition of a Code

A *code* is any set whose elements stand for the elements of another set. The Morse code is a familiar example. Dots and dashes stand for letters and numbers. Figure 9-1a illustrates the idea of the Morse code. The set on the left is the English alphabet and the set on the right is the Morse code. Notice that dot-dash stands for *A*, dash-dot-dot-dot for *B*, dash-dot-dash-dot for *C*, and so on. By memorizing the Morse code, you can use dots and dashes to represent letters of the alphabet.

One-to-one Code

A code is a *one-to-one* code if each element stands for one and only one element in another set, and vice-versa (see Fig. 9-1b). The Morse code is a one-to-one code because each element of dots and dashes represents a unique letter or number.

The color code used with resistors is another example of a one-to-one code (see Fig. 9-2a). The set on the left is the standard resistor values, and the set on the right is the color code. Each group of colored stripes on a resistor stands for a particular value of resistance.

Integers (whole numbers) are another one-to-one code (Fig. 9-2b). The set on the left is the integers (whole numbers), and the set on the right is groups of pebbles. Each integer (1, 2, 3, etc.) stands for a unique number of pebbles (o, oo, ooo, and so on). With the integer code, we learn to count, which leads to arithmetic and higher forms of mathematics.

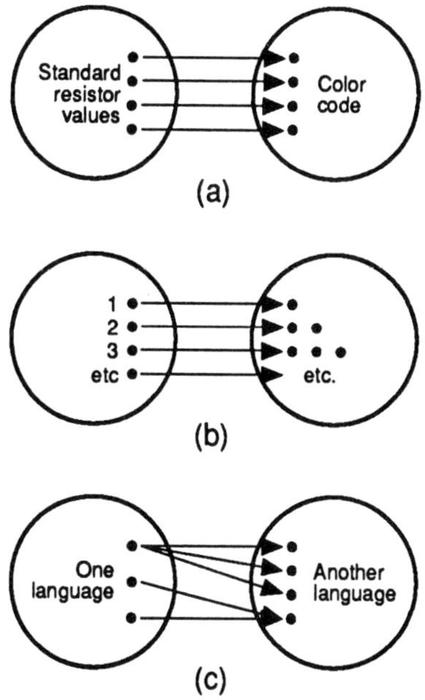

Fig. 9-2.

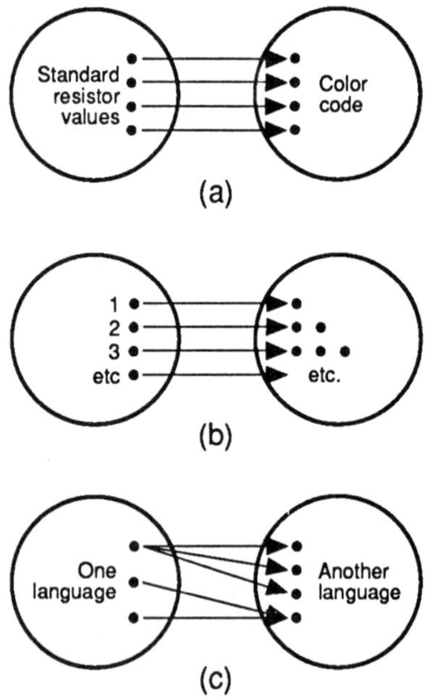

Fig. 9-2.

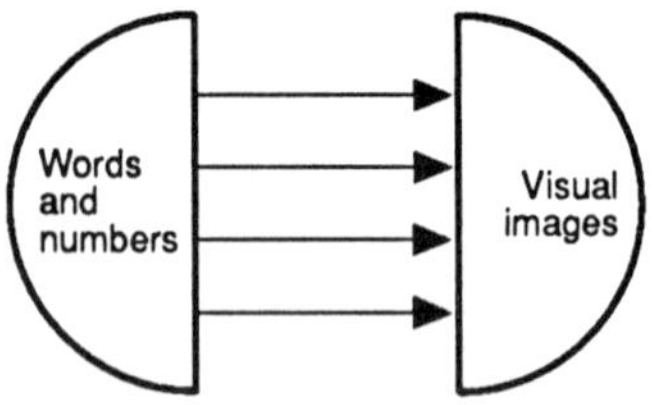

Fig. 9-3.

Coding and Decoding

Coding means representing the elements of a set by symbols in another set. For instance, with the Morse code the letter *B* is coded as dash-dot-dot-dot. *Decoding* is the opposite of coding. It refers to translating the code symbols back to their original meaning. For instance, dash-dot-dash-dot is decoded as the letter *C*.

Some Codes Are Not One-to-one

Not all codes are one-to-one. Languages, for example, are sets of words. Each word in one language can usually be translated into one or more words in another language, and vice-versa (Fig. 9-2c). For this reason, even though languages are codes, they do not have the one-to-one property. This lack of the one-to-one property is why computers have difficulty translating from one language to another. (An early English-to-Russian computer program translated the English sentence "The spirit is willing but the flesh is weak" to a Russian equivalent of "The vodka is good but the meat is rotten.")

Even the way we think is based on a code that exists between the left half of the brain and the right half of the brain as shown in Fig. 9-3. The left brain contains a set of words and numbers, while the right brain contains a set of visual images. A left-brain/right-brain code relates the two sets of data. Learning requires the creation of left-brain/right-brain codes that relate verbal and visual data in the two brain halves.

Codes Are Invented or Made Up

Codes are invented or made up by someone. There's no point asking why the inventor used a certain code symbol to represent the element in another set. For instance, it's pointless to ask why dot-dash represents the letter *A* in the Morse code. It just does because Samuel F. B. Morse defined it that way when he made up the Morse code. After a new code is created, people memorize its definitions and then they can communicate using the new code. Unless you know the code in a particular application, you cannot understand what is going on when someone uses the code.

In summary, it's important to be aware of codes, because they establish the rules of how things are related. Codes are used everywhere, especially in education. When you have difficulties with subjects like mathematics, it's often because you don't know a particular code that is being used. Watch for this in the study of complex numbers. All kinds of new codes are defined or made up for complex numbers. If you memorize the codes, you should be able to master the use of complex numbers with ac circuits.

9-2. NUMBERS ARE CODES

We are so familiar with numbers that we usually forget they are a code. At one time, numbers did not exist, and people could not count. Then, a creator got the idea of letting symbols represent physical quantities. This led to counting, arithmetic, and other forms of mathematics.

Real Numbers

Real numbers are a code for the points along a horizontal line. Figure 9-4a shows a horizontal axis. The number 0 stands for the origin, a reference point from which all other points are measured. Points to the right of the origin are labeled with positive numbers. Those to the left of the origin are labeled with negative numbers.

Real numbers are a one-to-one code, as shown in Fig. 9-4b. Each real number stands for a unique point along the horizontal axis. Because of this, we automatically know that 3 represents the point that is three units to the right of the origin, –2 stands for the point that is two units to the left of the origin, and so on.

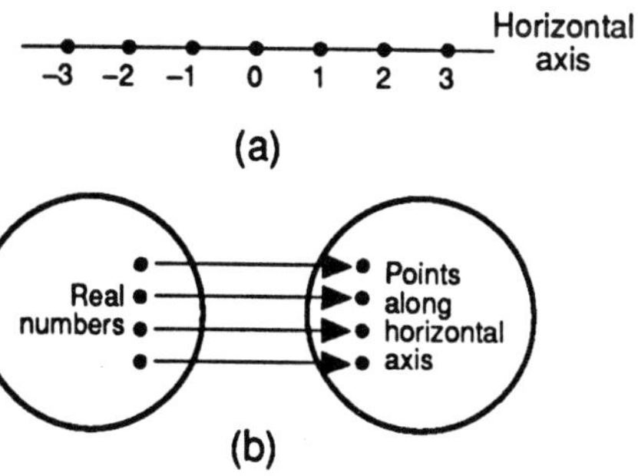

Fig. 9-4.

Imaginary Numbers

Imaginary numbers are numbers that contain the square root of a negative quantity. For instance, numbers like

$$\sqrt{-1}, \sqrt{-4}, \text{ and } \sqrt{-9}$$

are imaginary numbers. Why are they called imaginary? Because there is no way to get the square root of a negative number. No real number multiplied by itself can produce a negative number. For instance, the square of +2 is +4 and the square of –2 is +4; there is no way to get –4 by squaring real numbers. Because it's impossible to get the square root of a negative number, early mathematicians referred to numbers like

$$\sqrt{-1}, \sqrt{-4}, \text{ and } \sqrt{-9}$$

as imaginary numbers.

Imaginary numbers had no practical value until Gauss (1799) decided to use them as a code for the points along a vertical axis. In this code,

$$\sqrt{-1}$$

represents the point that is one unit above the origin in Fig. 9-5a. All other points on the vertical axis have proportional labels; this is why

$$2\sqrt{-1}$$

stands for the point that is two units above the origin. And so on for the other positive points. Similarly,

$$-\sqrt{-1}, -2\sqrt{-1}, \text{ and } -3\sqrt{-1}$$

represent the points that are one unit, two units, and three units below the origin.

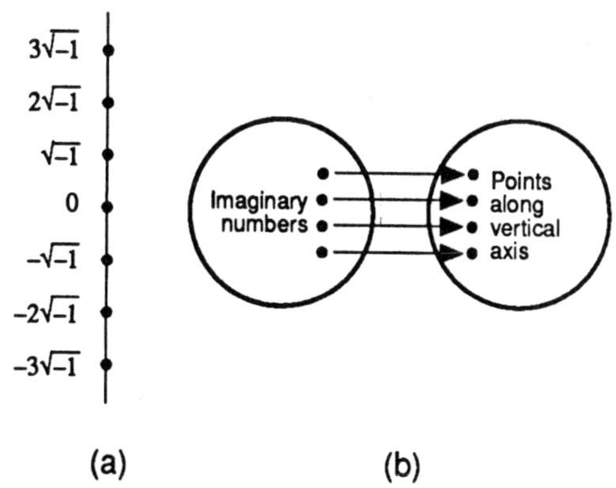

Fig. 9-5.

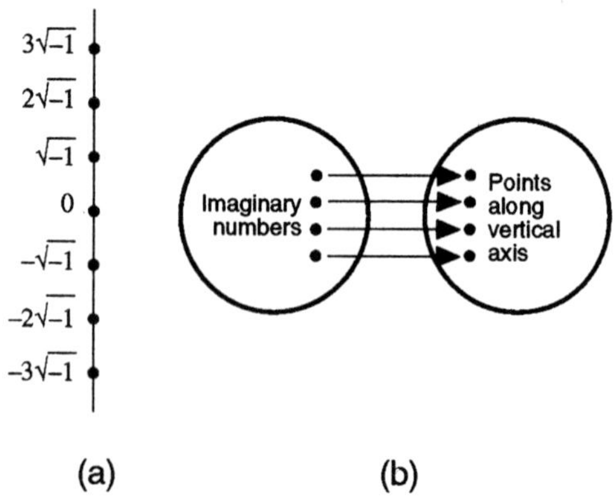

Fig. 9-5.

Definitions

It's important to understand that Gauss invented this code. He made it up. There's no point asking why he did it the way he did. Things like this are formally called *definitions*. It's important to know when something is a definition because then proof is not required. Asking for proof is like asking why dot-dash stands for the letter A in the Morse code.

A definition represents a starting point in a discussion. There is nothing that precedes it, so you don't have to ask where it came from. When something has been defined for the first time, all you have to do is memorize it just as you would memorize a code.

The code for imaginary numbers is a one-to-one code as illustrated in Fig. 9-5b. When you are using Gauss' code, each imaginary number represents a point along the vertical axis. Notice that the one-to-one property applies here. Each point is uniquely specified by an imaginary number, and vice-versa.

The *j* Operator

In Fig. 9-5a, the basic unit for the entire code is the square root of −1. All other code elements are multiples of this basic unit. But the quantity

$$\sqrt{-1}$$

is awkward to use in diagrams and equations, even ugly to look at. For this reason, it is usually recoded as the letter j. In other words, here is a new definition for you to memorize:

$$j = \sqrt{-1} \tag{9-1}$$

With this defining equation, a quantity like

$$3\sqrt{-1}$$

can be written as $3j$. Normally, the j is put in front as an immediate reminder that an imaginary number is involved. Therefore, $3j$ is usually written as $j3$.

Complex Numbers in Rectangular Form

A *complex number* is the algebraic sum of a real number and an imaginary number. Examples of complex numbers are

$$1 + j3$$

$$-2 + j1$$

$$-3 - j2$$

$$2 - j2$$

Numbers like these became important when Gauss decided to use them as a code for the points in the *x-y* plane.

Here is how Gauss defined a complex number in terms of a point in the plane. Points on the horizontal axis are coded with real numbers and those on the vertical axis with imaginary numbers as shown in Fig. 9-6a. A complex number like $1 + j3$ stands for the point that is one unit to the right of the origin and three units up (see Fig. 9-6a).

Similarly, $-2 + j1$ represents the point that is two units left of the origin and 1 unit up. The number $-3 - j2$ stands for the point that is three units left of the origin and 2 units down. The number $2 - j2$ represents the point that is two units right of the origin and two units down. In general, $x + jy$ represents the point that is x units to the right of the origin and y units up. (If x is negative, move left. If y is negative, move down.)

Complex numbers are a one-to-one code for the points in the plane because each number stands for one and only one point. This code is important because it creates a new metaphor: it makes complex numbers equivalent to points in the plane. The ultimate advantage of this is that it allows us to combine algebra and geometry when analyzing ac circuits.

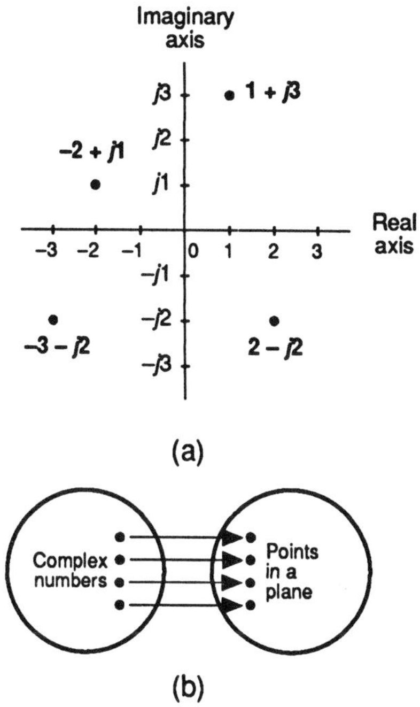

Fig. 9-6.

Complex Numbers in Polar Form

Figure 9-7a shows a point $x + jy$. It is located x units to the right of the origin and y units up. By drawing a hypotenuse from the origin to the point, we get a right triangle. With the Pythagorean theorem, the length of this hypotenuse is

$$M = \sqrt{x^2 + y^2}$$

So, given a complex number like $3 + j4$, the length of the hypotenuse is

$$M = \sqrt{3^2 + 4^2} = 5$$

as shown in Fig. 9-7b.

Look again at Fig. 9-7a. The tangent of an angle is the side opposite the angle divided by the adjacent side. Therefore,

$$\tan \phi = \frac{y}{x} \tag{9-2}$$

In terms of the inverse function, this equation may be rewritten as

$$\phi = \arctan \frac{y}{x}$$

For example, the point coded $3 + j4$ in Fig. 9-7b has an angle of

$$\phi = \arctan \frac{4}{3} = 53°$$

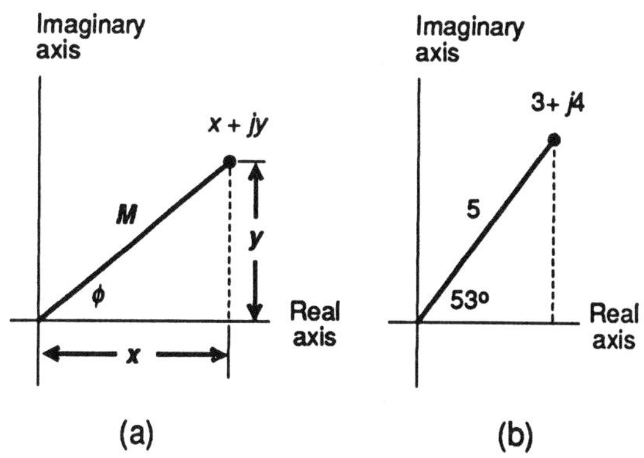

Fig. 9-7.

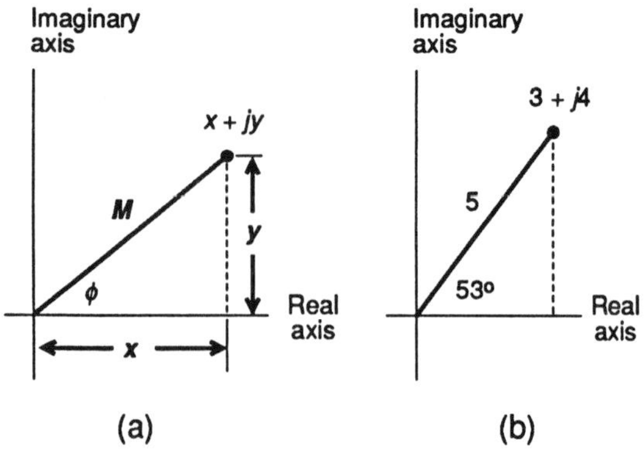

Fig. 9-7.

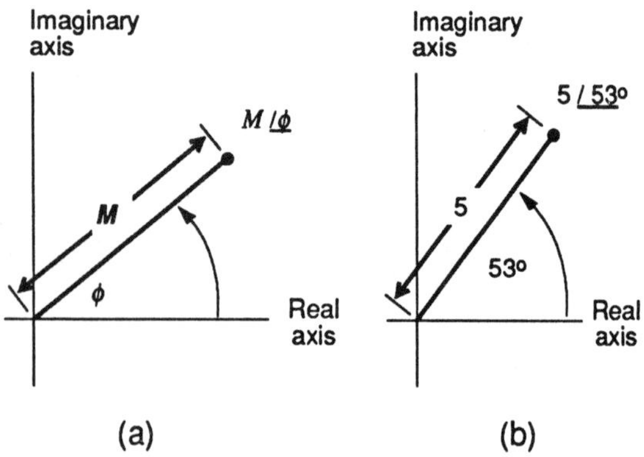

Fig. 9-8.

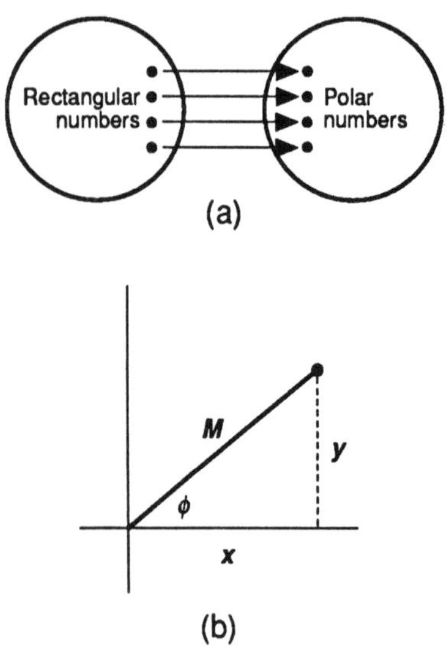

Fig. 9-9.

In Fig. 9-7b, the complex number $3 + j4$ has produced a hypotenuse of 5 and an angle of 53°. If the complex number is changed, either the hypotenuse or the angle or both will be changed. Because of this, we can create a new code by specifying the hypotenuse and the angle instead of the x and y values (see Fig. 9-8a). The standard notation for the point is to label it as

$$M \angle \phi$$

For instance, the point in Fig. 9-8b is

$$5 \angle 53^\circ$$

You read this as "5 at an angle of 53°." This label tells you the point is located 5 units from the origin at an angle of 53°.

Numbers in this new form are called complex numbers in *polar form,* or simply polar numbers. Numbers like $x + jy$ are called complex numbers in *rectangular form,* or simply rectangular numbers. A rectangular number and polar number are equal when they both represent the same point. For instance, the point in Fig. 9-7b can be expressed as $3 + j4$ or as 5 at an angle of 53°. Because of this, we can write

$$3 + j4 = 5 \angle 53^\circ$$

9-3. CONVERSIONS

Rectangular numbers are convenient in some problems, while polar numbers may be easier to use in other problems. Because of this, we often have to convert from one type of number to another. This allows us to use whichever type of complex number is more convenient.

Rectangular-to-polar Conversion

As shown in Fig. 9-9a, polar numbers are a one-to-one code for rectangular numbers. Each polar number represents a single rectangular number. You can see this by looking at Fig. 9-9b. Because of the Pythagorean theorem, the magnitude of the polar number is

$$M = \sqrt{x^2 + y^2} \tag{9-3}$$

and the angle is

$$\phi = \arctan \frac{y}{x} \tag{9-4}$$

With these two equations, you can convert any rectangular number into an equivalent polar number.

Polar-to-rectangular Conversion

In Fig. 9-9b, basic trigonometry gives these relations:

$$x = M \cos \phi \tag{9-5}$$

$$y = M \sin \phi \tag{9-6}$$

With these two equations, you can convert a polar number into an equivalent rectangular number.

EXAMPLE 9-1

Convert $3 + j2$ into an equivalent polar number.

SOLUTION

Figure 9-10a shows the point. The polar number has a magnitude of

$$M = \sqrt{3^2 + 2^2} = 3.61$$

and an angle of

$$\phi = \arctan \frac{2}{3} = 34^\circ$$

Therefore,

$$3 + j2 = 3.61 \angle 34^\circ$$

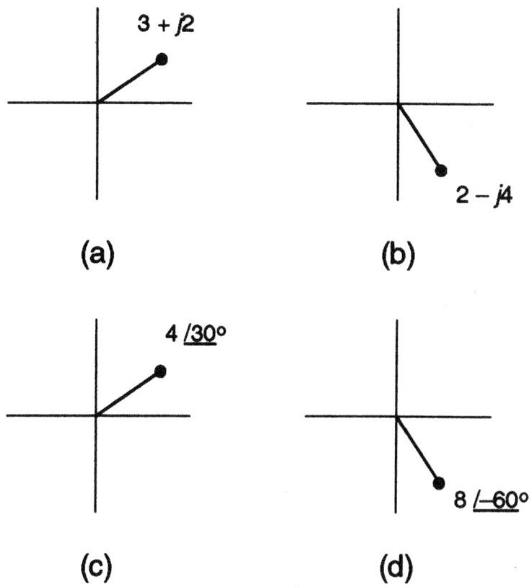

Fig. 9-10.

EXAMPLE 9-2

Convert $2 - j4$ into an equivalent polar number.

SOLUTION

Figure 9-10b shows the point. The polar number has a magnitude of

$$M = \sqrt{2^2 + (-4)^2} = 4.47$$

and an angle of

$$\phi = \arctan \frac{4}{-2} = -63^\circ$$

Therefore,

$$2 - j4 = 4.47 \angle -63^\circ$$

EXAMPLE 9-3

Convert $4 \angle 30^\circ$ into an equivalent rectangular number.

SOLUTION

Figure 9-10c shows the point. The equivalent rectangular number has an x coordinate of

$$x = M \cos \phi \quad = 4 \cos 30^\circ = 3.46$$

and a y coordinate of

$$y = M \sin \phi \quad = 4 \sin 30^\circ = 2$$

Therefore,

$$4 \angle 30^\circ = 3.46 + j2$$

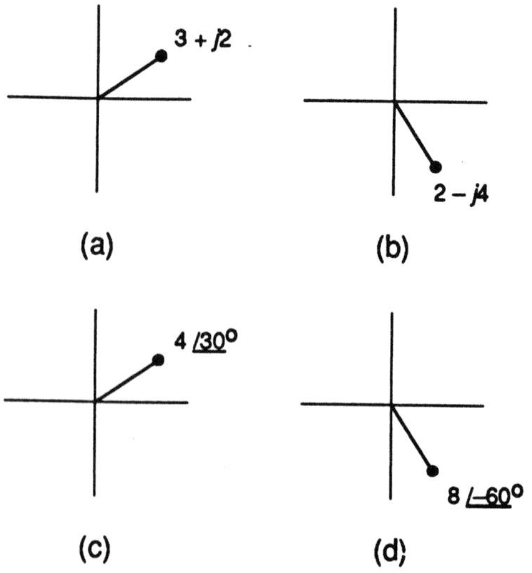

Fig. 9-10.

EXAMPLE 9-4

Convert $8 \angle -60^\circ$ into an equivalent rectangular number.

SOLUTION

Figure 9-10d shows the point. The equivalent rectangular number has coordinates of

$$x = M \cos \phi \quad = 8 \cos (-60^\circ) = 4$$

$$y = M \sin \phi \quad = 8 \sin (-60^\circ) = -6.93$$

The polar/rectangular equivalence is

$$8 \angle -60^\circ = 4 - j6.93$$

9-4. ARITHMETIC OF COMPLEX NUMBERS

You can add, subtract, multiply, and divide complex numbers. Addition and subtraction are usually done with complex numbers in rectangular form, while multiplication and division are usually done with complex numbers in polar form.

Rectangular Addition and Subtraction

You cannot numerically add complex numbers if they are in polar form. It is necessary to express complex numbers in rectangular form before you can add hem. For two complex numbers in the form of $x_1 + jy_1$ and $x_2 + jy_2$, the defining equation for their sum is

$$(x_1 + jy_1) + (x_2 + jy_2) = x_1 + x_2 + j(y_1 + y_2)$$

This says you add the real parts, then add the imaginary parts.

Subtraction is similar. You cannot numerically subtract complex numbers if they are in polar form. It is necessary to express complex numbers in rectangular form before you can subtract them. For two complex numbers in the form of $x_1 + jy_1$ and $x_2 + jy_2$, the defining equation for their difference is

$$(x_1 + jy_1) - (x_2 + jy_2) = x_1 - x_2 + j(y_1 - y_2)$$

This says you subtract the real parts, then subtract the imaginary parts.

Polar Multiplication and Division

When you want to multiply or divide complex numbers, it is usually best to express the complex numbers in polar form. Given two polar numbers like

$$M_1 \angle \phi_1$$

$$M_2 \angle \phi_2$$

the defining formula for multiplication is

$$M_1 \angle \phi_1 \text{ X } M_2 \angle \phi_2 = M_1 M_2 \angle \phi_1 + \phi_2$$

This says you multiply the magnitudes and add the angles.

Division is similar. If the numbers are not in polar form, convert them to polar form before attempting the division. This is usually the easiest way to divide complex numbers. The rule for polar division is defined as follows:

$$\frac{M_1 \angle \phi_1}{M_2 \angle \phi_2} = \frac{M_1}{M_2} \angle \phi_1 - \phi_2$$

This says you divide the magnitudes and subtract the angles.

EXAMPLE 9-5

What is the sum of $2 + j3$ and $4 + j7$?

SOLUTION

Add the real parts, then add the imaginary parts to get

$$(2 + j3) + (4 + j7) = 6 + j10$$

EXAMPLE 9-6

What is the sum of $-2 + j9$ and $6 - j3$?

SOLUTION

When we say sum, we mean algebraic sum. The sum of the two numbers is

$(-2 + j9) + (6 - j3) = 4 + j6$

In other words, adding the real parts means algebraic addition to get

$-2 + 6 = 4$

Similarly, adding the imaginary parts means algebraic addition to get

$9 - 3 = 6$

EXAMPLE 9-7

What is the difference of $5 - j6$ and $-8 + j5$?

SOLUTION

Subtract the real parts, then subtract the imaginary parts to get

$(5 - j6) - (-8 + j5) = 13 - j11$

EXAMPLE 9-8

What is the difference of $-3 + j6$ and $4 - j9$?

$(-3 + j6) - (4 - j9) = -7 + j15$

EXAMPLE 9-9

What is the product of $5 \angle 30^\circ \text{ X } 10 \angle 60^\circ$?

SOLUTION

Multiply the magnitudes and add the angles to get

$5 \angle 30^\circ \text{ X } 10 \angle 60^\circ = 50 \angle 90^\circ$

EXAMPLE 9-10

What is the product of $12 \angle 45^\circ \text{ X } 6 \angle -75^\circ$?

SOLUTION

$12 \angle 45^\circ \text{ X } 6 \angle -75^\circ = 72 \angle -30^\circ$

As you see, the algebraic sum of the angles is

$45^\circ + (-75^\circ) = -30^\circ$

EXAMPLE 9-11

What is the product of $8\angle -25^\circ \times 2\angle 65^\circ$?

SOLUTION

$$8\angle -25^\circ \times 2\angle 65^\circ = 16\angle 40^\circ$$

EXAMPLE 9-12

Divide $12\angle 50^\circ$ by $3\angle -15^\circ$.

SOLUTION

Divide the magnitudes and subtract the angles algebraically to get

$$\frac{12\angle 50^\circ}{3\angle -15^\circ} = 4\angle 65^\circ$$

EXAMPLE 9-13

Divide $50\angle -60^\circ$ by $4\angle -15^\circ$.

SOLUTION

$$\frac{50\angle -60^\circ}{4\angle -15^\circ} = 12.5\angle -45^\circ$$

9-5. COMPLEX IMPEDANCE

Each of the basic components (resistors, inductors, and capacitors) has an impedance to alternating current. This impedance may be expressed as a complex number.

Phasor Voltage and Phasor Current

Earlier chapters discussed phasor voltages and currents. Recall that the length of a phasor equals the peak or rms value of a sinusoidal voltage or current, and the angle of the phasor is the phase angle of the sinusoidal voltage or current.

Polar numbers can be used as a one-to-one code for phasors. The magnitude of the polar number is the same as the length of the phasor, and the angle of the polar number is the same as the angle of the phasor. What this means is that we can use polar numbers to represent sinusoidal voltages and currents. In turn, this allows us to analyze more complicated ac circuits than we could in earlier chapters.

Here are two examples of using polar numbers to represent phasors and sinusoidal quantities. Figure 9-11a shows a sinusoidal voltage with a peak of 20 mV and a phase angle of 60°. The corresponding phasor is 14.1 mV (rms value) at an angle of 60° as shown in Fig. 9-11b. The equivalent polar number is 14.1 mV $\angle 60°$. Similarly, the sinusoidal current of Fig. 9-11c has a peak value of 5 mA and a phase angle of -45°. The corresponding phasor has a magnitude of 3.54 mA (rms) and an angle of –45°.

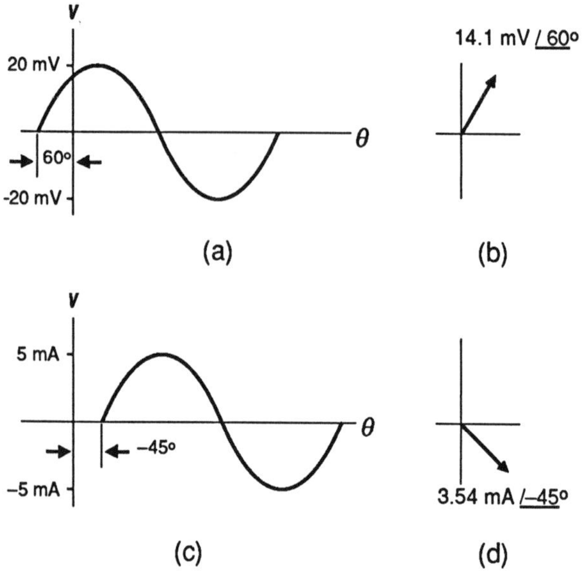

Fig. 9-11.

In the foregoing examples, we used rms values for the magnitudes of the phasors. If you prefer, you can use peak values for the phasors. As long as you are consistent in your usage, you can choose whichever suits the application. For instance, if you are measuring voltages with a voltmeter, you may prefer rms values because most voltmeters are calibrated in rms values. If you are measuring voltages on an oscilloscope, you may prefer peak values because they are easier to work with.

Impedance of a Basic Components

Impedance is the opposition to the flow of alternating current. Mathematically, it is defined as the phasor voltage divided by phasor current:

$$\mathbf{Z} = \frac{\mathbf{V}}{\mathbf{I}} \tag{9-7}$$

where boldface letters **Z**, **V**, and **I** are used to indicate complex numbers.

Figure 9-12a shows the phasor voltage and current for an ideal resistor. Since the voltage and current are in phase, they both have an angle of 0°. Therefore, the impedance of a resistor is

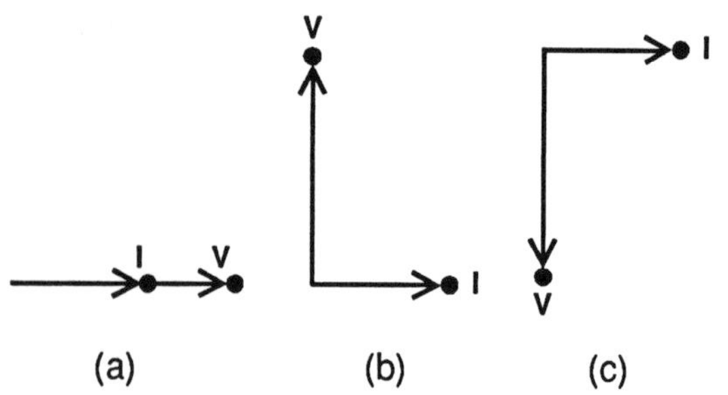

Fig. 9-12.

$$\mathbf{Z} = \frac{\mathbf{V}}{\mathbf{I}} = \frac{V \angle 0°}{I \angle 0°} = R \angle 0° = R$$

This says the impedance of an ideal resistor is equal to its resistance. Since the final expression has no imaginary part, the impedance of an ideal resistance is always a real number.

Notice the use of italic letters for V, I, and R in the foregoing equation. Italic letters are used for magnitudes of phasor quantities. So, V and I represents the rms voltage and current. R represents the magnitude of the resistance, equivalent to the real part because there is no imaginary part.

Figure 9-12b shows the phasor voltage and current for an ideal inductor. Since the voltage leads the current by an angle of 90°, the impedance of an inductor is

$$\mathbf{Z} = \frac{\mathbf{V}}{\mathbf{I}} = \frac{V \angle 90°}{I \angle 0°} = X_L \angle 90° = jX_L$$

This says the impedance of an ideal inductor is equal to j times its inductive reactance. The impedance of an ideal inductor is always a positive imaginary number.

Figure 9-12c shows the phasor voltage and current for an ideal capacitor. Since the voltage lags the current by an angle of 90°, the impedance of an ideal capacitor is

$$\mathbf{Z} = \frac{\mathbf{V}}{\mathbf{I}} = \frac{V \angle 0^\circ}{I \angle -90^\circ} = X_C \angle -90^\circ = -jX_C$$

This says the impedance of an ideal capacitor is equal to $-j$ times its capacitive reactance. The impedance of an ideal capacitor is always a negative imaginary number.

EXAMPLE 9-14

What is the impedance of an inductor with the phasors shown in Fig. 9-13a?

SOLUTION

Divide the phasor voltage by the phasor current to get the impedance:

$$\mathbf{Z} = \frac{j25\text{ V}}{10\text{ mA}} = j2.5\text{ k}\Omega = 2.5\text{ k}\Omega \angle 90^\circ$$

EXAMPLE 9-15

What is the impedance of a capacitor with the phasors shown in Fig. 9-13b?

SOLUTION

Divide the phasor voltage by the phasor current to get the impedance:

$$\mathbf{Z} = \frac{-j15\text{ V}}{9\text{ mA}} = -j1.67\text{ k}\Omega = 1.67\text{ k}\Omega \angle -90^\circ$$

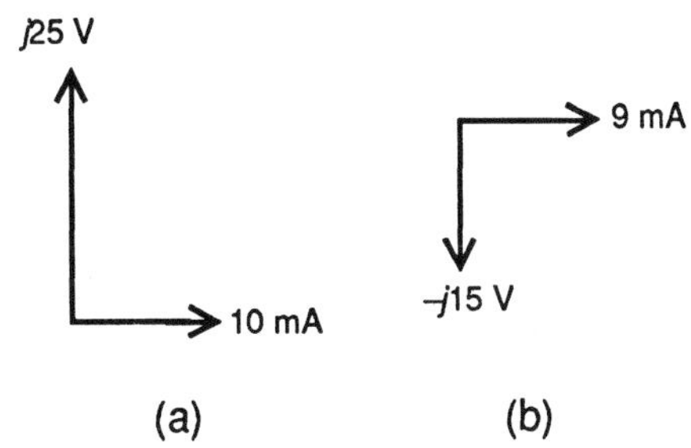

Fig. 9-13.

9-6. OHM'S LAW

Steinmetz (1865-1923) was the scientist who worked out the complete solution to ac circuits using complex numbers. One of the most important things he did was to prove that Ohm's law can be extended to ac circuits, provided that phasor voltage and current are used. Stated another way, Ohm's law holds for ac circuits if complex numbers are used to represent voltages, currents, and impedances.

For ac circuits, Ohm's law is written as

$$\mathbf{I} = \frac{\mathbf{V}}{\mathbf{Z}} \qquad (9\text{-}8)$$

where **I**, **V**, and **Z** are complex numbers. In this equation, **V** is the phasor voltage across a load, and **Z** is complex impedance of the load. Given the phasor voltage and the impedance, you can calculate the phasor current. Read the following examples to see how it is done.

EXAMPLE 9-16

If **V** = 10 V at an angle of 0° and **Z** = 3 kΩ + *j*4 kΩ, what is the phasor current?

SOLUTION

Divide the phasor voltage by the impedance as follows:

$$\mathbf{I} = \frac{\mathbf{V}}{\mathbf{Z}} = \frac{10\ \text{V}}{3\ \text{k}\Omega + j4\ \text{k}\Omega} = \frac{10\ \text{V}\ \angle 0^\circ}{5\ \text{k}\Omega\ \angle 53^\circ} = 2\ \text{mA}\ \angle -53^\circ$$

Notice how the 3 kΩ + *j*4 kΩ is converted to polar form. Then, it's relatively easy to finish the calculation. The final answer tells us the current lags the voltage by 53°.

EXAMPLE 9-17

A load has **I** = 6 mA at an angle of 45° and **Z** = 3 kΩ. What is the phasor voltage across the load?

SOLUTION

Multiply the phasor current by the impedance to get

$$\mathbf{V} = \mathbf{IZ} = 6\ \text{mA}\ \angle 45^\circ \times 3\ \text{k}\Omega\ \angle 0^\circ = 18\ \text{V}\ \angle 45^\circ$$

This says the voltage is in phase with the current because both have an angle of 45°.

EXAMPLE 9-18

If **V** = 15 V at an angle of 25° and **I** = 3 mA at angle of –28°, what is the impedance?

SOLUTION

$$\mathbf{Z} = \frac{\mathbf{V}}{\mathbf{I}} = \frac{15\ \text{V}\ \angle 25^\circ}{3\ \text{mA}\ \angle -28^\circ} = 5\ \text{k}\Omega\ \angle 53^\circ = 3\ \text{k}\Omega + j4\ \text{k}\Omega$$

9-7. POWER

The *phasor power* in a load equals the product of its phasor voltage and its phasor current:

$$\mathbf{P} = \mathbf{VI} \tag{9-9}$$

The real part of **P** is the real power, the imaginary part of **P** is the reactive power, and the magnitude of **P** is the apparent power. For instance, if a load has **V** = 12 V at an angle of 25° and **I** = 3 A at an angle of 35°, the apparent phasor power is

$$\mathbf{P} = 12\text{ V}\angle 25^\circ \times 3\text{ A}\angle 35^\circ = 36\text{ VA}\angle 60^\circ$$

Convert this polar answer to rectangular form and you get

$$\mathbf{P} = 18\text{ W} + j31.2\text{ VAR}$$

The real power is 18 W, the reactive power is 31.2 VAR, and the apparent power is 36 VA.

MULTIPLE-CHOICE TEST

The following questions may have more than one right answer. Select the best answer. This is the one that is always true, or covers more situations, or agrees with the material of this chapter, etc.

1. The quantity $j8$ kΩ is the same as
   a. $-j8$ kΩ
   b. 8 kΩ at -90°
   c. 8 kΩ at +90°
   d. 8 kΩ

2. Imaginary numbers stand for points
   a. along the horizontal axis
   b. along the vertical axis
   c. in the plane
   d. with phasor voltages

3. A complex number may
   a. have magnitude and angle
   b. have real and imaginary parts
   c. represent a phasor
   d. all of the above

4. The number $1 - j1$ is equivalent to
   a. $1.414\angle -45^\circ$
   b. $1.414\angle -90^\circ$
   c. $2\angle 45^\circ$
   d. $2\angle -45^\circ$

5. If $x = 3$ and $y = 7$, the angle is
   a. equal to 45°
   b. less than 45°
   c. more than 45°
   d. negative

6. The impedance of an ideal capacitor always has an angle of
   a. 0°
   b. 90°
   c. 180°
   d. –90°

7. Ohm's law is valid for ac circuits if voltage and current are
   a. ac values
   b. complex numbers
   c. dc values
   d. instantaneous values

8. If the phasor power has an angle of 90°, the real power is
   a. finite
   b. imaginary
   c. infinite
   d. zero

9. Addition and subtraction are easiest when the complex numbers are in
   a. rectangular form
   b. polar form
   c. real form
   d. reactive form

10. Multiplication is easiest when the complex numbers are in
   a. rectangular form
   b. polar form
   c. real form
   d. reactive form

11. Division is easiest when the complex numbers are in
   a. rectangular form
   b. polar form
   c. real form
   d. reactive form

12. A code is any set whose elements represent
   a. numbers along the horizontal axis
   b. letters of the alphabet
   c. symbols
   d. elements of another set

13. Real numbers are a code for the points
    a. along a horizontal axis
    b. along a vertical axis
    c. in the x-y plane
    d. on a sphere

14. Imaginary numbers are a code for the points
    a. along a horizontal axis
    b. along a vertical axis
    c. in the x-y plane
    d. on a sphere

15. Complex numbers are a code for the points
    a. along a horizontal axis
    b. along a vertical axis
    c. in the x-y plane
    d. on a sphere

16. Ohm's law is valid for ac circuits if voltages, currents, and impedances are represented by
    a. real numbers
    b. imaginary numbers
    c. complex numbers
    d. none of the above

PROBLEMS

Sec. 9-3: Conversions

9-1. Convert $15 + j12$ to polar form.
9-2. Convert $32 - j60$ to polar form.
9-3. Convert $-24 + j33$ to polar form.
9-4. Convert $-14 - j12$ to polar form.
9-5. Convert $12 \angle 55^\circ$ to rectangular form.
9-6. Convert $35 \angle -73^\circ$ to rectangular form.
9-7. Convert $42 \angle 135^\circ$ to rectangular form.
9-8. Convert $57 \angle -120^\circ$ to rectangular form.

Sec. 9-4: Arithmetic of Complex Numbers

Addition:
9-9. What is the sum of $12 + j8$ and $23 + j34$?
9-10. What is the sum of $15 + j18$ and $32 - j26$?
9-11. What is the sum of $21 - j28$ and $23 - j34$?
9-12. What is the sum of $-15 + j18$ and $32 - j2$?

Subtraction:

9-13. What is the difference of $12 + j8$ and $23 + j34$?
9-14. What is the difference of $15 + j18$ and $32 - j26$?
9-15. What is the difference of $12 - j8$ and $23 - j34$?
9-16. What is the difference of $-15 + j18$ and $32 - j26$?

Multiplication:

9-17. Multiply $12 \angle 50^\circ$ and $20 \angle 15^\circ$?
9-18. Multiply $3 \angle -60^\circ$ and $10 \angle 25^\circ$?
9-19. Multiply $20 \angle 20^\circ$ and $15 \angle -25^\circ$?
9-20. Multiply $17 \angle -35^\circ$ and $22 \angle -45^\circ$?

Division:

9-21. Divide $42 \angle 60^\circ$ by $20 \angle 25^\circ$?
9-22. Divide $3 \angle -50^\circ$ and $10 \angle 35^\circ$?
9-23. Divide $20 \angle 30^\circ$ and $5 \angle -15^\circ$?
9-24. Divide $25 \angle -25^\circ$ and $3 \angle -35^\circ$?

Sec. 9-5: Complex Impedance

9-25. What is the impedance of an inductor with the phasors shown in Fig. 9-14a?
9-26. What is the impedance of a capacitor with the phasors shown in Fig. 9-14b?

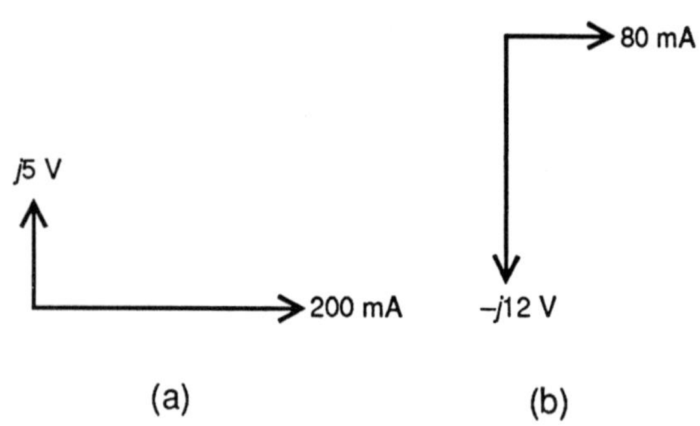

Fig. 9-14.

Sec. 9-6: Ohm's Law

9-27. If $\mathbf{V} = 27$ V at an angle of 30° and $\mathbf{Z} = 5\ \text{k}\Omega + j2\ \text{k}\Omega$, what is the phasor current?
9-28. A load has $\mathbf{I} = 12$ mA at an angle of –35° and $\mathbf{Z} = 5\ \text{k}\Omega$. What is the phasor voltage across the load?
9-29. A load has the phasors shown in Fig. 9-15a. What is the impedance of the load?
9-30. What is the impedance of the load with the phasors shown in Fig. 9-15b?

Sec. 9-7: Power

9-31. What is the phasor power in Fig. 9-15a? The real power? The reactive power?
9-32. What is the phasor power in Fig. 9-15b? The real power? The reactive power?

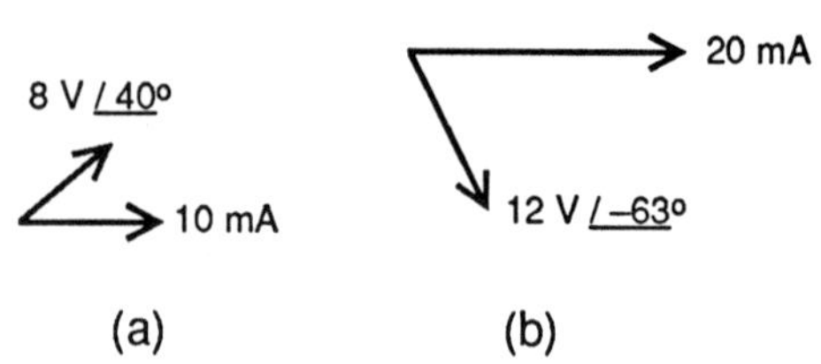

Fig. 9-15.

CHAPTER 10
ADVANCED ANALYSIS

In this chapter, we examine series, parallel, and series-parallel circuits with complex numbers. This will enhance your understanding of ac circuits by showing you how complex numbers are used in the analysis of ac circuits. We will also analyze *RC* coupling and bypass circuits because of their importance in electronics.

10-1. RESISTIVE CIRCUITS

Everything that you learned about dc circuits can be applied to ac circuits, provided you use complex numbers to calculate the voltages and currents. For instance, Ohm's law, Kirchhoff's laws, Thevenin's theorem, etc. are valid for ac circuits when you use complex numbers in your analysis.

When a circuit is purely resistive, every component is an ideal resistor with an impedance of

$$\mathbf{Z} = R \angle 0^\circ$$

This says the impedance of each resistor has a magnitude of R and an angle of 0°. The impedance may be expressed in rectangular form as

$$\mathbf{Z} = R + j0 = R$$

In other words, the impedance of an ideal resistor is a real number.

In series circuits, the total impedance of two resistors is the sum of the individual impedances:

$$\mathbf{Z} = \mathbf{Z}_1 + \mathbf{Z}_2 = R_1 + R_2 \qquad (10\text{-}1)$$

This means the total impedance is always a positive real number.

In parallel circuits, the total impedance of two resistors is the product over the sum:

$$\mathbf{Z} = \frac{\mathbf{Z}_1\mathbf{Z}_2}{\mathbf{Z}_1 + \mathbf{Z}_2} = \frac{R_1R_2}{R_1 + R_2} \qquad (10\text{-}2)$$

Again, the total impedance is always a positive real number.

As mentioned in earlier discussions, the phase angle of the total voltage may be defined as 0°. Then, the total current and all other quantities are referenced to the total voltage. In some problems, the phase angle of the total current may be defined as 0°. In this case, the total voltage and all other quantities are referenced to the total current. Read the following examples to see how to analyze purely resistive circuits with complex numbers.

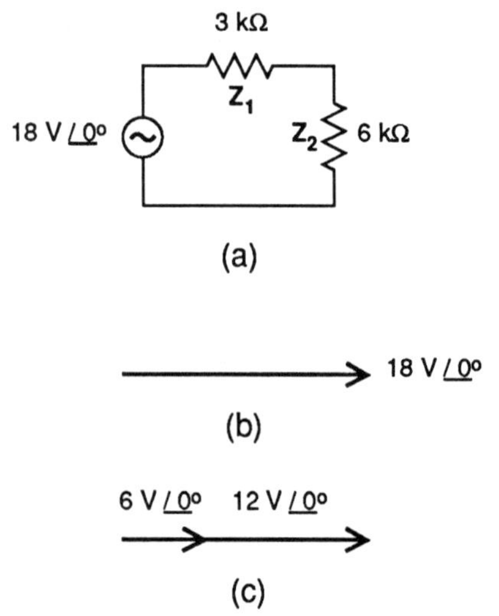

Fig. 10-1.

EXAMPLE 10-1

What is the phasor current in Fig. 10-1a? The phasor voltage across each resistor?

SOLUTION

The total impedance is the sum of the resistances:

$$\mathbf{Z} = 3\ \text{k}\Omega + 6\ \text{k}\Omega = 9\ \text{k}\Omega$$

This rectangular form has a polar equivalent of

$$\mathbf{Z} = 9\ \text{k}\Omega \angle 0^\circ$$

Notice that the applied voltage is given as $18\ \text{V} \angle 0^\circ$, whose phasor is shown in Fig. 10-1b. This means the total current and all other quantities are referenced to the applied or total voltage.

The total current is

$$\mathbf{I} = \frac{\mathbf{V}}{\mathbf{Z}} = \frac{18\ \text{V} \angle 0^\circ}{9\ \text{k}\Omega \angle 0^\circ} = 2\ \text{mA} \angle 0^\circ$$

Since the angle is zero, the total current is in phase with the total voltage. The individual voltages are

$$\mathbf{V}_1 = \mathbf{I}\mathbf{Z}_1 = (2\ \text{mA} \angle 0^\circ)(3\ \text{k}\Omega \angle 0^\circ) = 6\ \text{V} \angle 0^\circ$$

$$\mathbf{V}_2 = \mathbf{I}\mathbf{Z}_2 = (2\ \text{mA} \angle 0^\circ)(6\ \text{k}\Omega \angle 0^\circ) = 12\ \text{V} \angle 0^\circ$$

Figure 10-1c shows these individual voltages. Since both angles are zero, the individual voltages are in phase with the total voltage.

As a check on the work, Kirchhoff's voltage law can be used. In other words, the total voltage should equal the sum of the individual voltages:

$$18\ \text{V} \angle 0^\circ = 12\ \text{V} \angle 0^\circ + 6\ \text{V} \angle 0^\circ$$

As you see, the total voltage does equal the sum of the individual voltages. Figures 10-1b and c also show this equality.

EXAMPLE 10-2

What is the total phasor current in Fig. 10-2a?

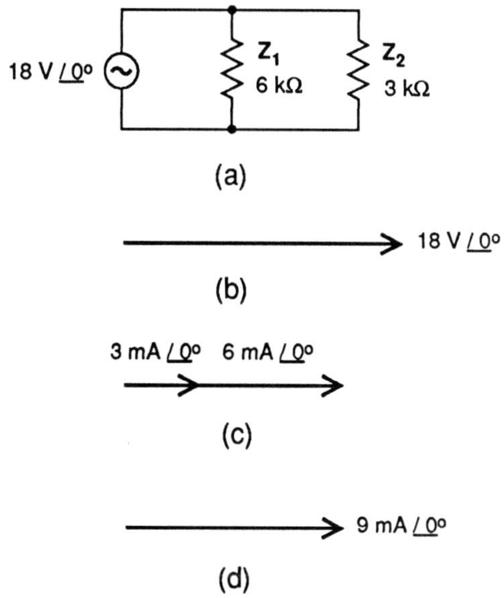

Fig. 10-2.

SOLUTION

Again, the applied voltage is given as $18\ \text{V} \angle 0^\circ$, whose phasor is shown in Fig. 10-2b. This means the total current and all other quantities are referenced to the total voltage.

The current through the first branch is

$$\mathbf{I}_1 = \frac{\mathbf{V}}{\mathbf{Z}_1} = \frac{18\ \text{V} \angle 0^\circ}{6\ \text{k}\Omega \angle 0^\circ} = 3\ \text{mA} \angle 0^\circ$$

The current through the second branch is

$$\mathbf{I}_2 = \frac{\mathbf{V}}{\mathbf{Z}_2} = \frac{18\ \text{V} \angle 0^\circ}{3\ \text{k}\Omega \angle 0^\circ} = 6\ \text{mA} \angle 0^\circ$$

Figure 10-2c shows these individual currents. Since both angles are zero, the individual currents are in phase with the total voltage.

The total current is

$$\mathbf{I} = \mathbf{I}_1 + \mathbf{I}_2 = 3\ \text{mA} \angle 0^\circ + 6\ \text{mA} \angle 0^\circ = 9\ \text{mA} \angle 0^\circ$$

Since the angle is zero, the total current is in phase with the total voltage as shown in Fig. 10-2d.

As a check on the work, the total impedance is

$$\mathbf{Z} = \frac{(6\ \text{k}\Omega)(3\ \text{k}\Omega)}{6\ \text{k}\Omega + 3\ \text{k}\Omega} = 2\ \text{k}\Omega$$

The total current is

$$\mathbf{I} = \frac{\mathbf{V}}{\mathbf{Z}} = \frac{18\ \text{V} \angle 0^\circ}{2\ \text{k}\Omega} = 9\ \text{mA} \angle 0^\circ$$

This agrees with the total current found earlier by adding the branch currents.

10-2. INDUCTIVE CIRCUITS

When a circuit is purely inductive, every component is an ideal inductor with an impedance of

$$\mathbf{Z} = X_L \angle 90^\circ$$

This says the impedance of each inductor has a magnitude of X_L and an angle of 90°. The impedance may be expressed in rectangular form as

$$\mathbf{Z} = 0 + jX_L = jX_L$$

In other words, the impedance of an ideal inductor is always a positive imaginary number.

In series circuits, the total impedance of two inductors is the sum of the individual impedances:

$$\mathbf{Z} = \mathbf{Z}_1 + \mathbf{Z}_2 = jX_{L1} + jX_{L2} = j(X_{L1} + X_{L2}) \quad (10\text{-}3)$$

This means the total impedance is a positive imaginary number.

In parallel circuits, the total impedance of two inductors is the product over the sum:

$$\mathbf{Z} = \frac{\mathbf{Z}_1\mathbf{Z}_2}{\mathbf{Z}_1 + \mathbf{Z}_2} = \frac{(jX_{L1})(jX_{L2})}{jX_{L1} + jX_{L2}} = \frac{j^2(X_{L1}X_{L2})}{j(X_{L1} + X_{L2})}$$

or

$$\mathbf{Z} = j\frac{X_{L1}X_{L2}}{X_{L1} + X_{L2}} \quad (10\text{-}4)$$

Again, the total impedance is a positive imaginary number.

Because the total impedance is always a positive imaginary number in purely inductive circuits, the total voltage always leads the total current by 90°. Furthermore, since the impedance of each inductor is always a positive imaginary number, the voltage across each inductor leads the current through the inductor by 90°.

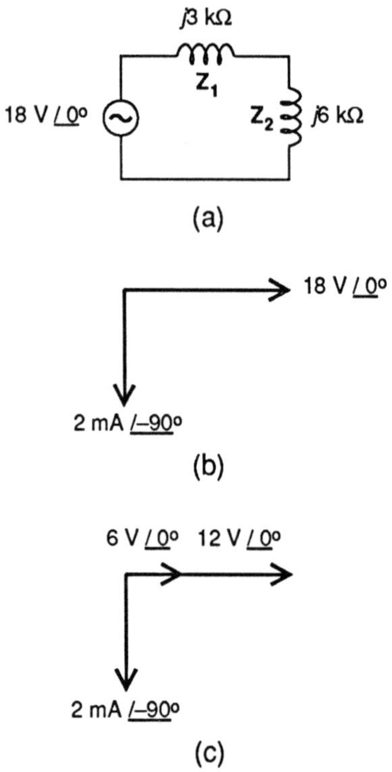

Fig. 10-3.

EXAMPLE 10-3

What is the phasor current in Fig. 10-3a? The phasor voltage across each inductor?

SOLUTION

The total impedance is

$$\mathbf{Z} = j3\ \text{k}\Omega + j6\ \text{k}\Omega = j9\ \text{k}\Omega$$

This rectangular form has a polar equivalent of

$$\mathbf{Z} = 9\ \text{k}\Omega\ \angle 90°$$

The total current is

$$\mathbf{I} = \frac{\mathbf{V}}{\mathbf{Z}} = \frac{18\ \text{V}\ \angle 0°}{9\ \text{k}\Omega\ \angle 90°} = 2\ \text{mA}\ \angle -90°$$

Since the angle is –90°, the total current lags the total voltage by 90° as shown in Fig. 10-3b. The individual voltages are

$$\mathbf{V}_1 = \mathbf{IZ}_1 = (2\text{ mA} \angle -90^\circ)(3\text{ k}\Omega \angle 90^\circ) = 6\text{ V} \angle 0^\circ$$

$$\mathbf{V}_2 = \mathbf{IZ}_2 = (2\text{ mA} \angle -90^\circ)(6\text{ k}\Omega \angle 90^\circ) = 12\text{ V} \angle 0^\circ$$

Figure 10-3c shows these individual voltages. Since both angles are zero, the individual voltages are in phase with the total voltage.

EXAMPLE 10-4

What is the total phasor current in Fig. 10-4a?

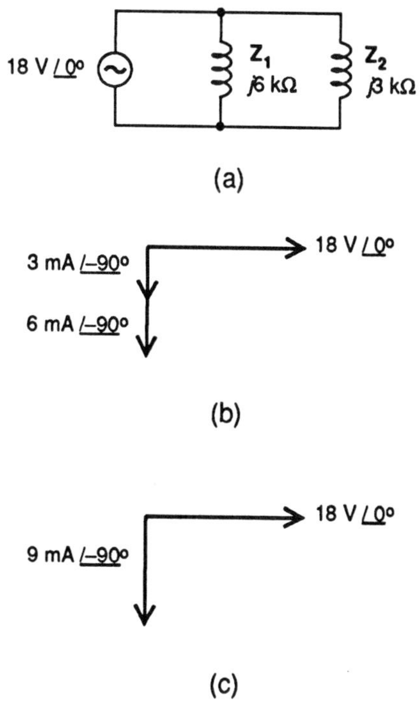

Fig. 10-4.

SOLUTION

The current through the first branch is

$$\mathbf{I}_1 = \frac{\mathbf{V}}{\mathbf{Z}_1} = \frac{18\text{ V} \angle 0^\circ}{6\text{ k}\Omega \angle 90^\circ} = 3\text{ mA} \angle -90^\circ$$

The current through the second branch is

$$\mathbf{I}_2 = \frac{\mathbf{V}}{\mathbf{Z}_2} = \frac{18\text{ V} \angle 0^\circ}{3\text{ k}\Omega \angle 90^\circ} = 6\text{ mA} \angle -90^\circ$$

Since both angles are –90°, these currents lag the total voltage by 90° as shown in Fig. 10-4b.

The total current is

$$\mathbf{I} = \mathbf{I}_1 + \mathbf{I}_2 = 3\text{ mA} \angle -90^\circ + 6\text{ mA} \angle -90^\circ = 9\text{ mA} \angle -90^\circ$$

Because the angle is –90°, the total current lags the total voltage by 90° as shown in Fig. 10-4b.

As a check on the work, the total impedance is

$$\mathbf{Z} = \frac{(j6\text{ k}\Omega)(j3\text{ k}\Omega)}{j6\text{ k}\Omega + j3\text{ k}\Omega} = \frac{j^2(18\text{ k}\Omega)}{j9\text{ k}\Omega} = j2\text{ k}\Omega$$

The total current is

$$\mathbf{I} = \frac{\mathbf{V}}{\mathbf{Z}} = \frac{18\text{ V} \angle 0^\circ}{j2\text{ k}\Omega} = \frac{18\text{ V} \angle 0^\circ}{2\text{ k}\Omega \angle 90^\circ} = 9\text{ mA} \angle -90^\circ$$

This agrees with the total current found earlier by adding the branch currents.

10-3. CAPACITIVE CIRCUITS

When a circuit is purely capacitive, every component is an ideal capacitor with an impedance of

$$\mathbf{Z} = X_C \angle -90^\circ$$

This says the impedance of each capacitor has a magnitude of X_C and an angle of -90°. The impedance may be expressed in rectangular form as

$$\mathbf{Z} = 0 - jX_C = -jX_C$$

In other words, the impedance of an ideal capacitor is always a negative imaginary number.

In series circuits, the total impedance of two capacitors is the sum of the individual impedances:

$$\mathbf{Z} = \mathbf{Z}_1 + \mathbf{Z}_2 = -jX_{C1} - jX_{C2} = -j(X_{C1} + X_{C2}) \tag{10-5}$$

This means the total impedance is a negative imaginary number.

In parallel circuits, the total impedance of two capacitors is the product over the sum:

$$\mathbf{Z} = \frac{\mathbf{Z}_1\mathbf{Z}_2}{\mathbf{Z}_1 + \mathbf{Z}_2} = \frac{(-jX_{C1})(-jX_{C2})}{-jX_{C1} - jX_{C2}} = \frac{j^2(X_{C1}X_{C2})}{-j(X_{C1} + X_{C2})}$$

or

$$\mathbf{Z} = -j\frac{X_{C1}X_{C2}}{X_{C1} + X_{C2}} \tag{10-6}$$

Again, the total impedance is a negative imaginary number.

Because the total impedance is always a negative imaginary number in purely capacitive circuits, the total current always leads the total voltage by 90°. Furthermore, since the impedance of each capacitor is always a negative imaginary number, the current into each capacitor leads the voltage across the capacitor by 90°.

EXAMPLE 10-5

What is the phasor current in Fig. 10-5a? The phasor voltage across each capacitor?

SOLUTION

The total impedance is

$$\mathbf{Z} = -j3\ \text{k}\Omega - j6\ \text{k}\Omega = -j9\ \text{k}\Omega$$

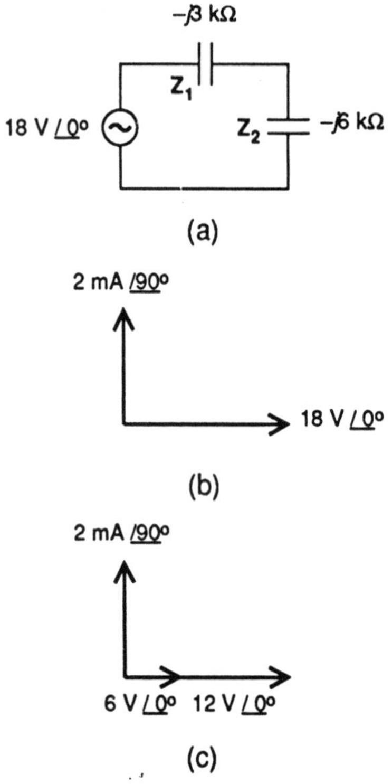

Fig. 10-5.

This rectangular form has a polar equivalent of

$$\mathbf{Z} = 9\ \text{k}\Omega \angle -90^\circ$$

The total current is

$$\mathbf{I} = \frac{\mathbf{V}}{\mathbf{Z}} = \frac{18\ \text{V} \angle 0^\circ}{9\ \text{k}\Omega \angle -90^\circ} = 2\ \text{mA} \angle 90^\circ$$

Since the angle is 90°, the total current leads the total voltage by 90° as shown in Fig. 10-5b. The individual voltages are

$$\mathbf{V}_1 = \mathbf{I}\mathbf{Z}_1 = (2\ \text{mA} \angle 90^\circ)(3\ \text{k}\Omega \angle -90^\circ) = 6\ \text{V} \angle 0^\circ$$

$$\mathbf{V}_2 = \mathbf{I}\mathbf{Z}_2 = (2\ \text{mA} \angle 90^\circ)(6\ \text{k}\Omega \angle -90^\circ) = 12\ \text{V} \angle 0^\circ$$

Figure 10-5c shows these individual voltages. Since both angles are zero, the individual voltages are in phase with the total voltage.

EXAMPLE 10-6

What is the total phasor current in Fig. 10-6a?

SOLUTION

The current through the first branch is

$$\mathbf{I}_1 = \frac{\mathbf{V}}{\mathbf{Z}_1} = \frac{18\ \text{V} \angle 0^\circ}{6\ \text{k}\Omega \angle -90^\circ} = 3\ \text{mA} \angle 90^\circ$$

The current through the second branch is

$$\mathbf{I}_2 = \frac{\mathbf{V}}{\mathbf{Z}_2} = \frac{18\ \text{V} \angle 0^\circ}{3\ \text{k}\Omega \angle -90^\circ} = 6\ \text{mA} \angle 90^\circ$$

Since both angles are 90°, these currents lead the total voltage by 90° as shown in Fig. 10-6b.

The total current is

$$\mathbf{I} = \mathbf{I}_1 + \mathbf{I}_2 = 3\ \text{mA} \angle 90^\circ + 6\ \text{mA} \angle 90^\circ = 9\ \text{mA} \angle 90^\circ$$

Because the angle is 90°, the total current leads the total voltage by 90° as shown in Fig. 10-6c.

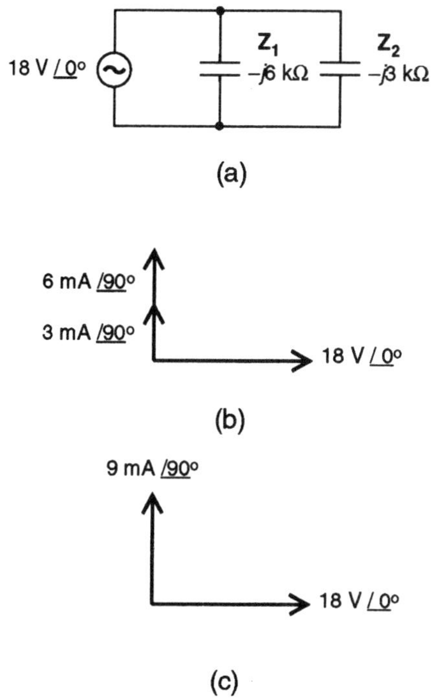

Fig. 10-6.

As a check on the work, the total impedance is

$$\mathbf{Z} = \frac{(-j6\ \text{k}\Omega)(-j3\ \text{k}\Omega)}{-j6\ \text{k}\Omega - j3\ \text{k}\Omega} = \frac{j^2(18\ \text{k}\Omega)}{-j9\ \text{k}\Omega} = -j2\ \text{k}\Omega$$

The total current is

$$\mathbf{I} = \frac{\mathbf{V}}{\mathbf{Z}} = \frac{18\ \text{V}\ \underline{/0^\circ}}{-j2\ \text{k}\Omega} = \frac{18\ \text{V}\ \underline{/0^\circ}}{2\ \text{k}\Omega\ \underline{/-90^\circ}} = 9\ \text{mA}\ \underline{/90^\circ}$$

This agrees with the total current found earlier by adding the branch currents.

10-4. *RL* CIRCUITS

Now, we are ready to analyze circuits with resistance and inductance. In series circuits, the total impedance of a resistor and an inductor is the sum of the individual impedances:

$$\mathbf{Z} = \mathbf{Z}_1 + \mathbf{Z}_2 = R + jX_L \tag{10-7}$$

This means the total impedance of a series *RL* circuit has a real part and an imaginary part.

In parallel circuits, the total impedance of a resistor and inductor is the product over the sum:

$$\mathbf{Z} = \frac{\mathbf{Z}_1\mathbf{Z}_2}{\mathbf{Z}_1 + \mathbf{Z}_2} = \frac{(R)(jX_L)}{R + jX_L} = \frac{jRX_L}{R + jX_L} \tag{10-8}$$

This shows that the total impedance of a parallel *RL* circuit is again a complex number. When circuit values are known, the foregoing can be converted to polar form and simplified (see Example 10-8).

Because the total impedance is always a complex number in *RL* circuits, the total current and total voltage will be out of phase by an angle between 0° and 90°. As usual, the resistive voltage and resistive current will be in phase, but the inductive voltage leads the inductive current by 90°.

EXAMPLE 10-7

What is the phasor current in Fig. 10-7a? The phasor voltage across each component?

SOLUTION

The total impedance is

$$\mathbf{Z} = 3\ \text{k}\Omega + j4\ \text{k}\Omega$$

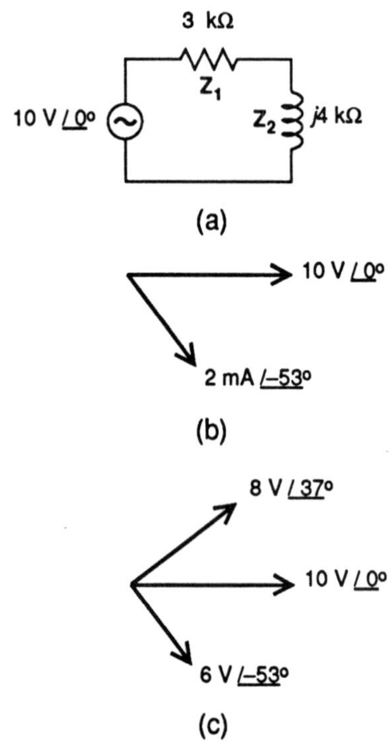

Fig. 10-7.

After converting this to polar form, we have

$$\mathbf{Z} = 5\ \text{k}\Omega \angle 53^\circ$$

Note: Here and elsewhere we will round the angle off to the nearest degree. This accuracy is adequate for our examples.

The total current is

$$\mathbf{I} = \frac{\mathbf{V}}{\mathbf{Z}} = \frac{10\ \text{V} \angle 0^\circ}{5\ \text{k}\Omega \angle 53^\circ} = 2\ \text{mA} \angle -53^\circ$$

Since the angle is -53°, the total current lags the total voltage as shown in Fig. 10-7b. The individual voltages are

$$\mathbf{V}_1 = \mathbf{I}\mathbf{Z}_1 = (2\ \text{mA} \angle -53^\circ)(3\ \text{k}\Omega) = 6\ \text{V} \angle -53^\circ$$

$$\mathbf{V}_2 = \mathbf{I}\mathbf{Z}_2 = (2\ \text{mA} \angle -53^\circ)(4\ \text{k}\Omega \angle 90^\circ) = 8\ \text{V} \angle 37^\circ$$

Figure 10-7c shows these individual voltages. As you see, the inductive voltage leads the total voltage by 37°, but the resistive voltage lags the total voltage by 53°.

EXAMPLE 10-8

What is the total phasor current in Fig. 10-8a?

SOLUTION

The current through the resistive branch is

$$\mathbf{I}_1 = \frac{\mathbf{V}}{\mathbf{Z}_1} = \frac{12\ \text{V} \angle 0^\circ}{3\ \text{k}\Omega \angle 0^\circ} = 4\ \text{mA} \angle 0^\circ$$

The current through the inductive branch is

$$\mathbf{I}_2 = \frac{\mathbf{V}}{\mathbf{Z}_2} = \frac{12\ \text{V} \angle 0^\circ}{4\ \text{k}\Omega \angle 90^\circ} = 3\ \text{mA} \angle -90^\circ$$

Figure 10-8b shows these currents.

The total current is

$$\mathbf{I} = \mathbf{I}_1 + \mathbf{I}_2 = 4\ \text{mA} \angle 0^\circ + 3\ \text{mA} \angle -90^\circ = 4\ \text{mA} - j3\ \text{mA}$$

Converting to polar form, we get

$$\mathbf{I} = 5\ \text{mA} \angle -37^\circ$$

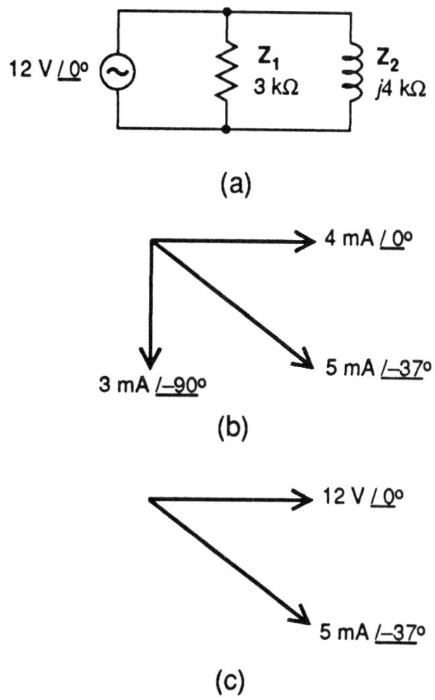

Fig. 10-8.

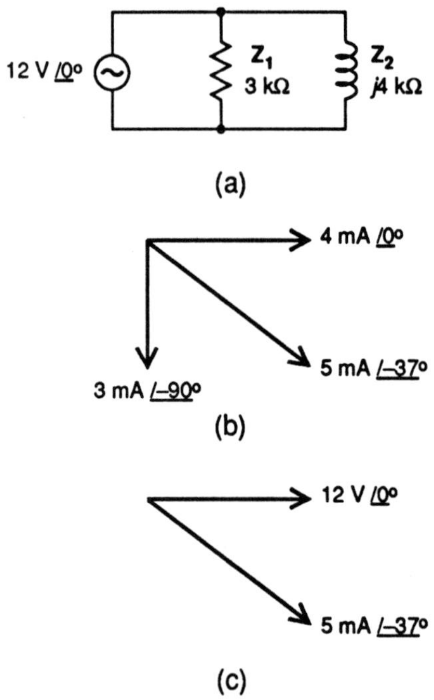

Fig. 10-8.

This total current is the phasor sum of the individual currents as you can see in Fig. 10-8b. Furthermore, the total current lags the total voltage by 37° as shown in Fig. 10-8c.

As a check on the work, the total impedance is

$$\mathbf{Z} = \frac{(3\ \text{k}\Omega)(j4\ \text{k}\Omega)}{3\ \text{k}\Omega + j4\ \text{k}\Omega} = \frac{j12\ \text{k}\Omega}{3\ \text{k}\Omega + j4\ \text{k}\Omega} = \frac{12\ \text{k}\Omega\ \angle 90^\circ}{5\ \text{k}\Omega\ \angle 53^\circ} = 2.4\ \text{k}\Omega\ \angle 37^\circ$$

The total current is

$$\mathbf{I} = \frac{\mathbf{V}}{\mathbf{Z}} = \frac{12\ \text{V}\ \angle 0^\circ}{2.4\ \text{k}\Omega\ \angle 37^\circ} = 5\ \text{mA}\ \angle -37^\circ$$

This agrees with the total current found earlier by adding the resistive and inductive currents.

10-5. *RC* CIRCUITS

Now, we are ready to analyze circuits with resistance and capacitance. In series circuits, the total impedance of a resistor and an capacitor is the algebraic sum of the individual impedances:

$$\mathbf{Z} = \mathbf{Z}_1 + \mathbf{Z}_2 = R - jX_C \qquad (10\text{-}9)$$

This means the total impedance of a series *RC* circuit has a real part and an imaginary part.

In parallel circuits, the total impedance of a resistor and capacitor is the product over the sum:

$$\mathbf{Z} = \frac{\mathbf{Z}_1\mathbf{Z}_2}{\mathbf{Z}_1 + \mathbf{Z}_2} = \frac{(R)(-jX_C)}{R - jX_C} = \frac{-jRX_C}{R - jX_C} \qquad (10\text{-}10)$$

This shows that the total impedance of a parallel *RC* circuit is again a complex number. When circuit values are known, the foregoing can be converted to polar form and simplified (see Example 10-10).

Because the total impedance is always a complex number in *RC* circuits, the total current and total voltage will be out of phase by an angle between 0° and 90°. As usual, the resistive voltage and resistive current will be in phase, but the capacitive current leads the capacitive voltage by 90°.

EXAMPLE 10-9

What is the phasor current in Fig. 10-9a? The phasor voltage across each component?

SOLUTION

The total impedance is

$$\mathbf{Z} = 3\ \text{k}\Omega - j4\ \text{k}\Omega$$

After converting this to polar form, we have

$$\mathbf{Z} = 5\ \text{k}\Omega \angle -53^\circ$$

The total current is

$$\mathbf{I} = \frac{\mathbf{V}}{\mathbf{Z}} = \frac{10\ \text{V} \angle 0^\circ}{5\ \text{k}\Omega \angle -53^\circ} = 2\ \text{mA} \angle 53^\circ$$

Since the angle is 53°, the total current leads the total voltage as shown in Fig. 10-9b. The individual voltages are

$$\mathbf{V}_1 = \mathbf{IZ}_1 = (2\ \text{mA} \angle 53^\circ)(3\ \text{k}\Omega) = 6\ \text{V} \angle 53^\circ$$

$$\mathbf{V}_2 = \mathbf{IZ}_2 = (2\ \text{mA} \angle 53^\circ)(4\ \text{k}\Omega \angle -90^\circ) = 8\ \text{V} \angle -37^\circ$$

Figure 10-9c shows these individual voltages. As you see, the capacitive voltage leads the total voltage by 53°, but the resistive voltage lags the total voltage by 37°.

EXAMPLE 10-10

What is the total phasor current in Fig. 10-10a?

SOLUTION

The current through the resistive branch is

$$\mathbf{I}_1 = \frac{\mathbf{V}}{\mathbf{Z}_1} = \frac{12\ \text{V} \angle 0^\circ}{3\ \text{k}\Omega \angle 0^\circ} = 4\ \text{mA} \angle 0^\circ$$

The current through the capacitive branch is

$$\mathbf{I}_2 = \frac{\mathbf{V}}{\mathbf{Z}_2} = \frac{12\ \text{V} \angle 0^\circ}{4\ \text{k}\Omega \angle -90^\circ} = 3\ \text{mA} \angle 90^\circ$$

Figure 10-10b shows these currents.

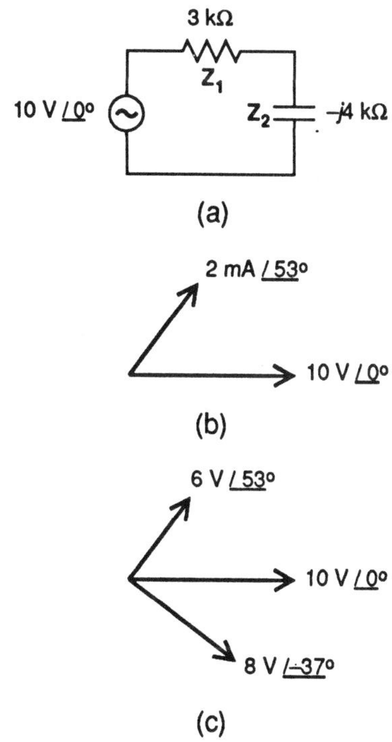

Fig. 10-9.

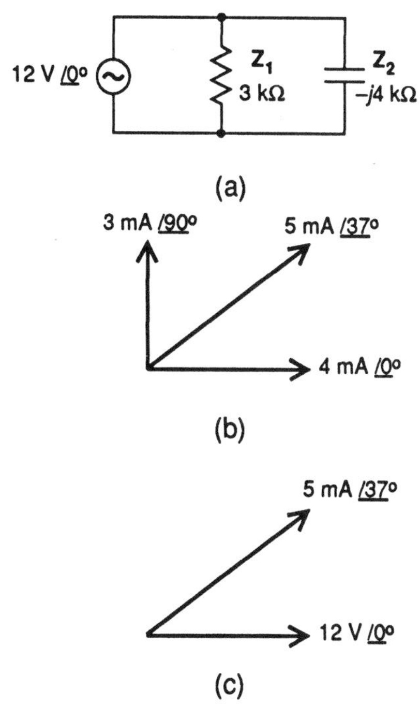

Fig. 10-10.

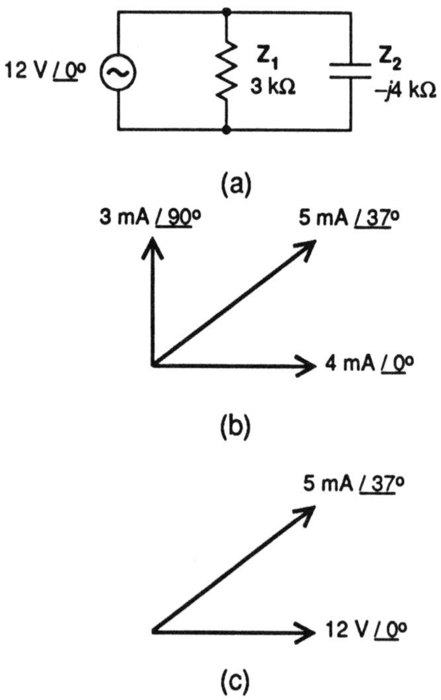

Fig. 10-10.

The total current is

$$\mathbf{I} = \mathbf{I}_1 + \mathbf{I}_2 = 4\text{ mA}\angle 0^\circ + 3\text{ mA}\angle 90^\circ = 4\text{ mA} + j3\text{ mA}$$

Converting to polar form, we get

$$\mathbf{I} = 5\text{ mA}\angle 37^\circ$$

This total current is the phasor sum of the individual currents as you can see in Fig. 10-10b. Furthermore, the total current leads the total voltage by 37° as shown in Fig. 10-10c.

As a check on the work, the total impedance is

$$\mathbf{Z} = \frac{(3\text{ k}\Omega)(-j4\text{ k}\Omega)}{3\text{ k}\Omega - j4\text{ k}\Omega} = \frac{-j12\text{ k}\Omega}{3\text{ k}\Omega - j4\text{ k}\Omega} = \frac{12\text{ k}\Omega\angle -90^\circ}{5\text{ k}\Omega\angle -53^\circ} = 2.4\text{ k}\Omega\angle -37^\circ$$

The total current is

$$\mathbf{I} = \frac{\mathbf{V}}{\mathbf{Z}} = \frac{12\text{ V}\angle 0^\circ}{2.4\text{ k}\Omega\angle -37^\circ} = 5\text{ mA}\angle 37^\circ$$

This agrees with the total current found earlier by adding the resistive and capacitive currents.

10-6. SERIES-TO-PARALLEL CONVERSION

Up to now, we have discussed series and parallel connections of basic components. The next step is series-parallel circuits. Before attempting the analysis of series-parallel circuits, however, we will discuss the conversion of series impedance to parallel impedance, and vice-versa. Then, we can use these conversions to simplify the analysis of series-parallel circuits.

RL Conversions

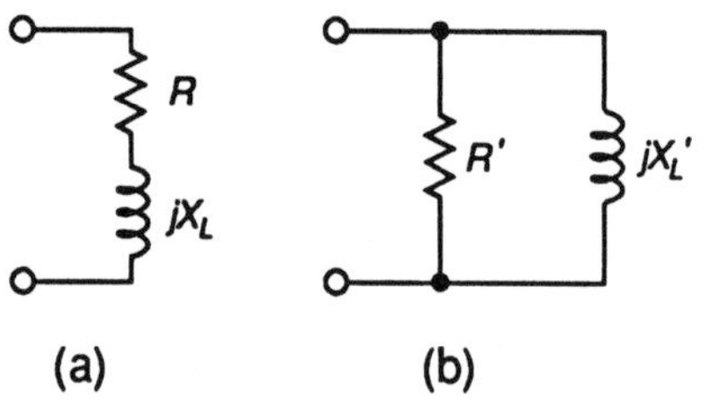

Fig. 10-11.

In the analysis of series-parallel ac circuits, you sometimes need to convert a series section like Fig. 10-11a to a parallel section like Fig. 10-11b. Using complex algebra, it is possible to derive these two conversion formulas:

$$R' = (1 + Q^2)R \qquad (10\text{-}11)$$

and

$$X_L' = (1 + 1/Q^2)X_L \qquad (10\text{-}12)$$

where $Q = X_L/R$. This is the Q of a series *RL* circuit. Given the values of R and X_L in the series section, you can use these two formulas to calculate the values of an equivalent parallel section. The new values are symbolized R' and X_L'.

RC Conversions

Figure 10-12a shows a series *RC* section. This time, the equivalent circuit has a resistance in parallel with a capacitive reactance (Fig. 10-12b). The conversion formulas are the same as before, except that X_C replaces X_L:

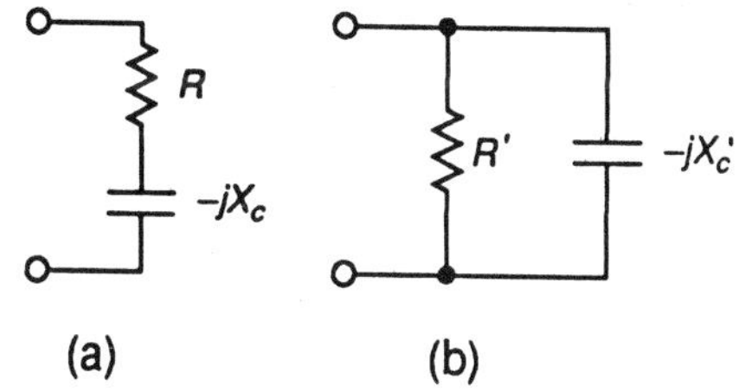

Fig. 10-12.

$$R' = (1 + Q^2)R \tag{10-13}$$

and

$$X_C' = (1 + 1/Q^2)X_C \tag{10-14}$$

where $Q = X_C/R$. This is the Q of a series *RC* circuit.

EXAMPLE 10-11

Convert the series section of Fig. 10-13a to an equivalent parallel section.

SOLUTION

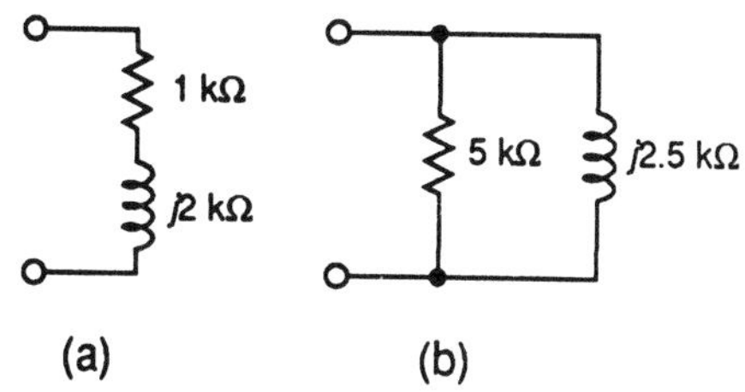

Fig. 10-13.

The first step is to calculate the Q of the circuit:

$$Q = \frac{X_L}{R} = \frac{2\text{ k}\Omega}{1\text{ k}\Omega} = 2$$

Then, Eq. (10-11) gives

$$R' = (1 + Q^2)R = (1 + 4)1\text{ k}\Omega = 5\text{ k}\Omega$$

and Eq. (10-12) gives

$$X_L' = (1 + 1/Q^2)X_L = (1 + 0.25)(2\text{ k}\Omega) = 2.5\text{ k}\Omega$$

Figure 10-13b shows the equivalent parallel circuit. This parallel equivalent section has the same total voltage and current as the original series section.

EXAMPLE 10-12

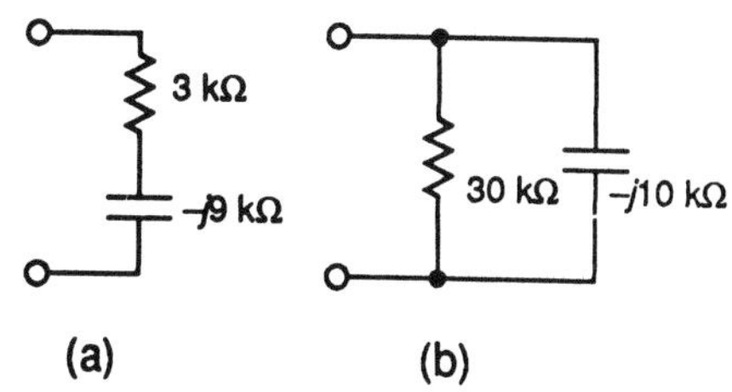

Fig. 10-14.

Convert the series section of Fig. 10-14a to an equivalent parallel section.

SOLUTION

The Q of the circuit is

$$Q = \frac{X_C}{R} = \frac{9\text{ k}\Omega}{3\text{ k}\Omega} = 3$$

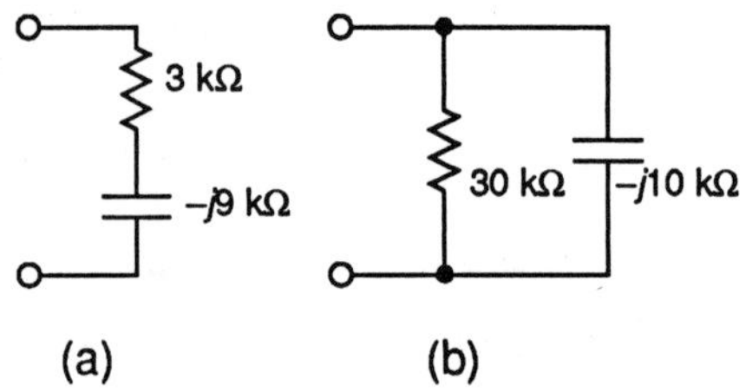

Fig. 10-14.

Then, Eqs. (10-13) and (10-14) give

$$R' = (1 + Q^2)R = (1 + 9)3 \text{ k}\Omega = 30 \text{ k}\Omega$$

and

$$X_C' = (1 + 1/Q^2)X_C = (1 + 0.111)9 \text{ k}\Omega = 10 \text{ k}\Omega$$

The equivalent circuit is a resistance of 30 kΩ in parallel with $-j10$ kΩ as shown in Fig. 10-14b.

10-7. PARALLEL-TO-SERIES CONVERSION

Sometimes, you will want to convert a parallel section like Fig. 10-15a to a series section like Fig. 10-15b. Using complex numbers, it is possible to derive these two conversion formulas:

$$R' = \frac{R}{1 + Q^2} \qquad (10\text{-}15)$$

and

$$X_L' = \frac{X_L}{1 + 1/Q^2} \qquad (10\text{-}16)$$

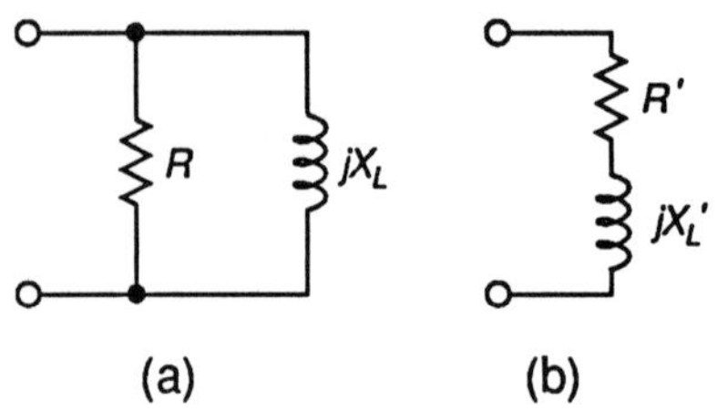

Fig. 10-15.

where $Q = R/X_L$. This is the Q of a parallel RL circuit. Given the values of R and X_L in the parallel section, you can use these two formulas to calculate the values of an equivalent series section. The new values are symbolized R' and X_L'.

Similarly, given a parallel RC circuit like Fig. 10-16a, we can convert it to an equivalent series RC section like Fig. 10-16b. The conversion formulas are the same as before, except that X_C replaces X_L:

$$R' = \frac{R}{1 + Q^2} \qquad (10\text{-}17)$$

and

$$X_C' = \frac{X_C}{1 + 1/Q^2} \qquad (10\text{-}18)$$

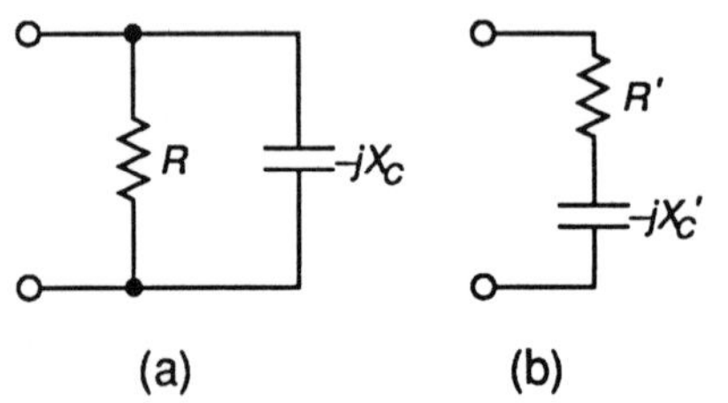

Fig. 10-16.

where $Q = R/X_C$. This is the Q of a parallel RC circuit.

EXAMPLE 10-13

Convert the parallel section of Fig. 10-17a to an equivalent series section.

SOLUTION

The first step is to calculate the Q of the circuit:

$$Q = \frac{R}{X_L} = \frac{2\ \text{k}\Omega}{1\ \text{k}\Omega} = 2$$

Then, Eqs. (10-15) and (10-16) give

$$R' = \frac{R}{1 + Q^2} = \frac{2\ \text{k}\Omega}{1 + 4} = 400\ \Omega$$

and

$$X_L' = \frac{X_L}{1 + 1/Q^2} = \frac{1\ \text{k}\Omega}{1 + 0.25} = 800\ \Omega$$

The equivalent circuit for Fig. 10-17a is a resistance of 400 Ω in series with j800 Ω as shown in Fig. 10-17b.

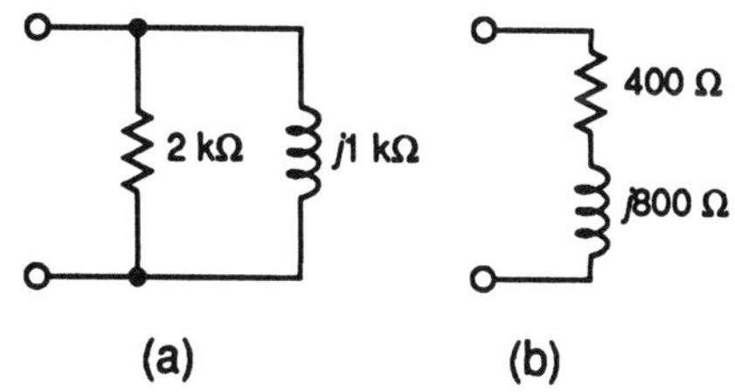

Fig. 10-17.

EXAMPLE 10-14

Convert the parallel section of Fig. 10-18a to an equivalent series section.

SOLUTION

The Q is

$$Q = \frac{R}{X_C} = \frac{2\ \text{k}\Omega}{2\ \text{k}\Omega} = 1$$

Then, Eqs. (10-17) and (10-18) give

$$R' = \frac{R}{1 + Q^2} = \frac{2\ \text{k}\Omega}{1 + 1} = 1\ \text{k}\Omega$$

and

$$X_C' = \frac{X_C}{1 + 1/Q^2} = \frac{2\ \text{k}\Omega}{1 + 1} = 1\ \text{k}\Omega$$

The equivalent circuit is a resistance of 1 kΩ in series with $-j1$ kΩ (Fig. 10-18b).

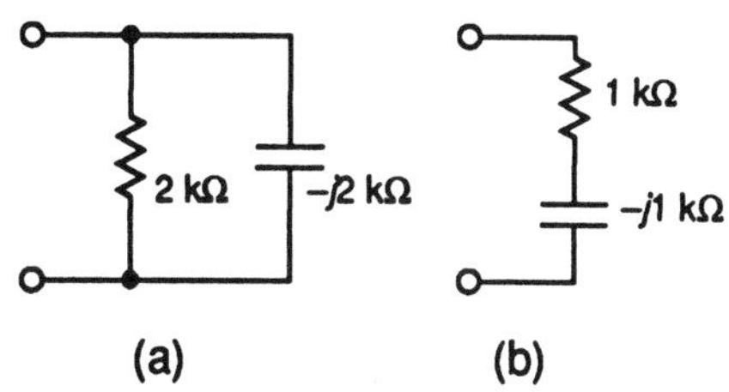

Fig. 10-18.

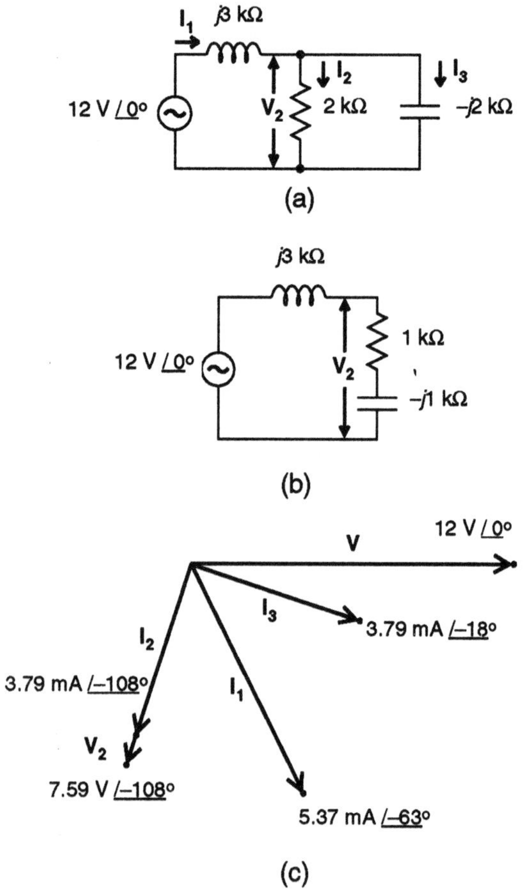

Fig. 10-19.

10-8. SERIES-PARALLEL CIRCUITS

The analysis of series-parallel ac circuits can become very complicated. Without a computer to do the calculations, it may be a nightmare. Let us take a look at a series-parallel circuit to get the idea of how the calculations are carried out.

Figure 10-19a shows a series-parallel circuit. How do we find the values of $\mathbf{I}_1$, $\mathbf{I}_2$, and $\mathbf{I}_3$? With Eqs. (10-17) and (10-18), we can convert the parallel *RC* section (2 kΩ and $-j2$ kΩ) to a series *RC* section (1 kΩ and $-j1$ kΩ) as shown in Fig. 10-19b. For details of the conversion, see Example 10-14 .

In Fig. 10-19b, the total impedance is

$$\mathbf{Z} = j3\ \text{k}\Omega + 1\ \text{k}\Omega - j1\ \text{k}\Omega = 1\ \text{k}\Omega + j2\ \text{k}\Omega$$

Next, we can calculate $\mathbf{I}_1$ as follows:

$$\mathbf{I}_1 = \frac{12\ \text{V}\ \angle 0^\circ}{1\ \text{k}\Omega + j2\ \text{k}\Omega} = 5.37\ \text{mA}\ \angle -63^\circ$$

With this current, we can calculate the voltage across the equivalent section as follows:

$$\mathbf{V}_2 = 5.37\ \text{mA}\ \angle -63^\circ \times (1\ \text{k}\Omega - j1\ \text{k}\Omega) = 7.59\ \text{V}\ \angle -108^\circ$$

Now, we can return to Fig. 10-19a to calculate the branch currents:

$$\mathbf{I}_2 = \frac{7.59\ \text{V}\ \angle -108^\circ}{2\ \text{k}\Omega} = 3.79\ \text{mA}\ \angle -108^\circ$$

and

$$\mathbf{I}_3 = \frac{7.59\ \text{V}\ \angle -108^\circ}{-j2\ \text{k}\Omega} = 3.79\ \text{mA}\ \angle -18^\circ$$

Figure 10-19c summarizes the calculations. This phasor diagram shows the magnitude and phase angles of the applied voltage and the branch currents. Notice that all currents lag the applied voltage. The foregoing analysis is straightforward, but the calculations with complex numbers are tedious when done by hand. This is why computers are often used to carry out the analysis of series-parallel circuits.

EXAMPLE 10-15

Find all currents and voltages in Fig. 10-20.

SOLUTION

The current through the first branch is

$$\mathbf{I}_1 = \frac{12\ \text{V}\ \angle 0^\circ}{1\ \text{k}\Omega + j2\ \text{k}\Omega} = \frac{12\ \text{V}\ \angle 0^\circ}{2.23\ \text{k}\Omega\ \angle 63^\circ} = 5.37\ \text{mA}\ \angle -63^\circ$$

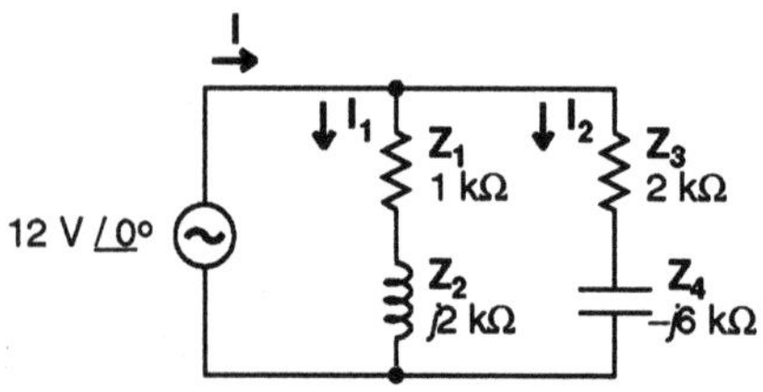

Fig. 10-20.

The current through the second branch is

$$\mathbf{I}_2 = \frac{12 \text{ V} \angle 0^\circ}{2 \text{ k}\Omega - j6 \text{ k}\Omega} = \frac{12 \text{ V} \angle 0^\circ}{6.32 \text{ k}\Omega \angle -72^\circ} = 1.9 \text{ mA} \angle 72^\circ$$

The total current is

$$\mathbf{I} = \mathbf{I}_1 + \mathbf{I}_2 = 5.37 \text{ mA} \angle -63^\circ + 1.9 \text{ mA} \angle 72^\circ$$

$$= 2.44 - j4.78 + 0.587 + j1.81 = 3.03 - j2.97 = 4.24 \angle -44^\circ$$

Next, we can calculate the individual voltages as follows:

$$\mathbf{V}_1 = \mathbf{I}_1\mathbf{Z}_1 = (5.37 \text{ mA} \angle -63^\circ)(1 \text{ k}\Omega) = 5.37 \text{ V} \angle -63^\circ$$

$$\mathbf{V}_2 = \mathbf{I}_1\mathbf{Z}_2 = (5.37 \text{ mA} \angle -63^\circ)(j2 \text{ k}\Omega) = 10.7 \text{ V} \angle 27^\circ$$

$$\mathbf{V}_3 = \mathbf{I}_2\mathbf{Z}_3 = (1.9 \text{ mA} \angle 72^\circ)(2 \text{ k}\Omega) = 3.8 \text{ V} \angle 72^\circ$$

$$\mathbf{V}_4 = \mathbf{I}_2\mathbf{Z}_4 = (1.9 \text{ mA} \angle 72^\circ)(-j6 \text{ k}\Omega) = 11.4 \text{ V} \angle -18^\circ$$

EXAMPLE 10-16

Find the total current in Fig. 10-21.

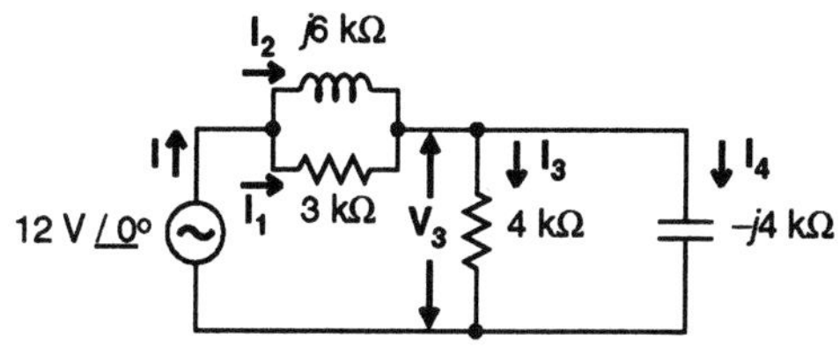

Fig. 10-21.

SOLUTION

The easiest way to do this is to convert each parallel section to a series section. Then, work out the total impedance and the total current. The first parallel section has a Q of

$$Q = \frac{R}{X_L} = \frac{3 \text{ k}\Omega}{6 \text{ k}\Omega} = 0.5$$

Then, Eqs. (10-15) and (10-16) give

$$R' = \frac{R}{1 + Q^2} = \frac{3 \text{ k}\Omega}{1 + 0.25} = 2.4 \text{ k}\Omega$$

and

$$X_L' = \frac{X_L}{1 + 1/Q^2} = \frac{6 \text{ k}\Omega}{1 + 4} = 1.2 \text{ k}\Omega$$

The equivalent series section is a resistance of 2.4 kΩ in series with $j1.2$ kΩ.

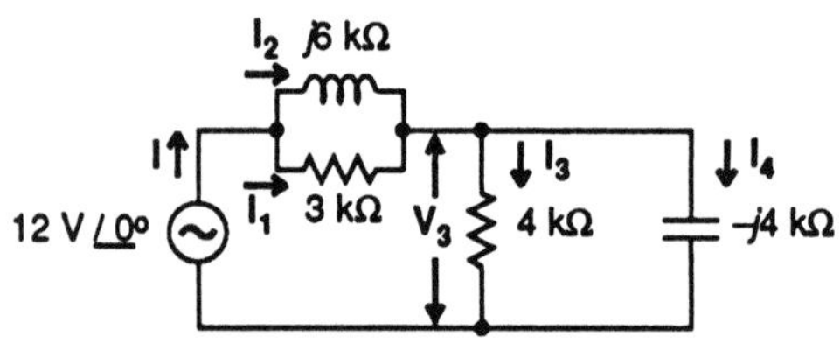

Fig. 10-21.

The second parallel section has a Q of

$$Q = \frac{R}{X_C} = \frac{4\ \text{k}\Omega}{4\ \text{k}\Omega} = 1$$

Then, Eqs. (10-17) and (10-18) give

$$R' = \frac{R}{1 + Q^2} = \frac{4\ \text{k}\Omega}{1 + 1} = 2\ \text{k}\Omega$$

and

$$X_C' = \frac{X_C}{1 + 1/Q^2} = \frac{4\ \text{k}\Omega}{1 + 1} = 2\ \text{k}\Omega$$

The equivalent series section is a resistance of 2 kΩ in series with $-j2$ kΩ.
The total impedance is

$$\mathbf{Z} = 2.4\ \text{k}\Omega + j1.2\ \text{k}\Omega + 2\ \text{k}\Omega - j2\ \text{k}\Omega = 4.4\ \text{k}\Omega - j800\ \Omega$$

and the total current is

$$\mathbf{I} = \frac{12\ \text{V}\ \angle 0^\circ}{4.4\ \text{k}\Omega - j800\ \Omega} = \frac{12\ \text{V}\ \angle 0^\circ}{4.47\ \angle -10^\circ} = 2.68\ \text{mA}\ \angle 10^\circ$$

10-9. COUPLING CIRCUIT

The circuit shown in Fig. 10-22a is often used to couple or transmit a signal from an ac source to a load resistor. The key to analysis of the circuit is the *critical frequency*, discussed in earlier chapters:

$$f_c = \frac{1}{2\pi RC} \tag{10-19}$$

As you recall, this is the frequency where the capacitive reactance equals the resistance. At this frequency, the output voltage equals 0.707 times the input voltage, and the lead angle of the output voltage is 45°. When the frequency is at least 10 times greater than the critical frequency, the capacitive reactance approaches zero and the output voltage equals the input voltage.

Figure 10-22b shows the phasors for input and output voltage at the critical frequency. The magnitude of the output voltage is 0.707 times the magnitude of the input voltage. Also, the output voltage leads the input voltage by 45°. Because the output voltage leads the input voltage, this circuit is sometimes called a *lead network*.

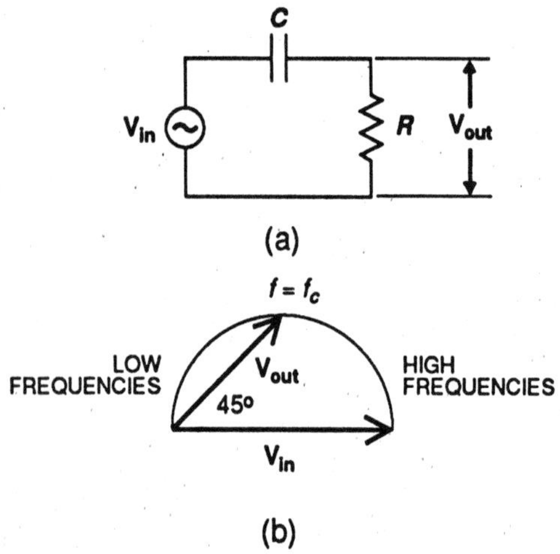

Fig. 10-22.

With an advanced derivation, we can prove that the tip of the output phasor lies on the half circle shown in Fig. 10-22b. In other words, as the frequency changes, the magnitude and angle of the phasor also change. But at all times, the tip of the phasor lies somewhere along the half circle. In fact, when the frequency is much lower than the critical frequency, the phasor moves into a vertical position. This results in a magnitude approaching zero and an angle approaching 90°. On the other hand, when the frequency is much greater than the critical frequency, the phasor moves into a horizontal position. In this case, the output magnitude approaches the input magnitude and the angle approaches 0°.

Here are some useful equations for analyzing a coupling circuit. The output voltage has a magnitude of

$$V_{out} = \frac{R}{\sqrt{R^2 + X_C^2}} V_{in} \qquad (10\text{-}20)$$

and an angle of

$$\phi = \arctan \frac{X_C}{R} \qquad (10\text{-}21)$$

In these equations, X_C plays an important role. When $X_C = R$, the output voltage equals 0.707 times the input voltage and the angle is 45°. At low frequencies, X_C approaches infinity. Therefore, the output voltage approaches zero and the angle approaches 90°. At high frequencies, X_C approaches zero. Therefore, the output voltage approximately equals the input voltage and the angle approaches 0°.

EXAMPLE 10-17

What is the critical frequency in Fig. 10-23a? Show the input and output phasor voltages.

SOLUTION

The critical frequency is

$$f_c = \frac{1}{2\pi(4.7\ \text{k}\Omega)(1\ \mu\text{F})} = 33.8\ \text{Hz}$$

At this frequency, the capacitive reactance equals the resistance. Then, Eqs. (10-20) and (10-21) give

$$V_{out} = \frac{4.7\ \text{k}\Omega}{\sqrt{(4.7\ \text{k}\Omega)^2 + (4.7\ \text{k}\Omega)^2}}\ 10\ \text{V} = 7.07\ \text{V}$$

and an angle of

$$\phi = \arctan \frac{4.7\ \text{k}\Omega}{4.7\ \text{k}\Omega} = 45°$$

Figure 10-23b shows these phasor voltages.

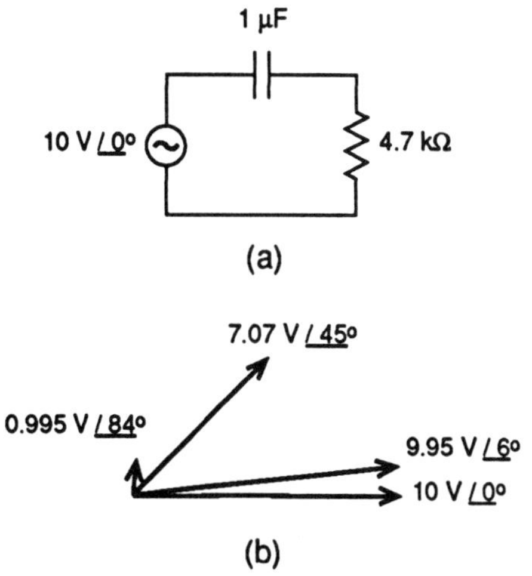

Fig. 10-23.

EXAMPLE 10-18

The critical frequency of Fig. 10-23a was calculated as 33.8 Hz in the preceding example. What are the phasor voltages at 3.38 Hz? At 338 Hz?

SOLUTION

When the frequency is 3.38 Hz (a factor of 10 decrease), the capacitive reactance increases by a factor of 10 to 47 kΩ. Therefore, the output voltage and angle become

$$V_{out} = \frac{4.7\text{ k}\Omega}{\sqrt{(4.7\text{ k}\Omega)^2 + (47\text{ k}\Omega)^2}}\,10\text{ V} = 0.995\text{ V}$$

and an angle of

$$\phi = \arctan\frac{47\text{ k}\Omega}{4.7\text{ k}\Omega} = 84^\circ$$

When the frequency is 338 Hz (a factor of 10 increase), the capacitive reactance decreases by a factor of 10 to 470 Ω. Therefore, the output voltage and angle become

$$V_{out} = \frac{4.7\text{ k}\Omega}{\sqrt{(4.7\text{ k}\Omega)^2 + (470\ \Omega)^2}}\,10\text{ V} = 9.95\text{ V}$$

and an angle of

$$\phi = \arctan\frac{470\ \Omega}{4.7\text{ k}\Omega} = 6^\circ$$

Figure 10-23b shows the phasors. This diagram tells the whole story. The input voltage is 10 V at an angle of 0°. At the critical frequency, the output phasor is 7.07 V at an angle of 45°. At one-tenth of the critical frequency, the output phasor is 0.995 V at an angle of 84°. At 10 times the critical frequency, the output phasor is 9.95 V at an angle of 6°. Therefore, good coupling of the signal occurs when the frequency is at least 10 times the critical frequency. In this case, the output voltage is approximately equal to the input voltage and is almost in phase with the input voltage.

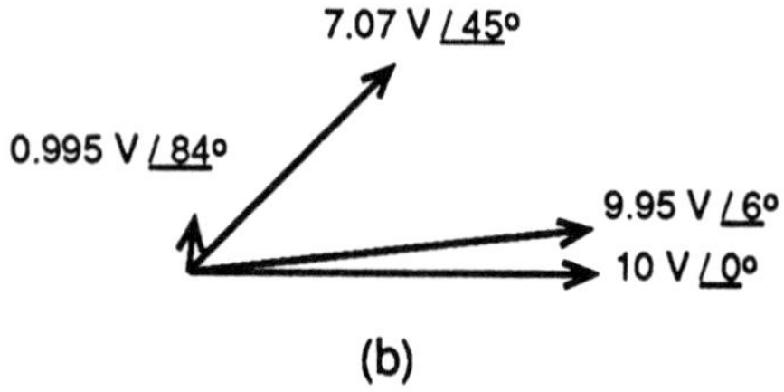

Fig. 10-23.

10-10. BYPASS CIRCUIT

The circuit shown in Fig. 10-24a is often used to bypass or block a signal from reaching the output. The key to analysis of the circuit is the critical frequency, again given by

$$f_c = \frac{1}{2\pi RC} \tag{10-22}$$

As you recall, this is the frequency where the capacitive reactance equals the resistance. At this frequency, the output voltage equals 0.707 times the input voltage, and the lag angle of the output voltage is 45°. When the frequency is at least 10 times greater than the critical frequency, the capacitive reactance approaches zero and the output voltage approaches zero.

Figure 10-24b shows the phasors for input and output voltage at the critical frequency. The magnitude of the output voltage is 0.707 times the magnitude of the input voltage. Also, the output voltage lags the input voltage by 45°. Because the output voltage lags the input voltage, this circuit is sometimes called a *lag network.*

Again, it is possible to prove that the tip of the output phasor lies on the half circle shown in Fig. 10-24b. When the frequency is much higher than the critical frequency, the phasor moves into a vertical position. This results in a magnitude approaching zero and an angle approaching –90°. On the other hand, when the frequency is much smaller than the critical frequency, the phasor moves into a horizontal position. In this case, the output magnitude approaches the input magnitude and the angle approaches 0°.

Here are the equations for analyzing a bypass circuit. The output voltage has a magnitude of

$$V_{out} = \frac{X_C}{\sqrt{R^2 + X_C^2}} V_{in} \tag{10-23}$$

and an angle of

$$\phi = -\arctan \frac{R}{X_C} \tag{10-24}$$

When $X_C = R$, the output voltage equals 0.707 times the input voltage and the angle is –45°. At high frequencies, X_C approaches zero. Therefore, the output voltage approaches zero and the angle approaches –90°. At low frequencies, X_C approaches infinity. Therefore, the output voltage approximately equals the input voltage and the angle approaches 0°.

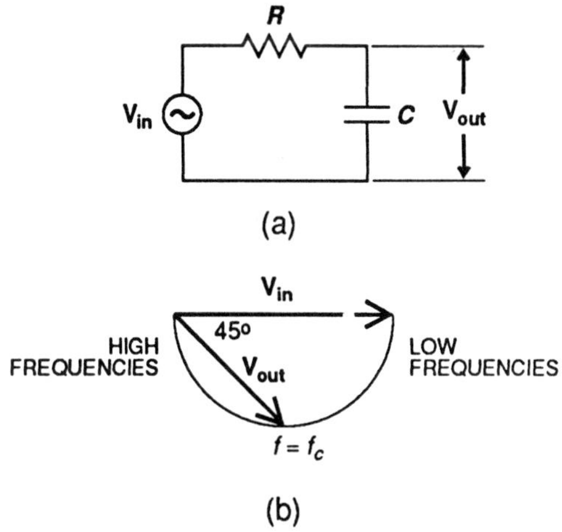

Fig. 10-24.

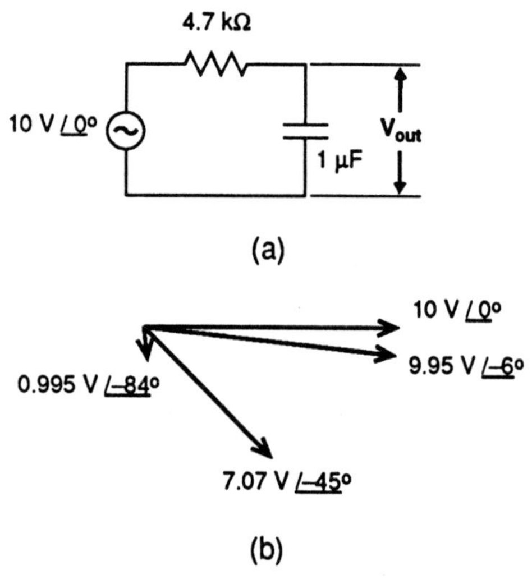

Fig. 10-25.

EXAMPLE 10-19

What is the critical frequency in Fig. 10-25a? Show the input and output phasor voltages.

SOLUTION

The critical frequency is

$$f_c = \frac{1}{2\pi(4.7\ \text{k}\Omega)(1\ \mu\text{F})} = 33.8\ \text{Hz}$$

At this frequency, the capacitive reactance equals the resistance. Then, Eqs. (10-22) and (10-23) give

$$V_{out} = \frac{4.7\ \text{k}\Omega}{\sqrt{(4.7\ \text{k}\Omega)^2 + (4.7\ \text{k}\Omega)^2}}\ 10\ \text{V} = 7.07\ \text{V}$$

and an angle of

$$\phi = \arctan \frac{4.7\ \text{k}\Omega}{4.7\ \text{k}\Omega} = -45^\circ$$

Figure 10-25b shows these phasor voltages.

EXAMPLE 10-20

The critical frequency of Fig. 10-25a was calculated as 33.8 Hz in the preceding example. What are the phasor voltages at 3.38 Hz? At 338 Hz?

SOLUTION

When the frequency is 3.38 Hz (a factor of 10 decrease), the capacitive reactance increases by a factor of 10 to 47 kΩ. Therefore, the output voltage and angle become

$$V_{out} = \frac{47\ \text{k}\Omega}{\sqrt{(4.7\ \text{k}\Omega)^2 + (47\ \text{k}\Omega)^2}}\ 10\ \text{V} = 9.95\ \text{V}$$

and an angle of

$$\phi = -\arctan \frac{4.7\ \text{k}\Omega}{47\ \text{k}\Omega} = -6^\circ$$

When the frequency is 338 Hz (a factor of 10 increase), the capacitive reactance decreases by a factor of 10 to 470 Ω. Therefore, the output voltage and angle become

$$V_{out} = \frac{470\ \Omega}{\sqrt{(4.7\ k\Omega)^2 + (470\ \Omega)^2}}\ 10\ V = 0.995\ V$$

and an angle of

$$\phi = -\arctan \frac{4.7\ k\Omega}{470\ \Omega} = -84^\circ$$

Figure 10-25b shows the phasors. The input voltage is 10 V at an angle of 0°. At the critical frequency, the output phasor is 7.07 V at an angle of −45°. At one-tenth of the critical frequency, the output phasor is 9.95 V at an angle of 6°. At ten times the critical frequency, the output phasor is 0.995 V at an angle of −84°. Therefore, good bypassing of the signal occurs when the frequency is at least 10 times the critical frequency. In this case, the output voltage is much smaller than the input voltage.

10-11. LEAD-LAG NETWORK

Figure 10-26a is called a *lead-lag network* because the output voltage leads the input voltage at lower frequencies and lags the input voltage at higher frequencies. At very low frequencies, the series capacitor appears open to the ac signal, and there is no output voltage. At very high frequencies, the shunt capacitor appears shorted, and there is no output voltage. In between these two extremes, the output voltage reaches a maximum value of one-third the input voltage.

What is the frequency where the output voltage reaches its maximum value? It is the critical frequency that we have been discussing for lead and lag circuits:

$$f_c = \frac{1}{2\pi RC} \qquad (10\text{-}25)$$

Given a circuit like Fig. 10-26a with specific values of R and C, you can calculate the critical frequency with Eq. (10-25). If you build the circuit, you will find that the output voltage reaches a maximum value at the critical frequency. Furthermore, the output voltage leads the input voltage at all frequencies below the critical frequency. Conversely, the output voltage lags the input voltage at all frequencies above the critical frequency.

The mathematical analysis of Fig. 10-26 is very complicated because complex numbers have to be applied to a series-parallel circuit. With considerable effort, it can be shown that the magnitude and angle are given by these equations:

$$V_{out} = \frac{V_{in}}{\sqrt{9 + (X_C/R - R/X_C)^2}} \qquad (10\text{-}26)$$

and

$$\phi = \arctan \frac{X_C/R - R/X_C}{3} \qquad (10\text{-}27)$$

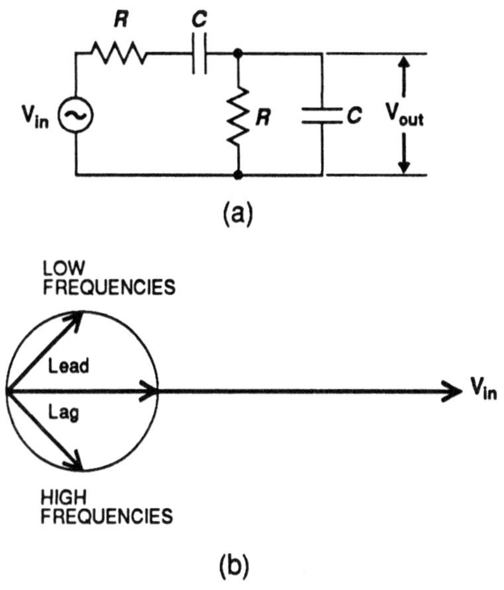

Fig. 10-26.

At the critical frequency, the denominator of Eq. (10-26) reduces to a value of 3. Therefore, the output voltage equals one-third of the input voltage. Above or below the critical frequency, the denominator is greater than 3, so the output voltage is less than one-third of the input. At extremely low or high frequencies, the output voltage approaches zero.

Another interesting feature of the lead-lag circuit is the phase angle. At the critical frequency, Eq. (10-27) produces an angle of 0°. Below the critical frequency, the angle becomes positive. Above the critical frequency, the angle becomes negative. If you study the equation, you will see that the phase angle varies from +90° to –90°.

In summary, Fig. 10-26b shows the phasor diagram. The tip of the output phasor always falls somewhere on the circle. At low frequencies, the output voltage is small at an angle of +90°. At high frequencies, the output voltage is small at an angle of –90°. At the critical frequency, the magnitude of the output voltage reaches a maximum value of $V_{in}/3$ at an angle of 0°.

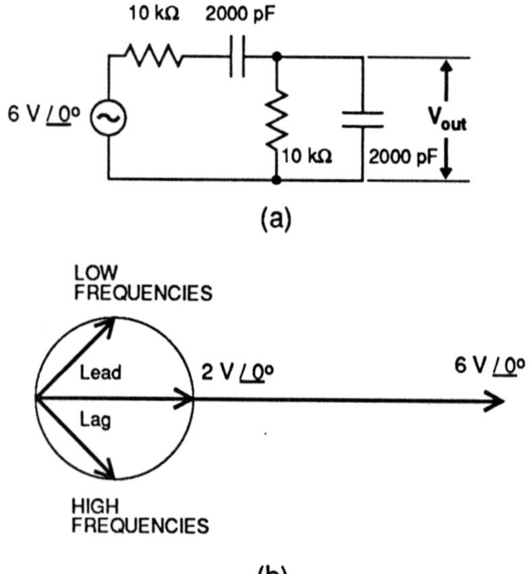

Fig. 10-27.

EXAMPLE 10-21

What is the critical frequency in Fig. 10-27a? Show the phasor diagram.

SOLUTION

The critical frequency is

$$f_c = \frac{1}{2\pi(10\ \text{k}\Omega)(2000\ \text{pF})} = 7.96\ \text{kHz}$$

At this frequency the output voltage is one-third of the input voltage and the angle is 0°. Figure 10-27b shows the phasor diagram. The phasor input voltage is 6 V $\angle 0°$ and the output voltage is 2 V $\angle 0°$.

MULTIPLE-CHOICE TEST

The following questions may have more than one right answer. Select the best answer. This is the one that is always true, or covers more situations, or agrees with the material of this chapter, etc.

1. The laws for dc circuits are still valid for ac circuits if you use
   a. real numbers
   c. imaginary numbers
   c. complex numbers
   d. negative numbers

2. The Q of a parallel RC circuit equals the
   a. inductive reactance divided by the resistance
   b. resistance divided by the inductive reactance
   c. capacitive reactance divided by the resistance
   d. resistance divided by the capacitive reactance

3. In a series *RL* circuit, the total voltage
   a. leads the total current
   b. lags the total current
   c. is in phase with the total current
   d. may be any of the above

4. The *Q* of a series *RL* circuit equals the
   a. inductive reactance divided by the resistance
   b. resistance divided by the inductive reactance
   c. capacitive reactance divided by the resistance
   d. resistance divided by the capacitive reactance

5. In a parallel *RC* circuit, the total voltage
   a. leads the total current
   b. lags the total current
   c. is in phase with the total current
   d. may be any of the above

6. At low frequencies, the angle of output voltage in a coupling circuit is near
   a. 0°
   b. 45°
   c. 90°
   d. 180°

7. The *Q* of a parallel *RL* circuit equals the
   a. inductive reactance divided by the resistance
   b. resistance divided by the inductive reactance
   c. capacitive reactance divided by the resistance
   d. resistance divided by the capacitive reactance

8. A parallel *RC* circuit may be converted to a
   a. parallel *RC* circuit
   b. parallel *RL* circuit
   c. series *RC* circuit
   d. series *RL* circuit

9. In a series *RC* circuit, the total voltage
   a. leads the total current
   b. lags the total current
   c. is in phase with the total current
   d. may be any of the above

10. The *Q* of a series *RC* circuit equals the
   a. inductive reactance divided by the resistance
   b. resistance divided by the inductive reactance
   c. capacitive reactance divided by the resistance
   d. resistance divided by the capacitive reactance

11. In a parallel *RL* circuit, the total current
    a. is in phase with the applied voltage
    b. is larger than the individual currents
    c. lags the applied voltage
    d. leads the applied voltage

12. In a series *RC* circuit, the total voltage
    a. leads the current
    b. lags the current
    c. is in phase with the current
    d. may be any of the above

13. In a series *RL* circuit, the total current
    a. leads the total voltage
    b. lags the total voltage
    c. is in phase with the total voltage
    d. may be any of the above

14. In a parallel *RC* circuit, an open capacitor means the
    a. applied voltage is in phase with the total current
    b. circuit is purely resistive
    c. total impedance has an angle of zero
    d. all of the above

15. In a series-parallel circuit, all sections are first converted to
    a. series sections
    b. parallel sections
    c. either type of section
    d. impossible to say

PROBLEMS

Sec. 10-1: Resistive Circuits

10-1. A resistive circuit has three series resistances: 2 kΩ, 4 kΩ, and 6 kΩ. If the applied voltage is 7 V at an angle of zero, what is the total current? The phasor voltage across each resistor?

10-2. A resistive circuit has four parallel resistances: 5 kΩ, 3 kΩ, 6 kΩ, and 4 kΩ. If the applied voltage is 14 V at an angle of zero, what is the total current?

Sec. 10-2: Inductive Circuits

10-3. An inductive circuit has four inductive reactances in series with values of 2.5 kΩ, 6.25 kΩ, 1.75 kΩ, and 3.2 kΩ. If the applied voltage is 9 V at an angle of zero, what is the total current? The phasor voltage across each inductor?

10-4. An inductive circuit has three inductive reactances in parallel with values of 5.2 kΩ, 6.7 kΩ, and 2.4 kΩ. If the applied voltage is 17 V at an angle of zero, what is the total current? The phasor voltage across each inductor?

Sec. 10-3: Capacitive Circuits

10-5. A capacitive circuit has three capacitive reactances in series with values of 31 kΩ, 42 kΩ, and 25 kΩ. If the applied voltage is 25 V at an angle of zero, what is the total current? The phasor voltage across each capacitor?

10-6. An capacitive circuit has five capacitive reactances in parallel with values of 2.5 kΩ, 6.4 kΩ, 3.6 kΩ, 5.8 kΩ, and 2.9 kΩ. If the applied voltage is 30 V at an angle of zero, what is the total current? The phasor voltage across each capacitor?

Sec. 10-4: *RL* Circuits

10-7. An *RL* circuit has a resistance in series with an inductor. The impedances are 3.5 kΩ and $j4.75$ kΩ. If the applied voltage is 250 mV at an angle of zero, what is the total current? The phasor voltage across each component?

10-8. An *RL* circuit has a resistance in parallel with an inductor. The impedances are 20 kΩ and $j15$ kΩ. If the applied voltage is 100 mV at an angle of zero, what is the total current?

10-9. The ac generator of Fig. 10-28 has a frequency of 5 kHz. What is the voltage across the inductor?

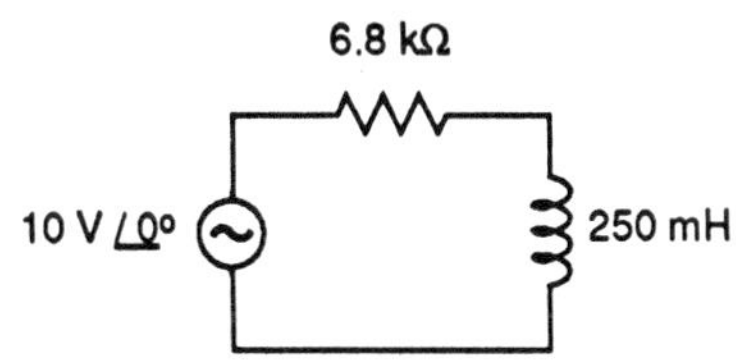

Fig. 10-28.

Sec. 10-5: *RC* Circuits

10-10. An *RC* circuit has a resistance in series with an capacitor. The impedances are 12 kΩ and $-j5$ kΩ. If the applied voltage is 375 mV at an angle of zero, what is the total current? The phasor voltage across each component?

10-11. An *RC* circuit has a resistance in parallel with an capacitor. The impedances are 4.7 kΩ and $-j3.3$ kΩ. If the applied voltage is 200 mV at an angle of zero, what is the total current?

10-12. The ac generator of Fig. 10-29 has a frequency of 200 kHz. What is the voltage across the capacitor?

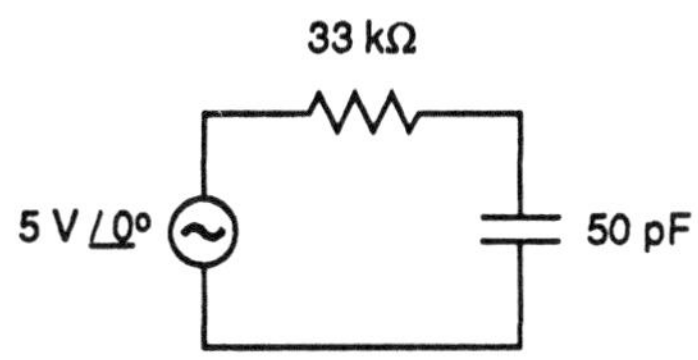

Fig. 10-29.

Sec. 10-6: Series-to-parallel Conversion

10-13. A series *RL* circuit has a total impedance of 2 kΩ + $j10$ kΩ. What is the equivalent parallel circuit?

10-14. A series *RC* circuit has a total impedance of 32 kΩ − $j86$ kΩ. What is the equivalent parallel circuit?

10-15. In Fig. 10-28, the frequency is 5 kHz. What is the parallel equivalent circuit at this frequency?

10-16. In Fig. 10-29, the frequency is 200 kHz. What is the parallel equivalent circuit at this frequency?

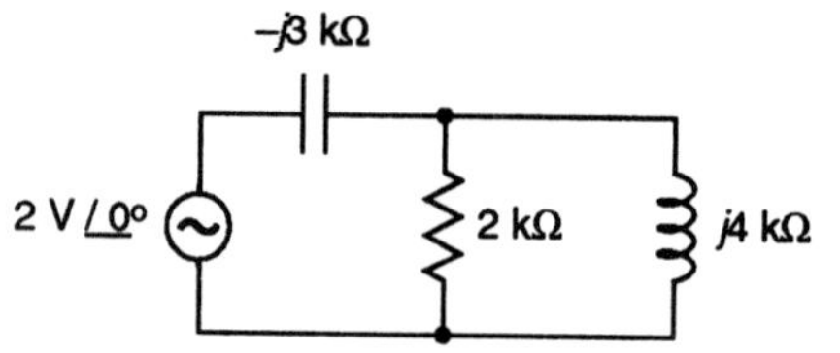

Fig. 10-30.

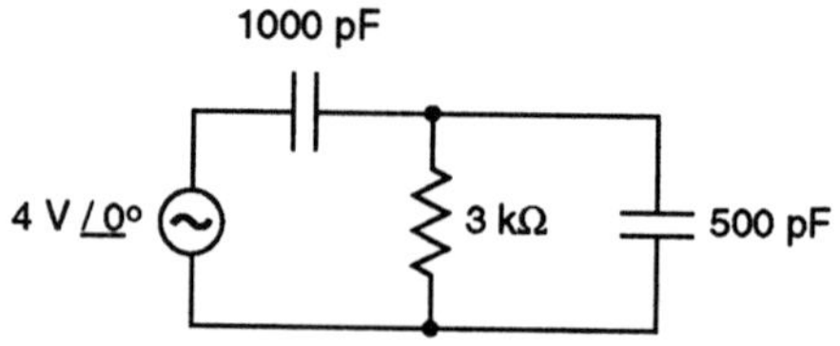

Fig. 10-31.

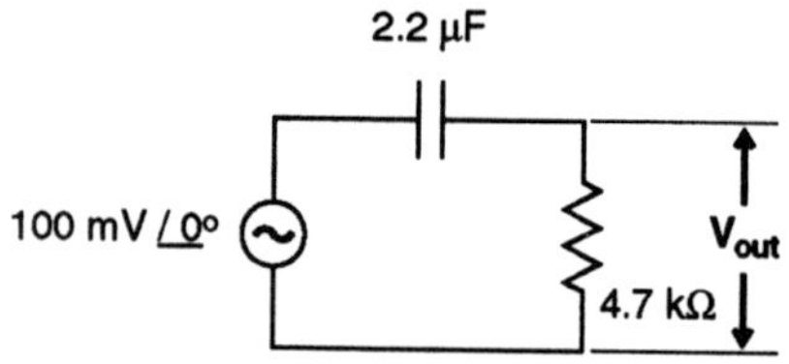

Fig. 10-32.

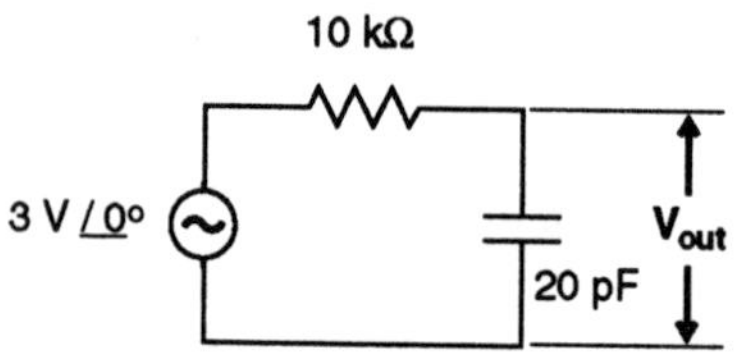

Fig. 10-33.

Sec. 10-7: Parallel-to-series Conversion

10-17. An *RL* circuit has 1 kΩ in parallel with *j*5 kΩ. What is the equivalent series circuit?

10-18. A parallel *RC* circuit has 12 kΩ in parallel with –*j*8 kΩ. What is the equivalent series circuit?

Sec. 10-8: Series-parallel Circuits

10-19. What is the total current in Fig. 10-30? The voltage across each component?

10-20. The frequency doubles in Fig. 10-30. What is the total current?

10-21. In Fig. 10-31, the ac generator has a frequency of 75 kHz. What is the total current?

10-22. A circuit has 2 kΩ + *j*3 kΩ in parallel with 1 kΩ – *j*2 kΩ. What is the total current if the applied voltage is 10 V?

Sec. 10-9. Coupling Circuit

10-23. What is the critical frequency in Fig. 10-32? The output phasor voltage at this frequency?

10-24. In Fig. 10-32, what is the output phasor voltage when the frequency is 10 times the critical frequency? When the frequency is one-tenth of the critical frequency?

Sec. 10-10. Bypass Circuit

10-25. What is the critical frequency in Fig. 10-33? The output phasor voltage at this frequency?

10-26. In Fig. 10-33, what is the output phasor voltage when the frequency is 10 times the critical frequency? When the frequency is one-tenth of the critical frequency?

Sec. 10-11. Lead-lag Network

10-27. A lead-lag circuit has a resistance of 3.3 kΩ and a capacitance of 470 pF. What is the critical frequency?

10-28. A lead-lag circuit has a resistance of 68 kΩ and a capacitance of 10,000 pF. What is the critical frequency?

CHAPTER 11
RESONANCE

Resonance occurs when the inductive reactance equals the capacitive reactance in a series or parallel *RLC* circuit. The two opposite types of reactance cancel, leaving a purely resistive equivalent circuit. Because resonance occurs at only one frequency, resonant circuits can filter out one frequency from many others that may be present. This is how a radio or TV receiver is able to tune to one particular station, even though the antenna simultaneously picks up all signals.

11-1. SERIES RESONANCE

Figure 11-1 shows a series *RLC* circuit. The applied voltage *V* produces a current *I*. At very low frequencies, the capacitive reactance is much greater than the inductive reactance, and the circuit acts like a series *RC* circuit. At very high frequencies, the inductive reactance is much greater than the capacitive reactance, and the circuit acts like a series *RL* circuit. In between, something unusual happens.

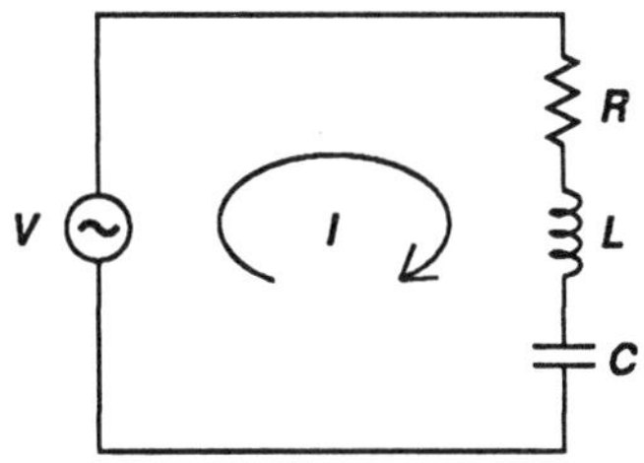

Fig. 11-1.

Resonance

There is an intermediate frequency where the inductive reactance cancels the capacitive reactance. At this frequency, the total impedance is purely resistive and the current is maximum. This condition is called *resonance.* The phenomenon of resonance makes it possible for a radio to tune in a specific station and for a TV receiver to select a particular channel.

Impedance

The total impedance in the series *RLC* circuit of Fig. 11-1 is

$$\mathbf{Z} = R + jX_L - jX_C$$

or

$$\mathbf{Z} = R + j(X_L - X_C) \qquad (11\text{-}1)$$

If we graph both reactances on the same set of axes, we get Fig. 11-2. At the intersection of the two graphs, $X_L = X_C$. The frequency at the point of intersection is called the *resonant frequency* (symbolized f_r).

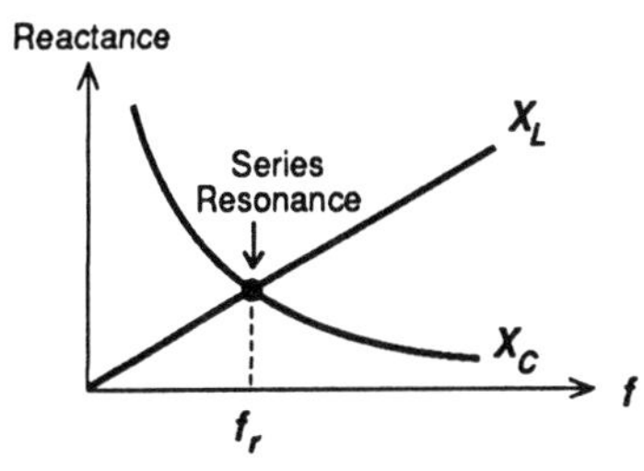

Fig. 11-2.

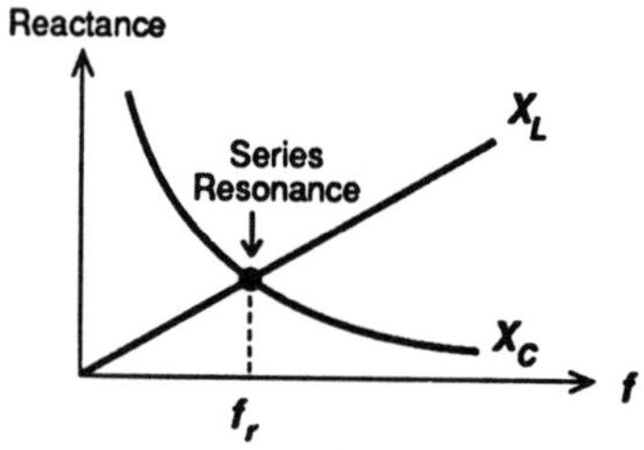

Fig. 11-2.

The current in a series *RLC* circuit is given by

$$\mathbf{I} = \frac{\mathbf{V}}{\mathbf{Z}} = \frac{\mathbf{V}}{R + j(X_L - X_C)}$$

At the resonant frequency, the imaginary component of the impedance is zero because X_L equals X_C. This means the impedance has only a real component. Because of this, the magnitude or rms current at resonance is maximum, given by

$$I_{max} = \frac{V}{R} \qquad (11\text{-}2)$$

In other words, the equivalent circuit is purely resistive because the effects of the inductor and capacitor cancel out. Since the resistance is the only opposition to flow, the current is maximum as shown in Fig. 11-3.

The magnitude or rms value of the current at any frequency is given by

$$I = \frac{V}{\sqrt{R^2 + (X_L - X_C)^2}}$$

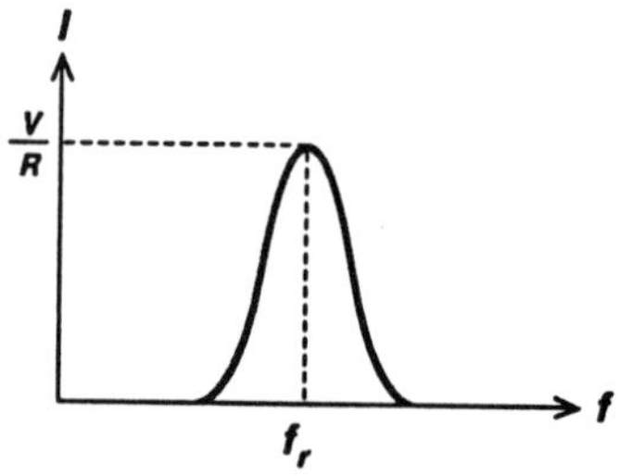

Fig. 11-3.

When the frequency is less than or greater than f_r, the denominator is greater than R and the current is less than maximum. At very low and very high frequencies, the current in a series *RLC* circuit approaches zero as shown in Fig. 11-3.

Incidentally, the graph of Fig. 11-3 is called the *resonance curve* for a series *RLC* circuit. As you see, the current reaches a maximum value at the resonant frequency. The current drops off on either side of the resonant frequency. The rate of drop off depends on the Q of the circuit, which is discussed later.

Series Resonant Frequency

We can derive a useful formula for the series resonant frequency as follows. At the resonant frequency, the inductive reactance equals the capacitive reactance:

$$X_L = X_C$$

which can be written as

$$2\pi f_r L = \frac{1}{2\pi f_r C}$$

By solving for f_r, we get

$$f_r = \frac{1}{2\pi\sqrt{LC}} \qquad (11\text{-}3)$$

MEMORIZE

$$f_r = \frac{1}{2\pi\sqrt{LC}}$$

At this frequency the inductive and capacitive effects cancel, leaving only a purely resistive effect. In other words, at the resonant frequency the equivalent circuit is an ac generator across a resistance.

EXAMPLE 11-1

What is the resonant frequency in Fig. 11-4a?

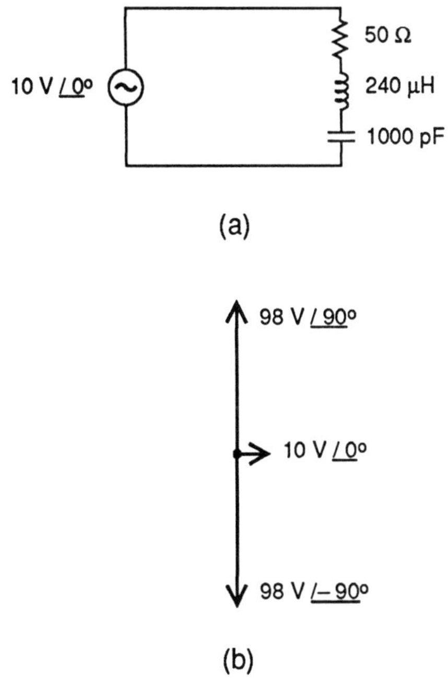

Fig. 11-4.

SOLUTION

$$f_r = \frac{1}{2\pi\sqrt{(240\ \mu\text{H})(1000\ \text{pF})}} = 325\ \text{kHz}$$

As a check on the work, the inductive and capacitive reactances should be equal at 325 kHz:

$$X_L = 2\pi fL = 2\pi(325\ \text{kHz})(240\ \mu\text{H}) = 490\ \Omega$$

and

$$X_C = \frac{1}{2\pi fC} = \frac{1}{2\pi(325\ \text{kHz})(1000\ \text{pF})} = 490\ \Omega$$

As you see, the reactances are equal at the resonant frequency.

EXAMPLE 11-2

What is the total phasor current in Fig. 11-4a at the resonant frequency? The phasor voltage across each component?

SOLUTION

The total impedance is

$$\mathbf{Z} = R + j(X_L - X_C)$$

At resonance, the inductive reactance equals the capacitive reactance, and the total impedance equals the resistance:

$$\mathbf{Z} = 50\ \Omega$$

The current at resonance equals the applied voltage divided by the total impedance:

$$\mathbf{I} = \frac{10\ \text{V}\ \angle 0^\circ}{50\ \Omega\ \angle 0^\circ} = 200\ \text{mA}\ \angle 0^\circ$$

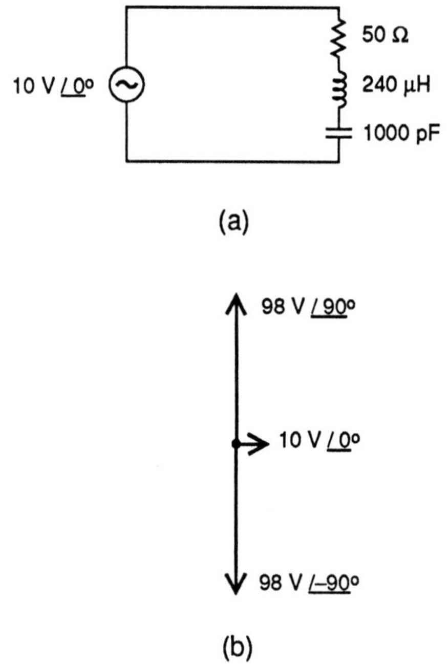

Fig. 11-4.

To calculate the phasor voltage across each component, multiply the total current by the impedance of each component as follows:

$\mathbf{V_R} = (200\text{ mA}\angle 0°)(50\ \Omega\angle 0°) = 10\text{ V}\angle 0°$

$\mathbf{V_L} = (200\text{ mA}\angle 0°)(490\ \Omega\angle 90°) = 98\text{ V}\angle 90°$

$\mathbf{V_C} = (200\text{ mA}\angle 0°)(490\ \Omega\angle -90°) = 98\text{ V}\angle -90°$

Figure 11-4b shows the phasor diagram. As shown, the inductive voltage has a magnitude of 98 V and a phase angle of 90°, and the capacitive voltage has a magnitude of 98 V and a phase angle of −90°. The resistive voltage has a magnitude of 10 V and a phase angle of 0°, the same as the applied voltage. This phasor diagram emphasizes how inductive and capacitive effects cancel each other, leaving only the resistive effect.

EXAMPLE 11-3

The capacitance of Fig. 11-4a doubles. What happens to the resonant frequency?

SOLUTION

The resonant frequency is

$$f_r = \frac{1}{2\pi\sqrt{(240\ \mu\text{H})(2000\text{ pF})}} = 230\text{ kHz}$$

As you see, doubling the capacitance does not reduce the frequency by a factor of 2. Because of the square root, the frequency is reduced by a factor of only 1.414.

11-2. MORE ABOUT SERIES RESONANCE

In this section, you will learn about the bandwidth of a series resonant circuit. This is quite important in communication circuits because the bandwidth indicates how well a radio or TV receiver tunes out the unwanted signals.

Critical Frequencies

Equation (11-1) is the formula for the impedance of a series *RLC* circuit:

$$\mathbf{Z} = R + j(X_L - X_C)$$

which has a magnitude of

$$Z = \sqrt{R^2 + (X_L - X_C)^2} \qquad (11\text{-}4)$$

and an angle of

$$\phi = \arctan\frac{X_L - X_C}{R} \qquad (11\text{-}5)$$

Both formulas contain the net reactance, $X_L - X_C$.

The critical frequencies of a series *RLC* circuit are the frequencies where the net reactance equals the resistance. There are two critical frequencies. The first critical frequency f_1 occurs when

$$R = X_C - X_L$$

In this case, Eq. (11-4) gives

$$Z = \sqrt{R^2 + (X_L - X_C)^2} = 1.414R$$

and the current is

$$I = \frac{V}{1.414R} = 0.707I_{max}$$

The second critical frequency f_2 occurs when

$$R = X_L - X_C$$

For this condition, Eq. (11-4) gives

$$Z = \sqrt{R^2 + (X_L - X_C)^2} = 1.414R$$

and the current has the same value as before:

$$I = \frac{V}{1.414R} = 0.707I_{max}$$

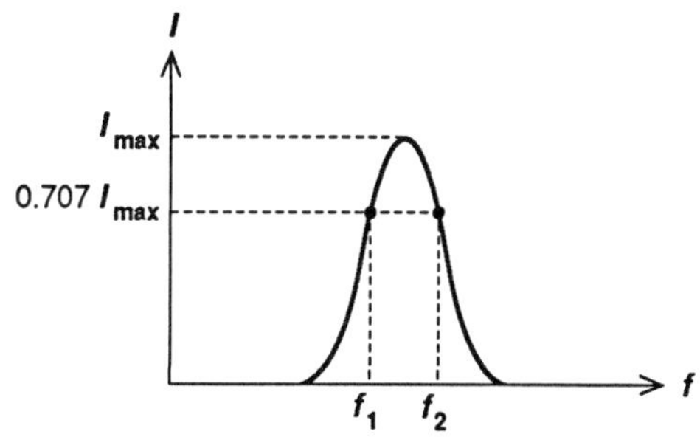

Fig. 11-5.

Figure 11-5 shows a graph of current versus frequency in a series *RLC* circuit. Notice how the current decreases to $0.707I_{max}$ at the two critical frequencies: f_1 and f_2. The points at the critical frequencies are called the *half-power* points. The reason is because the real power at either critical frequency is half of the real power at resonance. Here is the proof. At resonance, the real power is maximum and equals

$$P = (I_{max})^2R$$

At either critical frequency, the current is $0.707I_{max}$ and the real power decreases to

$$P = (0.707I_{max})^2R = 0.5(I_{max})^2R$$

which is half of the real power at resonance.

Bandwidth

The resonant frequency is one of the most important characteristics of a series *RLC* circuit because it is the frequency where the current reaches a maximum value. Another important characteristic of a series *RLC* circuit is its *bandwidth*, defined as the difference between the two critical frequencies. In symbols,

$$B = f_2 - f_1 \tag{11-6}$$

For instance, if $f_1 = 96$ kHz and $f_2 = 104$ kHz, the bandwidth is

$$B = 104\text{ kHz} - 96\text{ kHz} = 8\text{ kHz}$$

The resonant frequency and the bandwidth are like the finger prints of a series *RLC* circuit; they produce a unique resonant curve.

The *Q* of a Series Resonant Circuit

A third important characteristic of a series *RLC* circuit is its *Q*, defined as

$$Q = \frac{X_L}{R} \tag{11-7}$$

where X_L is the inductive reactance at the resonant frequency. Since $X_C = X_L$ at resonance, the *Q* is also given by X_C/R. For instance, if $X_L = 490\ \Omega$ and $R = 50\ \Omega$, then

$$Q = \frac{490\ \Omega}{50\ \Omega} = 9.8$$

Bandwidth and *Q*

By an advanced mathematical derivation, we can show this important relation between bandwidth and *Q*:

$$B = \frac{f_r}{Q} \tag{11-8}$$

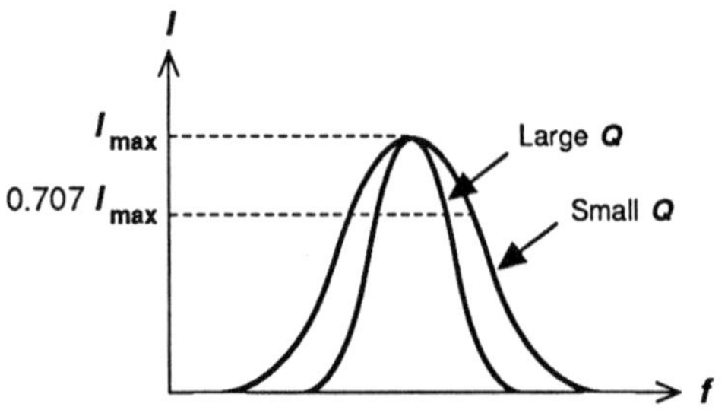

Fig. 11-6.

This is a neat equation. It relates the three most important characteristics of a series *RLC* circuit. Knowing any two of the quantities allows you to calculate the third.

Equation (11-8) says that the bandwidth is inversely proportional to the the *Q* of the circuit. The bigger the *Q*, the smaller the bandwidth. Figure 11-6 shows this idea visually. You can see that a large *Q* produces a small bandwidth.

EXAMPLE 11-4

A series *RLC* circuit has a resonant frequency of 5 MHz. What is the bandwidth for a Q of 10? For a Q of 50?

SOLUTION

When the Q is 10, the bandwidth is

$$B = \frac{f_r}{Q} = \frac{5\text{ MHz}}{10} = 500\text{ kHz}$$

When the Q is 50, the bandwidth is

$$B = \frac{f_r}{Q} = \frac{5\text{ MHz}}{50} = 100\text{ kHz}$$

Notice how the bandwidth shrinks when the Q increases.

EXAMPLE 11-5

In Fig. 11-7, what is the Q of the circuit at resonance? The bandwidth?

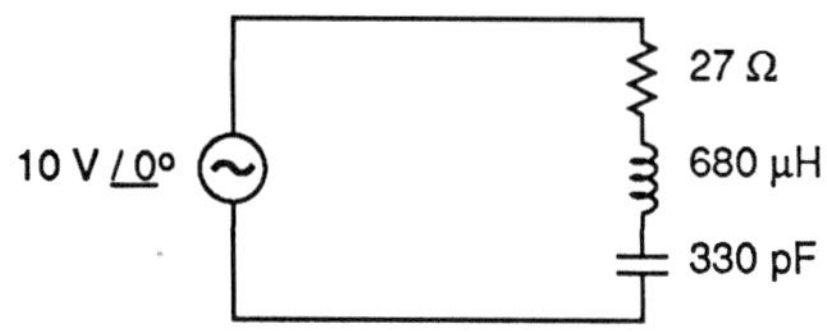

Fig. 11-7.

SOLUTION

First, calculate the resonant frequency:

$$f_r = \frac{1}{2\pi\sqrt{(680\ \mu\text{H})(330\text{ pF})}} = 336\text{ kHz}$$

Second, calculate the inductive reactance at resonance:

$$X_L = 2\pi(336\text{ kHz})(680\ \mu\text{H}) = 1.44\text{ k}\Omega$$

Now, you can calculate Q and the B:

$$Q = \frac{X_L}{R} = \frac{1.44\text{ k}\Omega}{27\ \Omega} = 53.3$$

and

$$B = \frac{f_r}{Q} = \frac{336\text{ kHz}}{53.3} = 6.3\text{ kHz}$$

EXAMPLE 11-6

A series *RLC* circuit has a resonant frequency of 455 kHz. If the bandwidth is 10 kHz, what is the Q?

SOLUTION

You can rearrange Eq. (11-8) and calculate a Q of

$$Q = \frac{f_r}{B} = \frac{455 \text{ kHz}}{10 \text{ kHz}} = 45.5$$

11-3. THE *Q* RISE IN REACTIVE VOLTAGES

Something unusual happens in a series resonant circuit like Fig. 11-8. When the Q is greater than one, the voltage across each reactive component is larger than the applied voltage. This phenomenon is known as the Q rise in reactive voltage. Here is the proof. At resonance, all of the applied voltage appears across the resistor. Therefore, the magnitude of current is

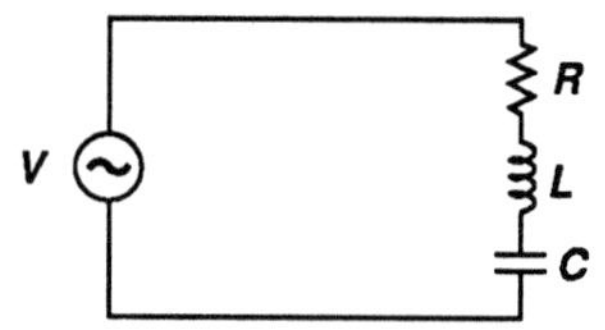

Fig. 11-8.

$$I = \frac{V}{R}$$

The magnitude of voltage across the inductor is

$$V_L = IX_L = \frac{V}{R} X_L$$

or

$$V_L = QV \tag{11-9}$$

This says the inductive voltage equals the Q of the circuit times the applied voltage.

Similarly, the capacitive reactance equals the inductive reactance at resonance. Therefore, the capacitor of Fig. 11-8 also experiences the Q rise in voltage because it has a voltage given by

$$V_C = IX_C = \frac{V}{R} X_C$$

or

$$V_C = QV \tag{11-10}$$

This capacitive voltage is 180° out of phase with the inductive voltage, which means the phasor sum of the two reactive voltages is zero.

EXAMPLE 11-7

In Fig. 11-9a, what is the inductive voltage at resonance? The capacitive voltage?

SOLUTION

First, calculate the resonant frequency:

$$f_r = \frac{1}{2\pi\sqrt{(20\ \mu\text{H})(500\ \text{pF})}} = 1.59\ \text{MHz}$$

Second, calculate the inductive reactance at resonance:

$$X_L = 2\pi(1.59\ \text{MHz})(20\ \mu\text{H}) = 200\ \Omega$$

Now, you can calculate Q and the B:

$$Q = \frac{X_L}{R} = \frac{200\ \Omega}{10\ \Omega} = 20$$

Since the applied voltage has a magnitude of 10 mV, the inductive voltage at resonance is

$$V_L = QV = 20(10\ \text{mV}) = 200\ \text{mV}$$

The capacitive voltage also equals 200 mV but it has the opposite phase angle as shown in Fig. 11-9b.

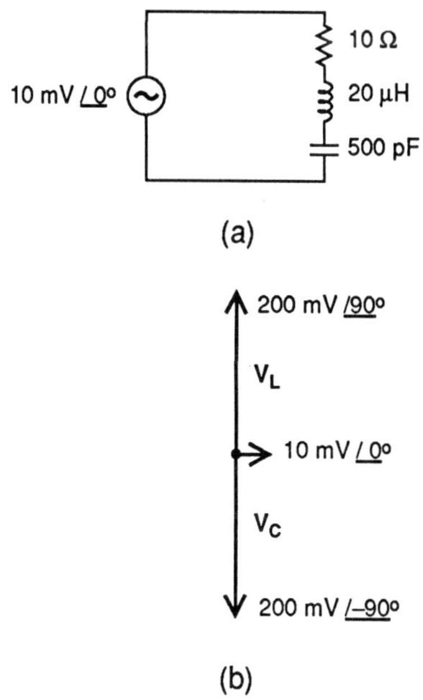

Fig. 11-9.

11-4. IDEAL PARALLEL RESONANCE

Figure 11-10a shows a parallel RLC circuit driven by a current source. In practical electronic circuits, the current source is usually a transistor. Since it is a parallel circuit, the same voltage appears across all branches.

Resonant Frequency

At low frequencies the capacitor of Fig. 11-10a appears open, and the equivalent circuit is a parallel RL circuit as shown in Fig. 11-10b. On the other hand, at high frequencies the inductor appears open, and the equivalent circuit is a parallel RC circuit like Fig. 11-10c. Between the extremes of low and high frequencies, there is a resonant frequency where the circuit acts purely resistive like Fig. 11-10d. When the inductor and capacitor are ideal, resonance occurs for this condition:

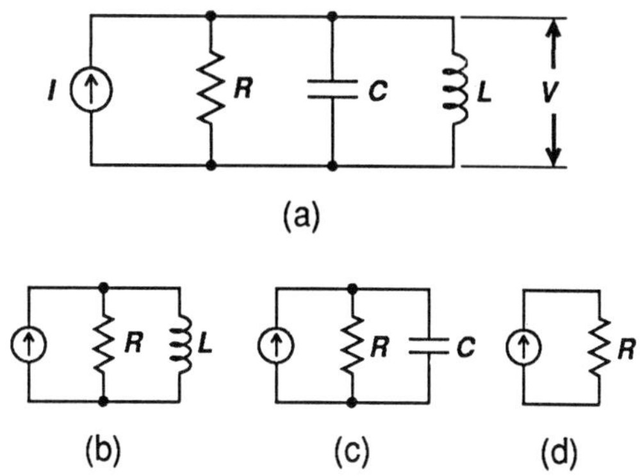

Fig. 11-10.

$$X_L = X_C$$

or

$$2\pi f_r L = \frac{1}{2\pi f_r C}$$

Solving for the resonant frequency, we get the same equation as before:

$$f_r = \frac{1}{2\pi\sqrt{LC}} \qquad (11\text{-}11)$$

Remember that this equation for the parallel resonant frequency assumes that all components are ideal. This means the resistor is purely resistive, the inductor is purely inductive, and the capacitor is purely capacitive.

Total Impedance

In Fig. 11-10a, the total impedance is

$$\mathbf{Z} = \frac{1}{1/R + 1/jX_L - 1/jX_C}$$

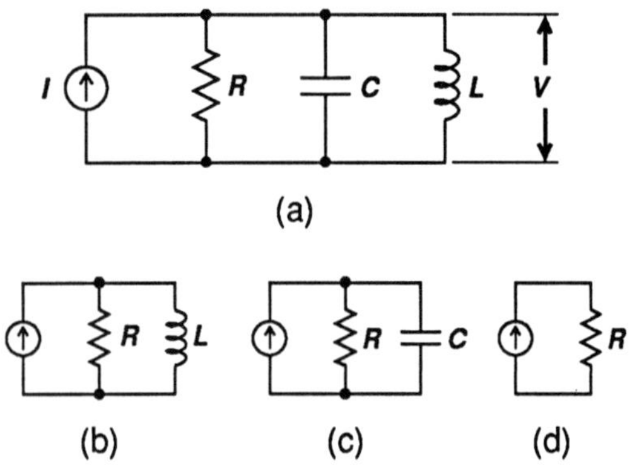

Fig. 11-10.

At resonance, the inductive and capacitive effects cancel, leaving a total impedance of

$$\mathbf{Z} = R\,\angle 0^\circ$$

This has a magnitude of R and an angle of 0°. Therefore, the voltage across the circuit at resonance is

$$V = IR \qquad (11\text{-}12)$$

Figure 11-11 shows the *resonance curve* (the graph of voltage versus frequency) for a parallel *RLC* circuit driven by a current source. As shown, the voltage reaches a maximum value at the resonant frequency. Below and above the resonant frequency, the voltage drops off. At very low and very high frequencies, the voltage approaches zero.

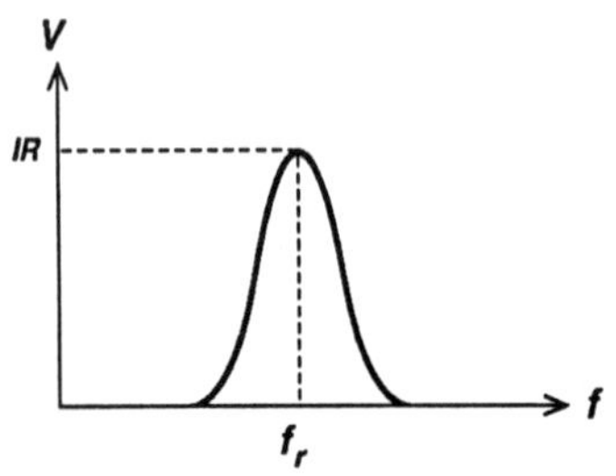

Fig. 11-11.

Effects of X_L and X_C Cancel

Why does the inductive reactance cancel the capacitive reactance in a parallel resonant circuit? We can use the product over the sum to see this as follows. The impedance of an inductor and capacitor in parallel is found with the product over the sum:

$$\mathbf{Z} = \frac{(jX_L)(-jX_C)}{jX_L - jX_C} = \frac{-j^2 X_L X_C}{j(X_L - X_C)} = -j\,\frac{X_L X_C}{X_L - X_C}$$

At the resonant frequency, $X_L = X_C$ and the denominator becomes zero, which implies infinite impedance for the parallel LC combination. Since no current can flow into an infinite impedance, the inductor and capacitor drops out of the equivalent circuit at resonance.

Here is another way to understand why an inductor and capacitor eliminate each other at resonance. The currents are equal in magnitude but opposite in angle. Therefore, the phasor sum of the inductive and capacitive currents is zero. Since the total current equals the resistive current at resonance, the parallel RLC circuit acts purely resistive.

EXAMPLE 11-8

In Fig. 11-12a, what is the resonant frequency? The voltage across the circuit at resonance? The branch currents at resonance?

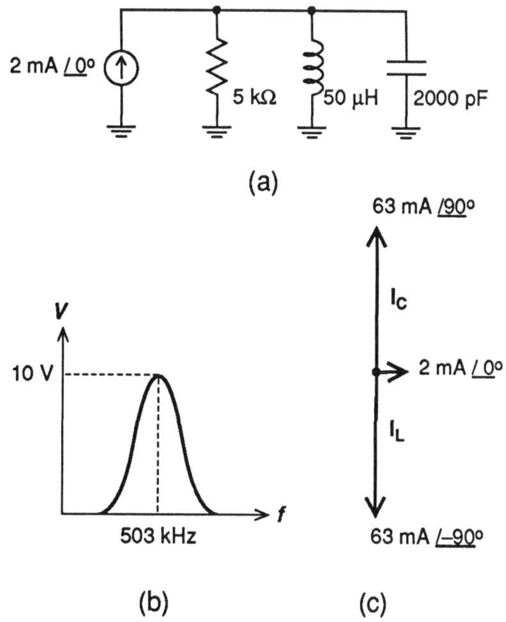

Fig. 11-12.

SOLUTION

$$f_r = \frac{1}{2\pi\sqrt{(50\ \mu\text{H})(2000\ \text{pF})}} = 503\ \text{kHz}$$

At this frequency the parallel RLC circuit appears purely resistive to the current source. Because of this, all of the source current passes through the 5 kΩ, producing a phasor voltage of

$$\mathbf{V} = \mathbf{IZ} = (2\ \text{mA}\ \angle 0^\circ)(5\ \text{k}\Omega) = 10\ \text{V}\ \angle 0^\circ$$

The magnitude of the voltage across the parallel resonant circuit equals 10 V. Figure 11-12b shows the resonance curve.

Since we have the voltage across the parallel RLC circuit, we can calculate the current into each reactance. At the resonant frequency, the inductive reactance is

$$X_L = 2\pi(503\ \text{kHz})(50\ \mu\text{H}) = 158\ \Omega$$

The capacitive reactance has the same value. The inductive current is

$$\mathbf{I}_L = \frac{10\ \text{V}\ \angle 0^\circ}{158\ \Omega\ \angle 90^\circ} = 63.3\ \text{mA}\ \angle -90^\circ$$

The capacitive current is

$$\mathbf{I}_C = \frac{10\ \text{V}\ \angle 0^\circ}{158\ \Omega\ \angle -90^\circ} = 63.3\ \text{mA}\ \angle 90^\circ$$

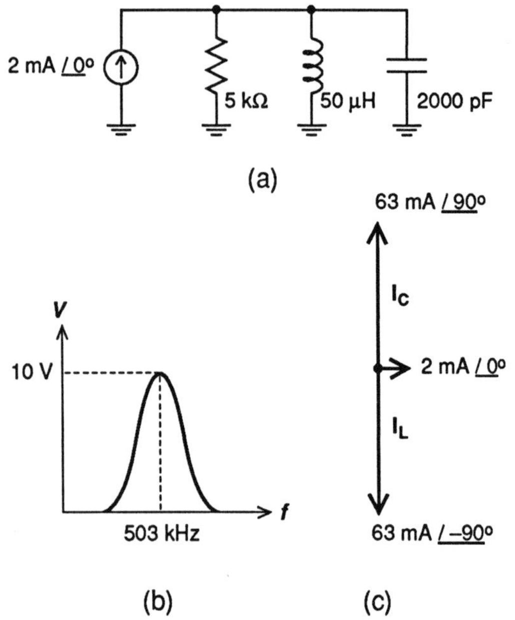

Fig. 11-12.

Figure 11-12c summarizes the currents. As you see, the resistive current is 2 mA at an angle of 0°, the inductive current is 63.3 mA at an angle of −90°, and the capacitive current is 63.3 mA at an angle of +90°. Since the reactive phasor currents are 180° out of phase, their phasor sum is zero. Therefore, the circuit is purely resistive at resonance, and the total current equals 2 mA at an angle of 0°.

11-5. PARALLEL Q AND BANDWIDTH

The quality factor or Q of a parallel resonant circuit is defined as the ratio of resistance to inductive reactance:

$$Q = \frac{R}{X_L} \tag{11-13}$$

This formula uses X_L, but X_C will work just as well because the calculation for Q is done at the resonant frequency.

The formula for bandwidth is the same as before:

$$B = \frac{f_r}{Q} \tag{11-14}$$

The bandwidth is the difference between the two critical frequencies, which are the frequencies at the half power points (see Fig. 11-13). Any operating frequency between f_1 and f_2 produces a voltage that is equal to or greater than 0.707 times the maximum voltage.

Q is Usually Large to Get Small Bandwidth

Equation (11-13) brings out an important point about parallel resonant circuits. To have a Q greater than unity, you need to use an R that is greater than X_L. Typically, electronic circuits have a Q that is greater than 10, which means the parallel resistance has to be more than 10 times the inductive reactance. This is exactly the opposite of a series resonant circuit where the series resistance has to be less than one-tenth of the inductive reactance to get a Q that is greater than 10.

Why do typical electronic circuits have Q's that are greater than 10? Because this produces a small bandwidth, less than one-tenth of the resonant frequency. Since resonant circuits are used to tune in radio and TV stations, the bandwidth has to be small to block unwanted signals. You can see this by looking at Fig. 11-13. When the bandwidth is small, f_1 and f_2 are close to the resonant frequency. This means unwanted signals, those outside the range of f_1 to f_2, produce much less voltage across the circuit. You will learn more about tuning if you study communication circuits. For now, all you need to know is that the Q of most resonant circuits is usually high, greater than 10.

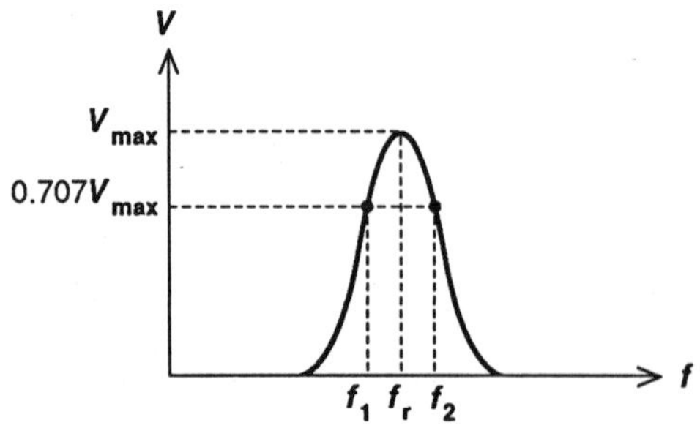

Fig. 11-13.

Critical Frequencies

When the Q is greater than 10, the critical frequencies can be calculated as follows. Divide the bandwidth by 2, then add and subtract this value from the resonant frequency to find the critical frequencies. In symbols,

$$f_1 = f_r - \frac{B}{2} \qquad (11\text{-}15)$$

and

$$f_2 = f_r + \frac{B}{2} \qquad (11\text{-}16)$$

These equations make the following concept clear: the smaller the bandwidth, the closer the critical frequencies are to the resonant frequency.

EXAMPLE 11-9

A parallel *RLC* circuit has a resonant frequency of 1.5 MHz. What is the bandwidth when the Q is 25? When the Q is 75?

SOLUTION

When the Q is 25, the bandwidth is

$$B = \frac{f_r}{Q} = \frac{1.5\text{ MHz}}{25} = 60\text{ kHz}$$

When the Q is 75, the bandwidth is

$$B = \frac{f_r}{Q} = \frac{1.5\text{ MHz}}{75} = 20\text{ kHz}$$

Notice how the bandwidth decreases when the Q increases.

EXAMPLE 11-10

In Fig. 11-14, what is the bandwidth?

SOLUTION

First, calculate the resonant frequency:

$$f_r = \frac{1}{2\pi\sqrt{(25\ \mu\text{H})(5000\text{ pF})}} = 450\text{ kHz}$$

Second, calculate the inductive reactance at resonance:

$$X_L = 2\pi(450\text{ kHz})(25\ \mu\text{H}) = 70.7\ \Omega$$

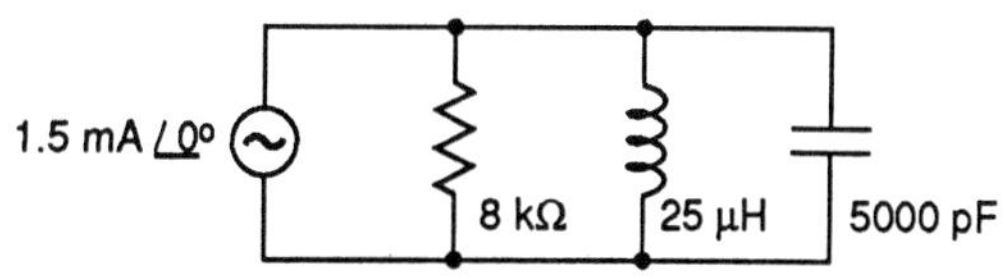

Fig. 11-14.

Now, you can calculate Q and the B:

$$Q = \frac{R}{X_L} = \frac{8\ \text{k}\Omega}{70.7\ \Omega} = 113$$

and

$$B = \frac{f_r}{Q} = \frac{450\ \text{kHz}}{113} = 4\ \text{kHz}$$

EXAMPLE 11-11

In Fig. 11-14, what are the critical frequencies?

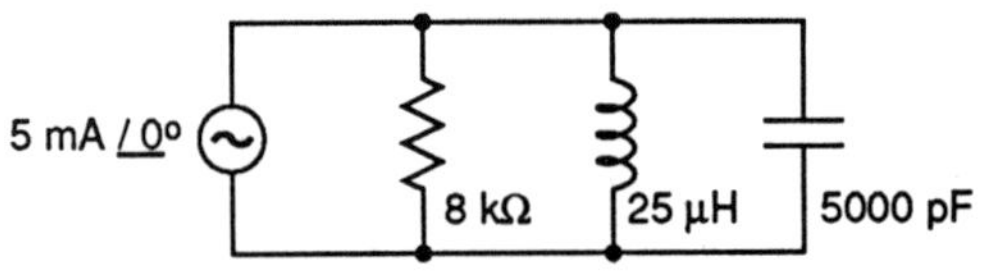

Fig. 11-14.

SOLUTION

In the previous example, we calculated a resonant frequency of 450 kHz and a bandwidth of 4 kHz. The two critical frequencies are

$$f_1 = f_r - \frac{B}{2} = 450\ \text{kHz} - \frac{4\ \text{kHz}}{2} = 448\ \text{kHz}$$

and

$$f_2 = f_r + \frac{B}{2} = 450\ \text{kHz} + \frac{4\ \text{kHz}}{2} = 452\ \text{kHz}$$

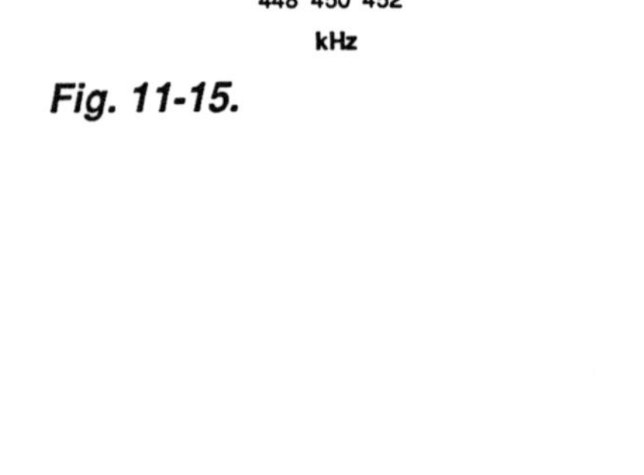

Fig. 11-15.

Figure 11-15 shows the resonance curve. As you can see, the critical frequencies are very close to the resonant frequency.

11-6. THE Q RISE IN REACTIVE CURRENTS

In a parallel resonant circuit like Fig. 11-16a, the magnitude of the inductive current is

$$I_L = \frac{V}{X_L} = \frac{IR}{X_L}$$

or

$$I_L = QI \qquad (11\text{-}17)$$

This says the inductive current equals the Q of the circuit times the source current.

Similarly, the capacitive reactance equals the inductive reactance at resonance. Therefore, the capacitor of Fig. 11-16a also experiences the Q rise in current because it has a current given by

$$I_C = \frac{V}{X_C} = \frac{IR}{X_C}$$

or

$$I_C = QI \qquad (11\text{-}18)$$

As shown in Fig. 11-16b, the capacitive current is 180° out of phase with the inductive current, which means the phasor sum of the two reactive currents is zero.

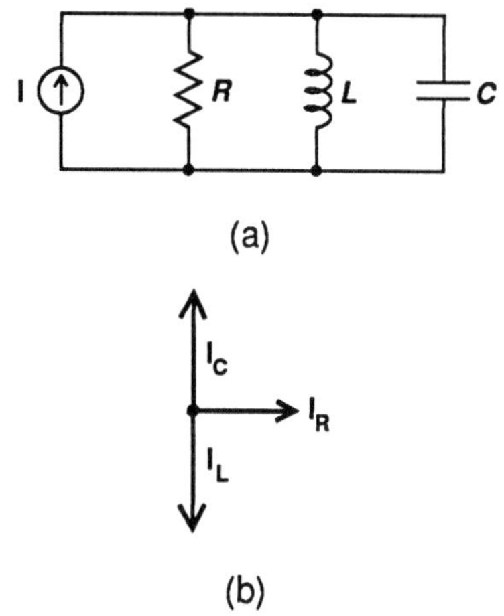

Fig. 11-16.

EXAMPLE 11-12

In Fig. 11-17a, what is the inductive current at resonance? The capacitive current?

SOLUTION

First, calculate the resonant frequency:

$$f_r = \frac{1}{2\pi\sqrt{(20\ \mu\text{H})(500\ \text{pF})}} = 1.59\ \text{MHz}$$

Second, calculate the inductive reactance at resonance:

$$X_L = 2\pi(1.59\ \text{MHz})(20\ \mu\text{H}) = 200\ \Omega$$

Now, you can calculate Q and the B:

$$Q = \frac{R}{X_L} = \frac{10\ \text{k}\Omega}{200\ \Omega} = 50$$

Since the source current has a magnitude of 2 mA, the inductive current at resonance is

$$I_L = QI = 50(2\ \text{mA}) = 100\ \text{mA}$$

The capacitive current also equals 100 mA but it has the opposite phase angle as shown in Fig. 11-17b.

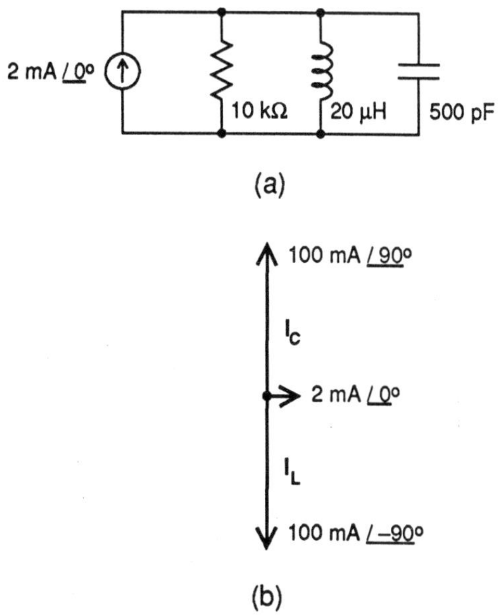

Fig. 11-17.

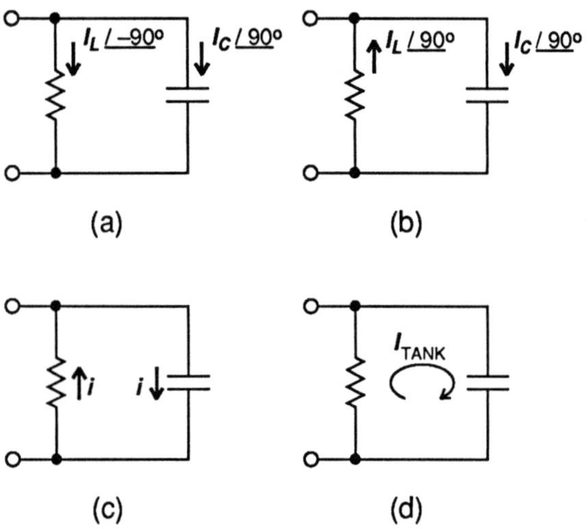

Fig. 11-18.

11-7. CIRCULATING OR TANK CURRENT

Figure 11-18a shows the phasor currents in the inductor and capacitor. These phasor currents represent sinusoidal currents with phase angles of −90° and +90°. Mathematically, we can reverse the direction of the inductive current and change its angle by 180° as shown in Fig. 11-18b. This is valid because reversing a current is equivalent to adding 180° to its polar value. Each of the reactive currents now represents a sinusoidal current with a phase angle of 90°.

Loop Current

In other words, we can visualize the instantaneous currents as shown in Fig. 11-18c. This emphasizes the idea that we are visualizing each current as sinusoidal. This means the currents increase to a maximum value, decrease to zero, reverse directions, increase to a maximum in the opposite direction, and so on.

Because the reactive currents are equal at resonance, you can draw a single *loop current* as shown in Fig. 11-18d. The idea is similar to what we did in dc circuits when we used loop currents with Kirchhoff's voltage law. This loop current is sinusoidal, so we can visualize it in one direction during the positive half cycle and in the opposite direction during the negative half cycle.

Stored Energy

The magnetic field around an inductor stores energy. With calculus, it can be shown that the stored energy is given by

$$W = \frac{LI^2}{2} \qquad (11\text{-}19)$$

Notice that the energy is directly proportional to the square of the current. If the current is sinusoidal, the inductor has maximum energy when the current is at a peak.

Similarly, the electric field of a capacitor stores energy. With calculus, it can be shown that the stored energy is given by

$$W = \frac{CV^2}{2} \qquad (11\text{-}20)$$

This says the energy is directly proportional to the square of the voltage. If the voltage is sinusoidal, the energy in the capacitor reaches a maximum when the voltage is at a peak.

Energy Transfer

In Fig. 11-18d, the loop current and voltage are sinusoidal quantities that are 90° out of phase. Because of this, the inductor energy is maximum when the capacitor energy is zero, and vice-versa. For instance, as the capacitor charges, the loop current decreases. This implies less energy in the magnetic field, but more energy in the electric field. Stated another way, energy is transferred from the inductor to the capacitor. This transfer continues until the loop current is zero and the voltage is maximum. At this instant, all of the energy is stored in the capacitor and none of it in the inductor.

As the cycle continues, the capacitor discharges and loses some of its energy. At the same time, the current increases in the opposite direction, and the energy increases in the magnetic field. Therefore, energy is transferred from the capacitor to the inductor. This transfer continues until the voltage is zero and the loop current is maximum. At this instant, all of the energy is stored in the inductor and none of it in the capacitor.

Subsequently, the energy is transferred back and forth between the inductor and the capacitor. If both components are ideal, this energy is trapped inside the *LC* part of the resonant circuit because neither component has any resistance to dissipate or convert the energy to heat.

Equation

The storage of energy in the *LC* part of the circuit is comparable to a tank storing water. This why the *LC* part of the circuit is called a *tank* circuit, and the loop current is often referred to as the *tank current*. Another name for the loop current is the *circulating* current because the flow is confined to the closed path formed by the inductor and capacitor. The circulating current has a magnitude equal to the magnitude of either reactive current. Therefore, its formula is

$$I_{\text{tank}} = QI \qquad (11\text{-}21)$$

This phenomenon is known as the Q rise in tank current. In other words, the tank current is Q times the source current.

EXAMPLE 11-13

What is the tank current in Fig. 11-19?

SOLUTION

This is similar to finding the reactive currents. Start by calculating the resonant frequency:

$$f_r = \frac{1}{2\pi\sqrt{(100\ \mu\text{H})(250\ \text{pF})}} = 1\ \text{MHz}$$

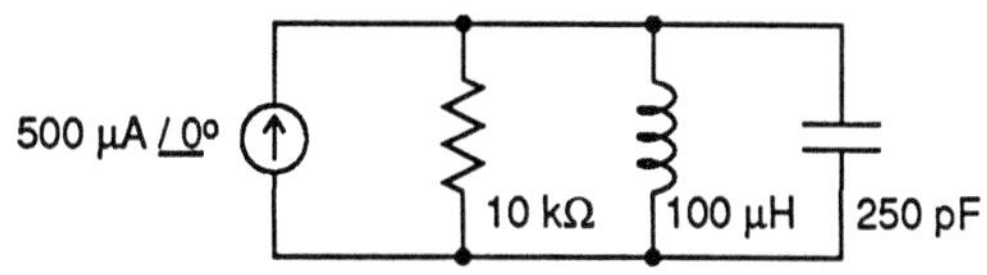

Fig. 11-19.

Next, calculate the inductive reactance at resonance:

$$X_L = 2\pi(1\ \text{MHz})(100\ \mu\text{H}) = 628\ \Omega$$

Now, you can calculate Q and the tank current:

$$Q = \frac{R}{X_L} = \frac{10\ \text{k}\Omega}{628\ \Omega} = 15.9$$

and

$$I_{\text{tank}} = QI = 15.9(500\ \mu\text{A}) = 7.95\ \text{mA}$$

11-8. Q OF AN INDUCTOR

Up to now, we have assumed an ideal inductor or coil. This is one that has no resistance. Every inductor has some resistance, usually small. In many applications, we can ignore this small resistance. But occasionally, we want to take the resistance into account. Here is how it is done.

Equivalent Circuit

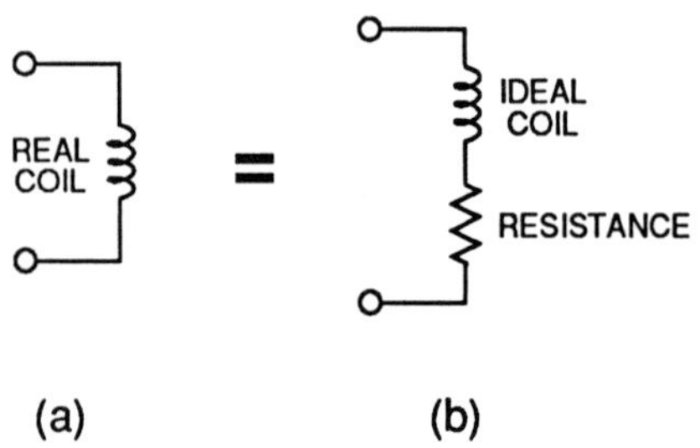

Fig. 11-20.

First, we visualize a real inductor or coil as shown in Fig. 11-20. Since the current through a real inductor has to flow through the inductance and the resistance in series, a real coil is equivalent to an ideal coil in series with the resistance. The Q of an inductor is is defined as

$$Q = \frac{X_L}{R} \tag{11-22}$$

Since an ideal coil has no resistance, its Q is infinite. Therefore, the higher the Q, the better the inductor. At lower frequencies (from 0 to 1 MHz), the resistance of the inductor is constant. Therefore, the Q of an inductor is directly proportional to frequency. The higher the frequency, the higher the Q.

Skin Effect

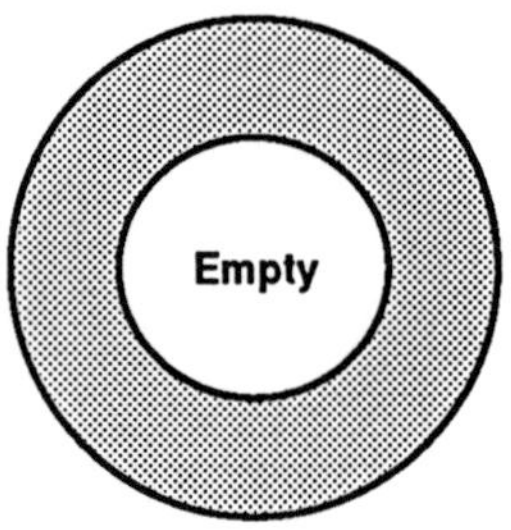

Fig. 11-21.

At higher frequencies (greater than 1 MHz), something unusual happens. The resistance of the coil begins to increase. This phenomenon is called the *skin effect*. It is caused by the self-inductance of a wire. As the frequency increases beyond 1 MHz, electrons tend to flow in the shaded region of the wire shown in Fig. 11-21. This occurs because there is more back emf in the center of the wire. Since the center of the wire no longer is a conducting path in Fig. 11-21, there is less cross-sectional area, which implies more resistance. The higher the frequency, the thinner the conducting layer becomes.

The increase in resistance caused by skin effect depends on the wire diameter, the frequency, and other factors. For instance, engineering books show the following for AWG-30 copper wire. The increase in resistance is approximately 1 percent at 400 kHz, 20 percent at 1 MHz, 200 percent at 10 MHz, and 900 percent at 100 MHz. If a coil of AWG-30 copper wire has 10 Ω of resistance at zero frequency (dc), then it has approximately 10.1 Ω at 400 kHz, 12 Ω at 1 MHz, 30 Ω at 10 MHz, and 100 Ω at 100 MHz.

Because of skin effect above 1 MHz, the Q is not as high as it would be without the skin effect. For this reason, there is a limit on the Q of manufactured inductors. Typical inductors have a Q from 10 to more than 300, depending on the frequency of operation.

EXAMPLE 11-14

An inductor has an inductance of 50 μH and a resistance of 2 Ω. What is its Q at 500 kHz?

SOLUTION

$$Q = \frac{X_L}{R} = \frac{2\pi(500\ \text{kHz})(50\ \mu\text{H})}{2\ \Omega} = \frac{157\ \Omega}{2\ \Omega} = 78.5$$

11-9. RESONANCE WITH NONIDEAL INDUCTOR

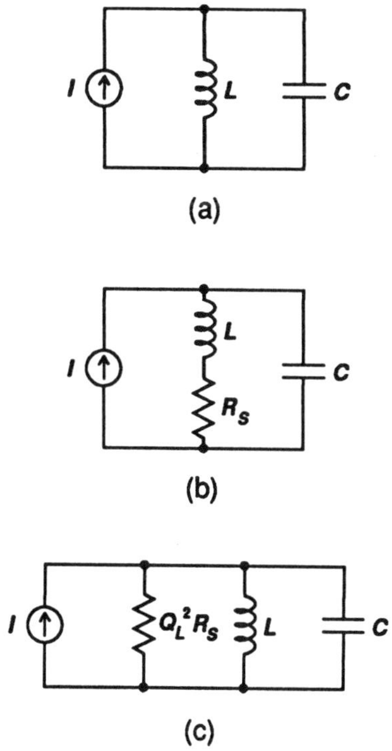

Fig. 11-22.

Figure 11-22a shows a tank circuit with a nonideal inductor whose resistance is R_s. Therefore, the Q of the coil may be written as

$$Q_L = \frac{X_L}{R_s} \qquad (11\text{-}23)$$

A Q_L equal to or greater than 10 is called a *high Q*, while one that is less than 10 is a *low Q*. Figure 11-22b shows the equivalent circuit where the nonideal coil has been replaced by an ideal coil and a series resistance.

In Chap. 10, you learned how to convert a series *RL* section to a parallel *RL* section. Here are the conversion formulas, Eqs. (10-11) and (10-12), rewritten for a nonideal inductor:

$$R' = (1 + Q_L^2)R_s$$

$$X_L' = (1 + 1/Q_L^2)X_L$$

For the high-Q case, Q_L is equal to or greater than 10, which means Q_L^2 is equal to or greater than 100.

When the Q is greater than 10, the conversion formulas simplify to these approximations:

$$R' = Q_L^2 R_s \qquad (11\text{-}24)$$

and

$$X_L' = X_L \qquad (11\text{-}25)$$

Notice that the equivalent parallel inductive reactance equals the original series inductive reactance. This means the parallel inductance equals the series inductance. In other words, the circuit of Fig. 11-22b is converted to the parallel *RLC* circuit of Fig. 11-22c.

We already know how to analyze the circuit. That is, we can calculate its resonant frequency, bandwidth, etc. The resonant frequency is given by

$$f_r = \frac{1}{2\pi\sqrt{LC}} \tag{11-26}$$

The bandwidth is

$$B = \frac{f_r}{Q} \tag{11-27}$$

Similarly, the critical frequencies, tank current, etc. are found by earlier methods.

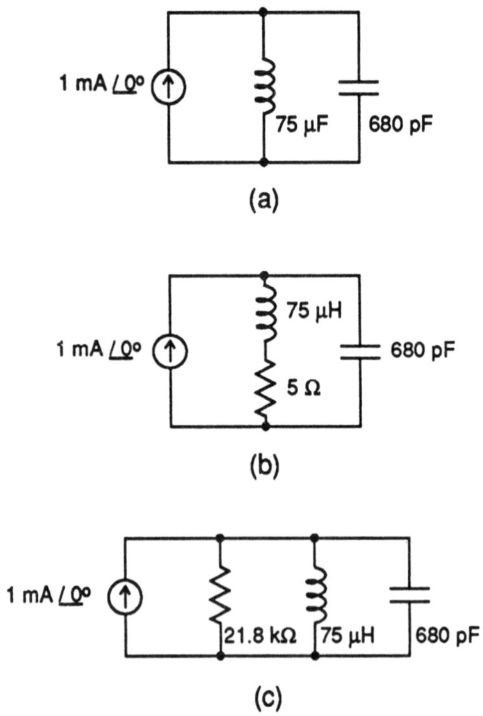

Fig. 11-23.

EXAMPLE 11-15

In Fig. 11-23a, the inductor is nonideal with a Q_L of 66. What is the resonant frequency, the voltage, the bandwidth, and the tank current?

SOLUTION

The resonant frequency is

$$f_r = \frac{1}{2\pi\sqrt{(75\ \mu\text{H})(680\ \text{pF})}} = 705\ \text{kHz}$$

At this frequency, the inductive reactance is

$$X_L = 2\pi f_r L = 2\pi(705\ \text{kHz})(75\ \mu\text{H}) = 332\ \Omega$$

and the R_s of the coil is

$$R_s = \frac{X_L}{Q_L} = \frac{332\ \Omega}{66} = 5\ \Omega$$

This gives an equivalent parallel resistance of

$$R' = Q_L{}^2 R_s = (66)^2(5\ \Omega) = 21.8\ \text{k}\Omega$$

Figure 11-23c shows the equivalent circuit.

At resonance, all of the 1 mA of source current passes through the equivalent parallel resistance to produce a voltage of

$$V = IR = (1\ \text{mA})\ (21.8\ \text{k}\Omega) = 21.8\ \text{V}$$

For the circuit of Fig. 11-23c, the Q of the circuit equals

$$Q = \frac{R}{X_L} = \frac{21.8\ \text{k}\Omega}{332\ \Omega} = 66$$

Notice that the Q of the circuit has the same value as Q_L, the Q of the coil. The bandwidth is

$$B = \frac{f_r}{Q} = \frac{705 \text{ kHz}}{66} = 10.7 \text{ kHz}$$

The tank current is

$$I_{\text{tank}} = QI = 66(1 \text{ mA}) = 66 \text{ mA}$$

11-10. PARALLEL RESONANCE WITH A LOAD

Figure 11-24a shows the type of resonant circuit usually encountered in electronics work. The inductor is nonideal with a Q_L of X_L/R_s. The inductor and the capacitor are the LC tank. There is a separate load resistance of R_L in parallel with the LC tank circuit. Figure 11-24b is the equivalent circuit where the nonideal inductor is replaced by an ideal inductor and a series resistance.

How do you analyze a circuit like this? Here is the easiest way. Assume the high-Q case, which is almost always valid in practical circuits. Then, convert the nonideal inductor into its parallel equivalent section as shown in Fig. 11-24c. The total parallel resistance is given by the product over the sum of the parallel resistances:

$$R = \frac{R_L Q_L^2 R_s}{R_L + Q_L^2 R_s} \tag{11-28}$$

After you have the value of R, the rest of the analysis reverts to the methods previously discussed.

In the previous section, there was no separate load resistor R_L across the LC tank (Fig. 11-22a). When there is no R_L, $Q = Q_L$ because the only power losses occur in the inductor. When a load resistance R_L is connected across an LC tank (Fig. 11-24a), the Q of the circuit is less than Q_L because of the additional power loss in R_L.

The Q of the circuit (same as Q_T, the total Q) equals the total parallel resistance divided by the inductive reactance at resonance:

$$Q = \frac{R}{X_L}$$

This Q is less than Q_L when there is a separate load resistance of R_L.

The main points to remember are these. Given a high-Q coil, convert the nonideal inductor into its parallel equivalent section. Then, combine the load resistance and the equivalent parallel resistance of the inductor into a single parallel resistance. Use this resistance to calculate the maximum voltage and the total Q. The remaining calculations are straightforward.

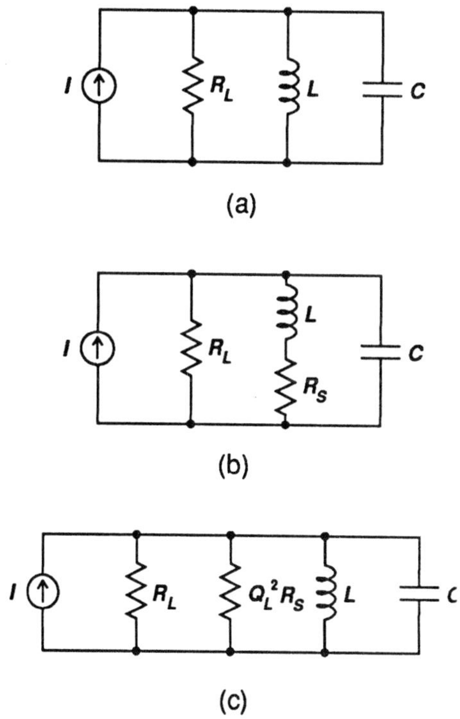

Fig. 11-24.

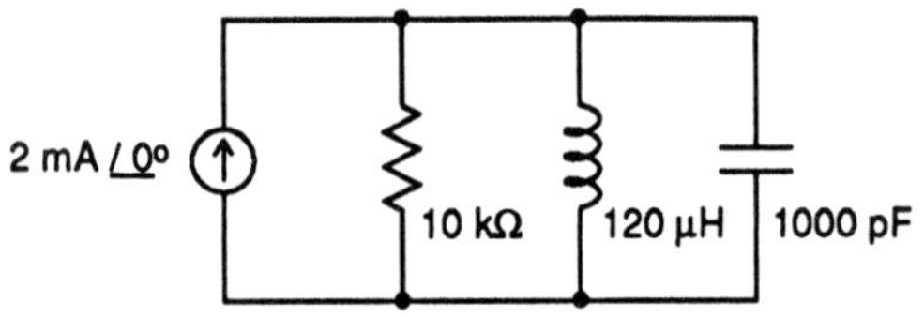

(a)

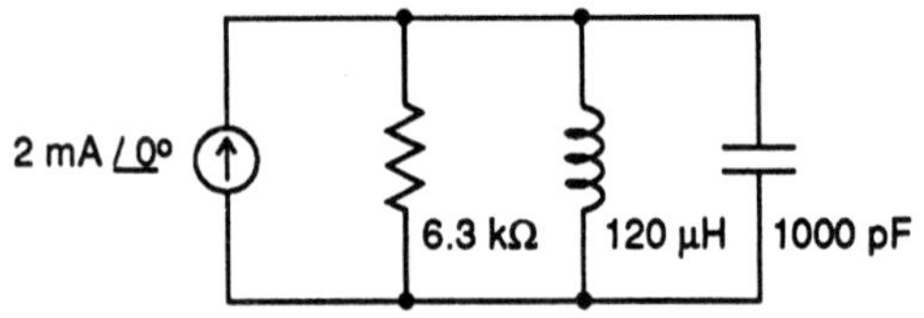

(b)

Fig. 11-25.

EXAMPLE 11-16

In Fig. 11-25a, the inductor has a Q_L of 49.4. What is the resonant frequency, the voltage, and the bandwidth?

SOLUTION

The resonant frequency is

$$f_r = \frac{1}{2\pi\sqrt{(120\ \mu\text{H})(1000\ \text{pF})}} = 459\ \text{kHz}$$

At this frequency, the inductive reactance is

$$X_L = 2\pi f_r L = 2\pi(459\ \text{kHz})(120\ \mu\text{H}) = 346\ \Omega$$

and the R_s of the coil is

$$R_s = \frac{X_L}{Q_L} = \frac{346\ \Omega}{49.4} = 7\ \Omega$$

The equivalent parallel resistance of the inductor is

$$Q_L{}^2 R_s = (49.4^2)(7\ \Omega) = 17\ \text{k}\Omega$$

At this point, you have to use the product over the sum to combine the equivalent parallel resistance of the inductor and the load resistance:

$$R = \frac{(10\ \text{k}\Omega)(17\ \text{k}\Omega)}{10\ \text{k}\Omega + 17\ \text{k}\Omega} = 6.3\ \text{k}\Omega$$

Figure 11-25b shows the equivalent circuit.

At resonance, all of the 1 mA of source current passes through the equivalent parallel resistance to produce a voltage of

$$V = IR = (1\ \text{mA})\ (6.3\ \text{k}\Omega) = 6.3\ \text{V}$$

For the circuit of Fig. 11-25b, the total Q equals

$$Q = \frac{R}{X_L} = \frac{6.3\ \text{k}\Omega}{346\ \Omega} = 18.2$$

Notice that the total Q (18.2) is less than Q of the coil (49.4). The bandwidth equals the resonant frequency divided by the total Q:

$$B = \frac{f_r}{Q} = \frac{459\ \text{kHz}}{18.2} = 25.2\ \text{kHz}$$

MULTIPLE-CHOICE TEST

The following questions may have more than one right answer. Select the best answer. This is the one that is always true, or covers more situations, or agrees with the material of this chapter, etc.

1. When a circuit is series resonant, which of the following is maximum?
   a. total current
   b. total voltage
   c. total impedance
   d. none of the above

2. At series resonance, the inductive reactance is
   a. less than the capacitive reactance
   b. equal to the capacitive reactance
   c. greater than the capacitive reactance
   d. none of the above

3. When a circuit is series resonant, which of the following is constant?
   a. total current
   b. total voltage
   c. total impedance
   d. none of the above

4. Below the resonant frequency, a series *RLC* circuit appears
   a. inductive
   b. capacitive
   c. resistive
   d. none of the above

5. If the inductance increases, the series resonant frequency will
   a. decrease
   b. stay the same
   c. increase
   d. none of the above

6. The total impedance of a series *RLC* circuit at resonance equals
   a. R
   b. jX_L
   c. $-jX_C$
   d. $jX_L - jX_C$

7. If the capacitance doubles, the series resonant frequency will
   a. decrease by a factor of 2
   b. increase by a factor of 2
   c. decrease by a factor of 1.414
   d. none of the above

8. If the Q increases, the bandwidth will
   a. decrease
   b. stay the same
   c. increase
   d. none of the above

9. When a circuit is parallel resonant, which of the following is maximum?
   a. capacitive reactance
   b. total voltage
   c. inductive reactance
   d. none of the above

10. At parallel resonance with a high Q, the inductive reactance is
   a. less than the capacitive reactance
   b. equal to the capacitive reactance
   c. greater than the capacitive reactance
   d. none of the above

11. When a circuit is parallel resonant with a high Q, which of the following is maximum?
   a. tank current
   b. total voltage
   c. total impedance
   d. all of the above

12. Below the resonant frequency, a parallel *RLC* circuit appears
   a. inductive
   b. capacitive
   c. resistive
   d. none of the above

13. If the inductance increases, the parallel resonant frequency will
   a. decrease
   b. stay the same
   c. increase
   d. none of the above

14. The total impedance of a parallel *RLC* circuit at resonance equals
   a. R
   b. jX_L
   c. $-jX_C$
   d. $jX_L - jX_C$

15. If the capacitance doubles, the parallel resonant frequency will
   a. decrease by a factor of 2
   b. increase by a factor of 2
   c. decrease by a factor of 1.414
   d. none of the above

16. If the Q decreases, the bandwidth of a parallel *RLC* circuit will
   a. decrease
   b. stay the same
   c. increase
   d. none of the above

PROBLEMS

Sec. 11-1: Series Resonance

11-1. What is the resonant frequency in Fig. 11-26?

11-2. What is the total phasor current in Fig. 11-26? The phasor voltage across each component at the resonant frequency?

11-3. The inductance doubles in Fig. 11-26. What happens to the resonant frequency?

11-4. To get a total current of 500 mA in Fig. 11-26, what value of resistance should we use?

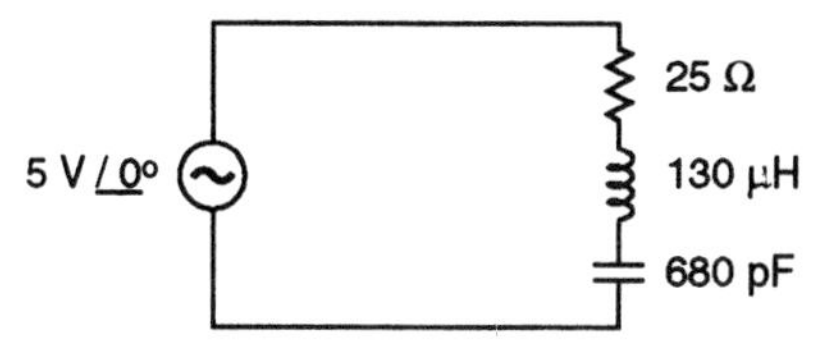

Fig. 11-26.

Sec. 11-2: More About Series Resonance

11-5. A series *RLC* circuit has a resonant frequency of 2 MHz. What is the bandwidth for a Q of 20? For a Q of 90?

11-6. In Fig. 11-26, what is the Q of the circuit at resonance? The bandwidth?

11-7. A series *RLC* circuit has a resonant frequency of 750 kHz. If the bandwidth is 12 kHz, what is the Q?

Sec. 11-3: The Q Rise in Reactive Voltages

11-8. In Fig. 11-26, what is the inductive voltage at resonance? The capacitive voltage?

11-9. If the inductance doubles in Fig. 11-26, what is the inductive voltage at resonance?

Sec. 11-4: Ideal Parallel Resonance

11-10. In Fig. 11-27, the inductor is ideal. What is the resonant frequency? The voltage across the circuit at resonance? The branch currents at resonance?

11-11. If the capacitance doubles in Fig. 11-27, what is the resonant frequency? (The inductor is ideal.)

11-12. To change the resonant frequency of Fig. 11-27 to 1 MHz, what value should the capacitance be? (Assume the coil is ideal.)

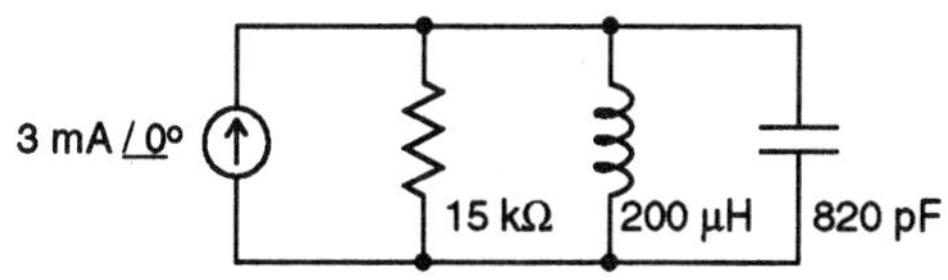

Fig. 11-27.

Sec. 11-5: Parallel Q and Bandwidth

11-13. A parallel *RLC* circuit has a resonant frequency of 250 kHz. What is the bandwidth when the Q is 45? When the Q is 125?

11-14. In Fig. 11-27, the inductor is ideal. What is the bandwidth? The critical frequencies?

11-15. If the capacitance of Fig. 11-27 changes to 680 pF, what is the bandwidth? The critical frequencies? (Assume the coil is ideal.)

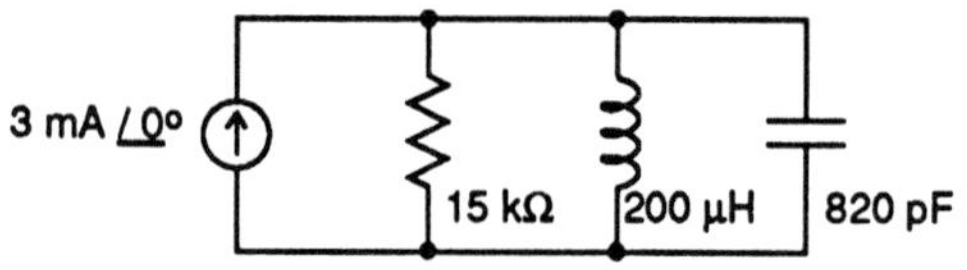

Fig. 11-27.

Sec. 11-6: The Q Rise in Reactive Currents

11-16. In Fig. 11-27, the inductor is ideal. What is the inductive current at resonance? The capacitive current?

11-17. If the capacitance is decreased by a factor of 2 in Fig. 11-27, what is the capacitive current at resonance? (Assume the coil is ideal.)

Sec. 11-7: Circulating or Tank Current

11-18. What is the tank current at resonance in Fig. 11-27 if the inductor is ideal?

11-19. If the capacitance is tripled in Fig. 11-27, what is the tank current at resonance? (Assume an ideal inductor.)

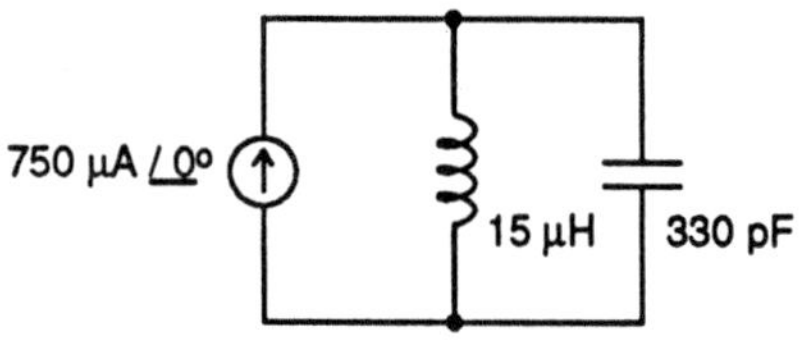

Fig. 11-28.

Sec. 11-8: Q of an Inductor

11-20. An inductor has an inductance of 250 μH and a resistance of 12 Ω. What is its Q at 130 kHz?

11-21. An inductor has a Q of 150 at 1 MHz. If its resistance is 15 Ω, what is its inductance?

11-22. An inductor has a Q of 150 at 1 MHz. If its inductance is 100 μH, what is its resistance?

Sec. 11-9: Resonance with Nonideal Inductor

11-23. In Fig. 11-28, the inductor has a Q_L of 106. What is the resonant frequency, the voltage, the bandwidth, and the tank current?

11-24. In Fig. 11-28, the capacitance is changed to 1000 pF. At the new resonant frequency, the inductor has a Q_L of 75. What is the resonant frequency, the voltage, the bandwidth, and the tank current?

Fig. 11-29.

Sec. 11-10: Parallel Resonance with a Load

11-25. In Fig. 11-29, the inductor has a Q_L of 66.4. What is the resonant frequency, the voltage, and the bandwidth?

11-26. The capacitance is increased to 1500 pF in Fig. 11-29. At the new resonant frequency, the inductor has a Q_L of 50. What is the resonant frequency, the voltage, the bandwidth, and the tank current?

CHAPTER 12
TRANSIENTS

When a switch closes or opens in a circuit, the voltages and currents around the circuit usually change. In purely resistive circuits, the voltages and currents change instantaneously from one value to another. But in *RL* and *RC* circuits, it takes time for the changes to occur because of inductive and capacitive effects. These changing voltages and currents are called *transients*.

12-1. RESISTIVE CIRCUITS

With the exception of Chap. 1, no switches have been used in any of the circuits we have discussed. This is why no transients appeared in the output of our circuits. But now, we will include switches to see what effects they produce.

Figure 12-1a shows a dc source connected to a purely resistive circuit that has a switch. When the switch closes or opens, all the currents and voltages in the circuit will change. In other words, we can always redraw the circuit with one of the resistors selected as the load resistor. When the switch is either closed or open, the current and voltage of the load resistor will change.

In a purely resistive circuit, the change in load current is instantaneous as shown in Fig. 12-1b. In discussing transients, it is customary to measure time from the instant a switch closes or opens. This is done by using $t = 0$ to represent the instant a switch closes or opens.

The initial value of current, symbolized by i_i, is the current at the instant the switch closes. The final value of current, symbolized by i_f, is the current slightly after the switch has closed. The transient is the change in current between the initial and final values. As you can see, the current of Fig. 12-1b steps between the initial and final value.

Similarly, the voltage across the load resistor is a transient between an initial value of v_i and a final value of v_f as shown in Fig. 12-1c. Again, the transient is a step or instantaneous change in value. Notice that a step in current takes no time at all. Mathematically, it means the change in time between the beginning and end of the step is zero. In symbols,

$$\Delta t = 0$$

No matter how complicated the circuit of Fig. 12-1a, you can always select one of the resistors and redraw the circuit with this resistor as the load resistor. As long as the circuit is purely resistive, the load current and load voltage will always be steps between the initial and final values.

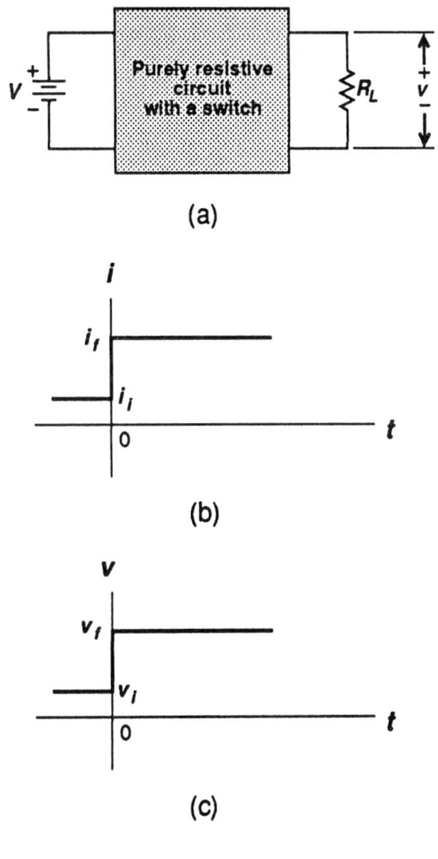

Fig. 12-1.

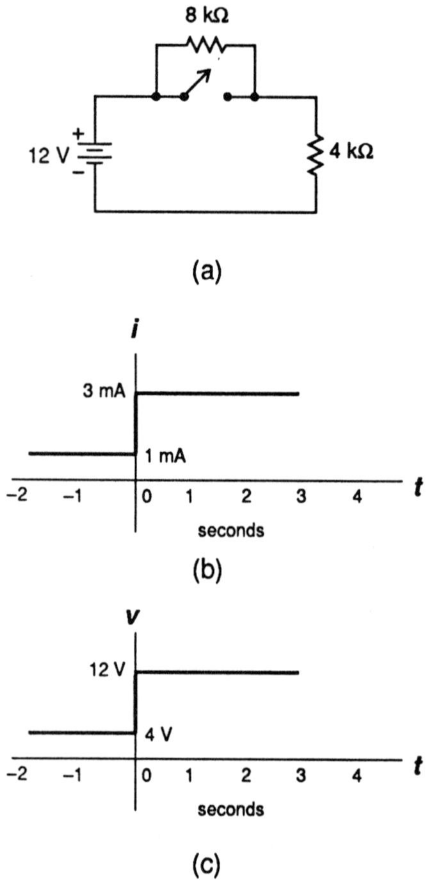

Fig. 12-2.

EXAMPLE 12-1

Figure 12-2a shows a purely resistive circuit with a dc source voltage of 12 V. What is current transient for the 4 kΩ when the switch closes? The voltage transient?

SOLUTION

With the switch open, the total resistance is 12 kΩ and the current is

$$I = \frac{12 \text{ V}}{12 \text{ k}\Omega} = 1 \text{ mA}$$

At the instant the switch closes, the load current is 1 mA. This is the initial value of load current.

With the switch closed, the total resistance is 4 kΩ and the current is

$$I = \frac{12 \text{ V}}{4 \text{ k}\Omega} = 3 \text{ mA}$$

Just after the switch closes, the current is 3 mA. This is the final value of load current.

The transient is the change between the initial and final values. In this example, the current transient is a step between 1 and 3 mA, as shown in Fig. 12-2b.

Similarly, the initial voltage is

$$v_i = (1 \text{ mA})(4 \text{ k}\Omega) = 4 \text{ V}$$

and the final voltage is

$$v_f = (3 \text{ mA})(4 \text{ k}\Omega) = 12 \text{ V}$$

Figure 12-2c shows the transient of voltage, a step between 4 V to 12 V.

12-2. *RL* CIRCUITS

The current through an inductor cannot change instantaneously. Why? Because this would produce infinite voltage across the inductor. You can see this by recalling this basic relation:

$$v = L\frac{di}{dt} \tag{12-1}$$

If the current were able to change instantaneously, dt (equivalent to Δt) would be zero because of the sudden step in current. This means di/dt would be infinite. In turn, this would produce infinite voltage. Since infinite voltage is impossible, a step in current is impossible. Therefore, no matter what the circuit, the current through an inductor cannot change instantaneously. This is why you will never see a step in current when inductance is present in a circuit.

Transients are Exponential

Figure 12-3a shows a circuit with an inductor and a resistor. With the switch open, the current is zero. Just after the switch is closed, the current is still zero because the current through an inductor cannot change instantaneously. As time begins to pass, the current starts to increase toward its final value.

Figure 12-3b shows the transient current. The switch closes at $t = 0$. Before this instant, the current is zero. Just after this instant, it is still zero. But then, the current begins to increase. Notice how it changes rapidly at first, then gradually approaches the final value. The final value is given by Ohm's law:

$$i_f = \frac{V}{R} \tag{12-2}$$

In other words, after the switch has been closed for a long time, the back emf across the inductor disappears and all of the source voltage appears across the resistor.

The instantaneous rate of change in current is initially large, then becomes smaller as the current approaches the final value. A transient like this is called an *exponential wave*, or simply an exponential. Since the instantaneous rate of change is positive at any point on this exponential, we refer to this waveform as a *positive* exponential.

The larger the inductance, the longer it takes for the positive exponential to reach the final value. For instance, if L is increased by a factor of 10 in Fig. 12-3a, it would take 10 times as long for the transient to reach the final value in Fig. 12-3b. On the other hand, the smaller the inductance, the faster the final current is reached. If L is decreased by a factor of 10 in Fig. 12-3a, the transient would reach its final value in 1/10 of the time shown in Fig. 12-3b.

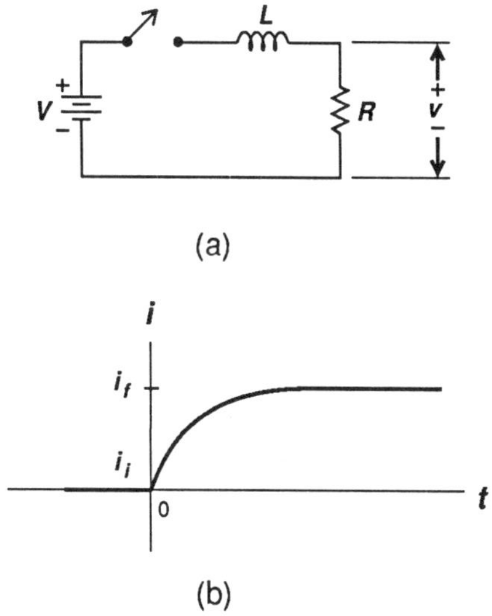

Fig. 12-3.

Definition of Time Constant

The *time constant* of a circuit is defined as the amount of time it takes for a transient to undergo a change of 63.2 percent of the total change between the initial and final values. For instance, after the switch closes in Fig. 12-3a, the transient current reaches 63.2 percent of its final value after one time constant.

We will use the Greek letter "tau" to represent the time constant. With a calculus derivation, it can be shown that the time constant of an *RL* circuit is given by

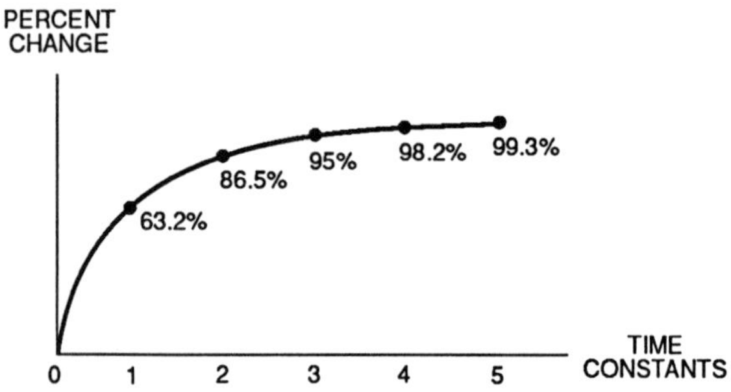

Fig. 12-4.

$$\tau = \frac{L}{R} \tag{12-3}$$

where τ = time constant in seconds
L = inductance in henries
R = resistance in ohms

For instance, if $L = 50$ H and $R = 20\ \Omega$ in Fig. 12-3a, the time constant is

$$\tau = \frac{L}{R} = \frac{50\text{ H}}{25\ \Omega} = 2\text{ s}$$

This means it takes 2 s for the current to increase from 0 to 63.2 percent of its final value.

Finished After Five Time Constants

Figure 12-4 shows a positive exponential. After one time constant, the current has reached 63.2 percent of its final value. After two time constants, the current is at 86.5 percent. After three time constants, it is at 95 percent. After four time constants, it is at 98.2 percent. After five time constants, the current has reached 99.3 percent of its final value. As an approximation, most people consider the transient *finished after five time constants* because the exponential is then within 1 percent of the final value.

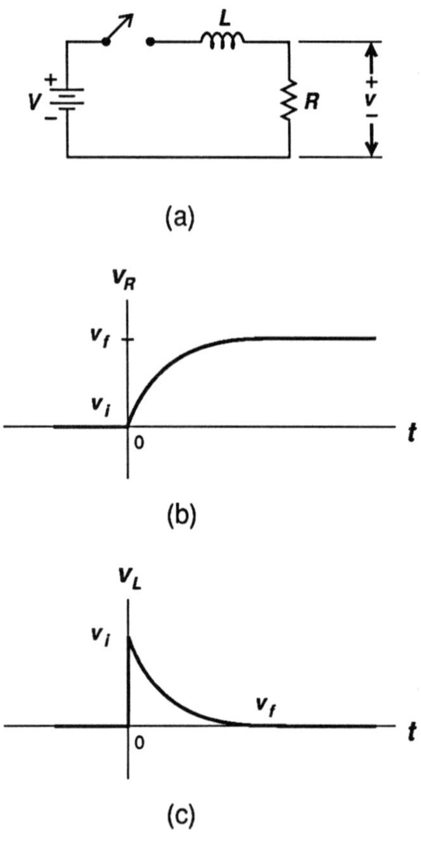

Fig. 12-5.

Transient Voltage Across Resistor

In Fig. 12-5a, the voltage across the resistor is zero before the switch closes. Because of Ohm's law, the voltage across the resistor of Fig. 12-5a is a positive exponential like Fig. 12-5b. In other words, we can multiply each instantaneous current in this circuit by the resistance to get the transient shown in Fig. 12-5b. The shape is identical to the current transient. Because of this, the voltage across the resistor reaches 63.2 percent of its final value after one time constant.

Transient Voltage Across Inductor

Because of Kirchhoff's voltage law, the voltage across the inductor of Fig. 12-5a after the switch closes is given by

$$v_L = V - v_R \tag{12-4}$$

This says the inductive voltage equals the source voltage minus the voltage across the resistor. Since the voltage across the resistor is a positive exponential, the voltage across the inductor is a negative exponential shown in Fig. 12-5c. This is called a negative exponential because the voltage is decreasing between the initial value and the final value.

At the instant the switch closes in Fig. 12-5a, all the source voltage appears across the inductor. This voltage is the back emf discussed in earlier chapters. As time passes, the current increases exponentially toward its final value. As this happens, the voltage across the inductor decreases exponentially toward zero as shown in Fig. 12-5c. Because the negative exponential is an inverted form of the positive exponential, we can subtract each of the values in Fig. 12-4 from 100 percent to get the graph of Fig. 12-6.

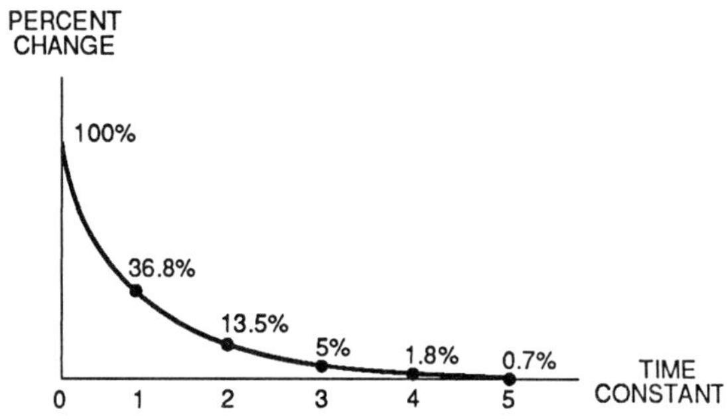

Fig. 12-6.

Summary of Analysis

When you are analyzing the transients of a circuit, the three most important quantities to calculate are the initial value, the final value, and the time constant. Once you have these three values, you can sketch the waveform of the transient. This sketch usually answers most of the practical questions that arise when transients are being taken into account. The following example will show you how it is done.

EXAMPLE 12-2

What is the current transient in Fig. 12-7a when the switch closes? The voltage transient for the resistor? The voltage transient for the inductor?

SOLUTION

At the instant the switch closes, the initial current is zero because the current through an inductor can never change instantaneously. In symbols,

$$i_i = 0$$

After the switch has been closed for a long time, the back emf across the inductor is zero. Since all of the source voltage is then across the resistor, the final current is

$$i_f = \frac{15\ \text{V}}{3\ \text{k}\Omega} = 5\ \text{mA}$$

The time constant equals the inductance divided by the resistance:

$$\tau = \frac{L}{R} = \frac{12\ \text{mH}}{3\ \text{k}\Omega} = 4\ \mu\text{s}$$

The current transient is the change between the initial and the final value. In this example, the current transient is a positive exponential between 0 and 5 mA, as shown in Fig. 12-7b. After 4 μs, the current equals 63.2 percent of the final value:

$$i = 0.632(5\ \text{mA}) = 3.16\ \text{mA}$$

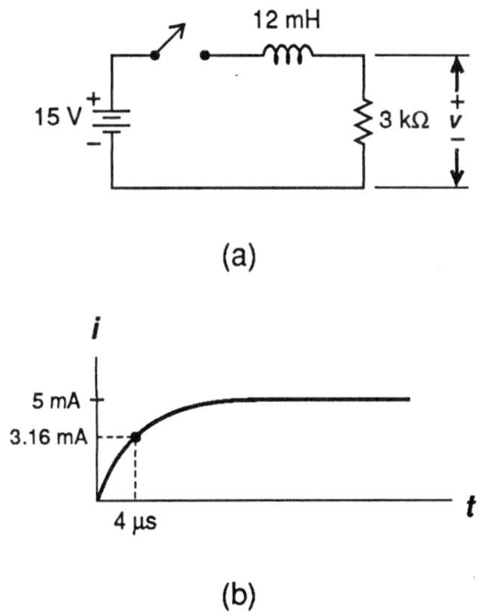

Fig. 12-7.

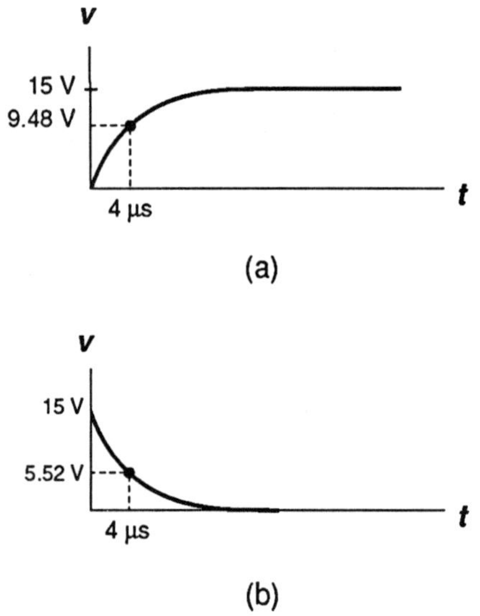

Fig. 12-8.

Similarly, the initial voltage across the resistor is

$$v_i = 0$$

and the final voltage is

$$v_f = (5\ \text{mA})(3\ \text{k}\Omega) = 15\ \text{V}$$

You can also get the final voltage by remembering that all of the source voltage appears across the resistor when the switch has been closed for a long time. Figure 12-8a shows the transient of voltage across the resistor, a positive exponential between 0 to 15 V.

After the switch closes, the voltage across the inductor equals the source voltage minus the voltage across the resistor. This is why the initial voltage across the inductor is

$$v_i = 15\ \text{V} - 0 = 15\ \text{V}$$

and the final voltage is

$$v_f = 15\ \text{V} - 15\ \text{V} = 0$$

Figure 12-8b shows the transient of voltage across the inductor, a negative exponential between 15 V and 0.

12-3. CHARGING TRANSIENTS IN *RC* CIRCUITS

Capacitor are widely used in electronics and are far more important than inductors. Inductors are seldom used in everyday electronics, except in special applications. For this reason, our emphasis will now shift to *R C* circuits and their transients.

Capacitor Voltage Cannot Change Instantaneously

Recall this important idea from Chap. 5: the voltage across a capacitor cannot change instantaneously. Why? Because this would produce infinite current into the capacitor. You can see this by recalling this basic relation:

$$i = C \frac{dv}{dt} \qquad (12\text{-}5)$$

If the voltage were able to change instantaneously, dt (equivalent to Δt) would be zero because of the sudden step in voltage. This means dv/dt would be infinite. In turn, this would produce infinite current. Since infinite current is impossible, a step in voltage is impossible. Therefore, no matter what the circuit, the voltage across a capacitor cannot change instantaneously. This is why you will never see a step in voltage when capacitance is present in a circuit.

Transients are Exponential

Figure 12-9a shows a circuit with a resistor and a capacitor. Assume the capacitor is initially uncharged. When the switch is closed, the voltage across the capacitor is zero because the voltage across a capacitor cannot change instantaneously. In symbols,

$$v_i = 0$$

As time passes, the capacitor begins to charge and its voltage increases. Eventually, the capacitor becomes fully charged.

Figure 12-9b shows the transient voltage. Before the switch is closed, the voltage is zero. After the switch is closed, the voltage increases exponentially toward a final value of

$$v_f = V$$

In other words, after the switch has been closed for a long time, all of the source voltage appears across the capacitor.

The larger the capacitance, the longer it takes for the positive exponential to reach the final value. For instance, if C is increased by a factor of 10 in Fig. 12-9a, it would take 10 times as long for the transient to reach the final value in Fig. 12-9b. On the other hand, the smaller the capacitance, the faster the final voltage is reached. If C is decreased by a factor of 10 in Fig. 12-9a, the transient would reach its final value in 1/10 of the time shown in Fig. 12-9b.

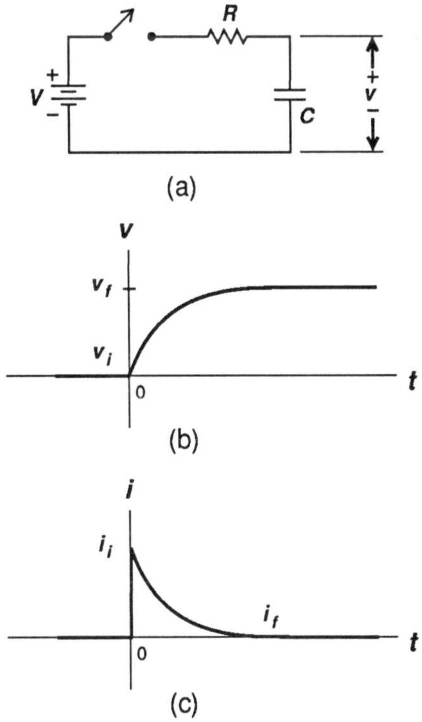

Fig. 12-9.

Charging Time Constant

The time constant of an RC circuit is still defined as the time it takes for the transients to undergo a change of 63.2 percent of the total change between the initial and final value. With a calculus derivation, it can be shown that the time constant of an RC circuit is given by

$$\tau = RC \qquad (12\text{-}6)$$

This says that the time constant of a series RC circuit equals the resistance times the capacitance. For instance, if R = 2.7 kΩ and C = 1 μF, the time constant is

$$\tau = (2.7\ \text{k}\Omega)(1\ \mu\text{F}) = 2.7\ \text{ms}$$

This means it takes 2.7 ms for the voltage to increase from 0 to 63.2 percent of its final value.

Transient Current

Before the switch closes in Fig. 12-9a, the current is zero. At the instant the switch is closed, the capacitor voltage is zero. This means that the capacitor

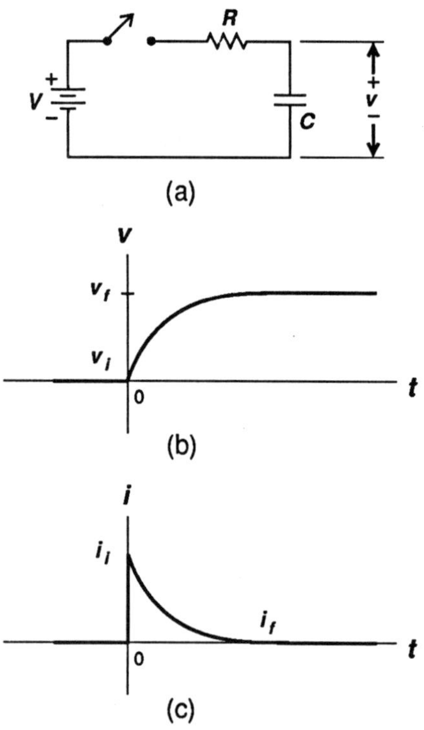

Fig. 12-9.

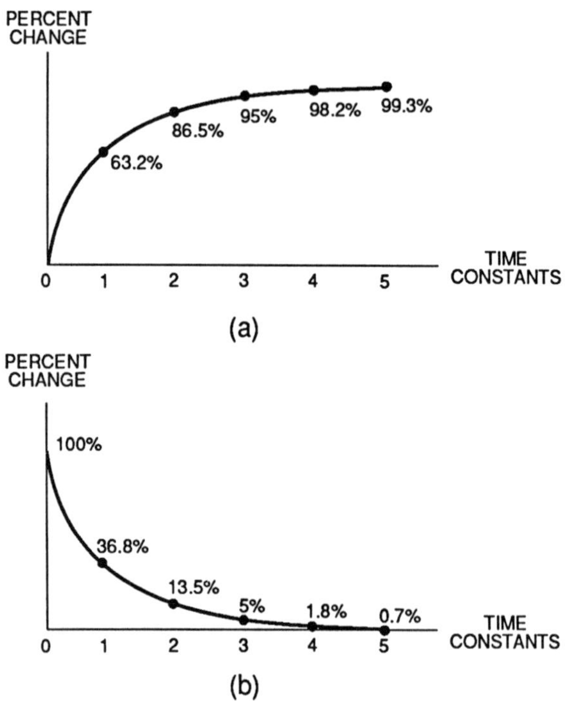

Fig. 12-10.

is equivalent to a short circuit at the instant the switch is closed. Because of this, all of the source voltage appears across the resistor and the initial current is

$$i_i = \frac{V}{R} \tag{12-7}$$

As the capacitor begins to charge, its voltage increases and the current decreases. When the capacitor is fully charged, the current is zero. Figure 12-9c shows the transient current, a negative exponential between the initial value and the final value.

Finished After Five Time Constants

Figure 12-10a shows the positive exponential of capacitor voltage. As indicated, the voltage equals 63.2 percent of the final voltage after one time constant. Again, we consider the transient voltage to be finished after five time constants because the voltage is then within 1 percent of its final value.

Similarly, Fig. 12-10b shows the negative exponential of current. As you can see, the current drops to 36.8 percent of its initial value after one time constant. After five time constants, the transient current is considered to be finished because the current is less than 1 percent of its initial value.

Transient Voltage Across Resistor

Because of Kirchhoff's voltage law, the voltage across the resistor of Fig. 12-9a after the switch closes is given by

$$v_R = V - v_C \tag{12-8}$$

This says the resistive voltage equals the source voltage minus the voltage across the capacitor. Since the voltage across the capacitor is a positive exponential, the voltage across the resistor is a negative exponential as shown in Fig. 12-11.

An alternative way to get the transient voltage across the resistor is by using Ohm's law:

$$v_R = iR \tag{12-9}$$

This says the voltage equals the current times the resistance. Since the current is a negative exponential (see Fig. 12-9c), the voltage is also a negative exponential (Fig. 12-11).

EXAMPLE 12-3

The capacitor of Fig. 12-12a is initially uncharged. What is the transient voltage across the capacitor when the switch closes? The transient current? The transient voltage for the resistor?

SOLUTION

At the instant the switch closes, the initial capacitor voltage is zero. In symbols,

$v_i = 0$

After the switch has been closed for a long time, the capacitor voltage equals the source voltage:

$v_i = 18$ V

The time constant equals the resistance times the capacitance:

$\tau = (3\ \text{k}\Omega)(2000\ \text{pF}) = 6\ \mu\text{s}$

The transient voltage across the capacitor is the change between the initial and the final value. In this example, the voltage transient is a positive exponential between 0 and 18 V, as shown in Fig. 12-12b. After 6 μs, the voltage equals 63.2 percent of the final value:

$v = 0.632(18\ \text{V}) = 11.4\ \text{V}$

The initial current equals the source voltage divided by the resistance:

$$i_i = \frac{18\ \text{V}}{3\ \text{k}\Omega} = 6\ \text{mA}$$

The final current is

$i_i = 0$

After one time constant the current is 0.36.8 percent of the initial value:

$i = 0.368(6\ \text{mA}) = 2.28\ \text{mA}$

Figure 12-13a shows the transient current.

The initial voltage across the resistor equals the source voltage:

$v_i = 18$ V

and the final voltage is

$v_f = 0$

After one time constant the voltage is 36.8 percent of the initial value:

$i = 0.368(18\ \text{V}) = 6.62\ \text{V}$

Figure 12-13b shows the transient voltage across the resistor.

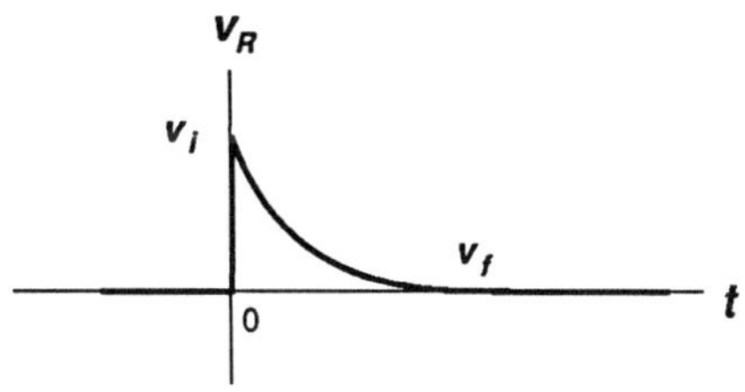

Fig. 12-11.

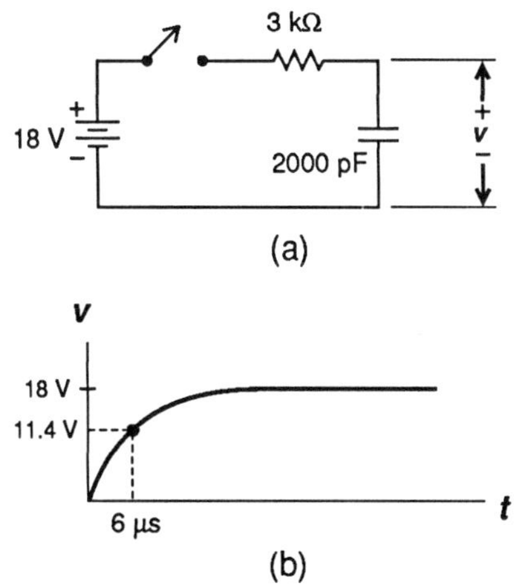

Fig. 12-12.

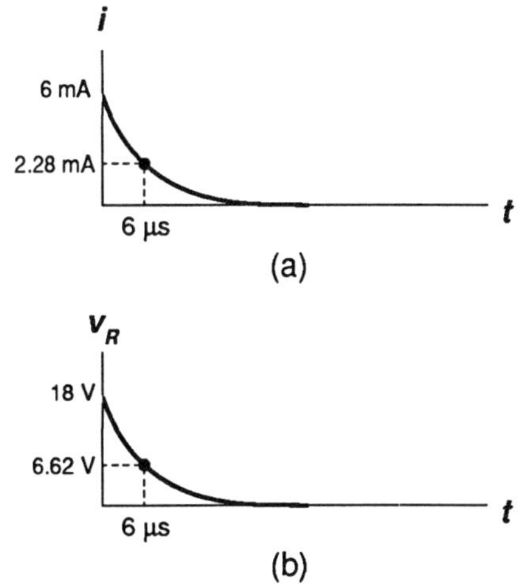

Fig. 12-13.

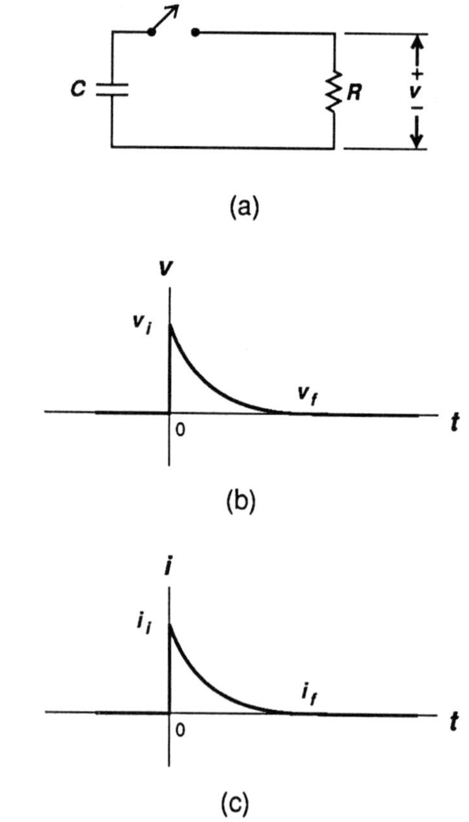

Fig. 12-14.

12-4. DISCHARGING TRANSIENTS IN *RC* CIRCUITS

Assume the capacitor of Fig. 12-14a was previously charged, so that it now has an initial voltage of v_i. With the switch open, the voltage across the resistor is zero. When the switch is closed, the output voltage will step from 0 to v_i.. The capacitor then begins to discharge through the resistor. As it does, the voltage decreases exponentially as shown in Fig. 12-14b. Since the current equals the voltage divided by the resistance, the current also decreases exponentially as shown in Fig. 12-14c.

The time constant of the circuit is still given by

$$\tau = RC \tag{12-10}$$

In other words, the discharging time constant of this circuit has the same equation as the charging time constant. In more complicated circuits, it is possible for the discharging time constant to be different from the charging time because the value of R may be different when the capacitor is discharging.

The discharging time constant tells you when the voltage equals 36.8 percent of the initial voltage. Also, we consider the discharging to be finished after five time constants.

EXAMPLE 12-4

The capacitor of Fig. 12-15a is initially uncharged. Switch S_1 is closed. After the capacitor is fully charged, switch S_1 is opened and S_2 is closed. What is the transient voltage across the capacitor?

SOLUTION

When S_1 is closed, the capacitor charges. When S_1 is opened and S_2 is closed, the capacitor discharges. Therefore, the voltage across the capacitor is a positive exponential from 0 to 12 V, followed by a negative exponential from 12 V back to zero.

When S_1 is closed, the capacitor is charging with a time constant of

$$\tau = (4.7\ \text{k}\Omega)(0.22\ \mu\text{F}) = 1.03\ \text{ms}$$

On the other hand, when S_1 is open and S_2 is closed, the capacitor is discharging with a time constant of

$$\tau = (6.8\ \text{k}\Omega)(0.22\ \mu\text{F}) = 1.49\ \text{ms}$$

Figure 12-15b shows the transient voltage across the capacitor. As you see, it charges exponentially from 0 to 12 V. After one time constant (1.03 ms), it equals 63.2 percent of 12 V, or 7.58 V. Later, the capacitor discharges exponentially from 12 V back to zero. After it has been discharging for 1.49 ms, its voltage equals 36.8 percent of 12 V, or 4.42 V.

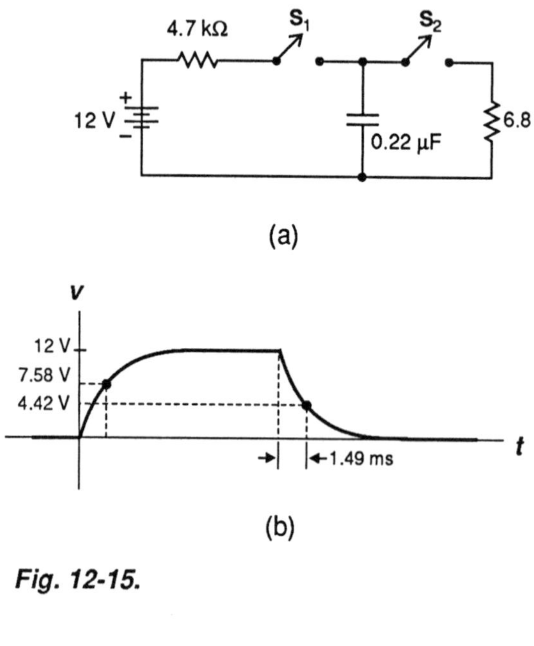

Fig. 12-15.

12-5. TRANSIENTS IN A VOLTAGE DIVIDER

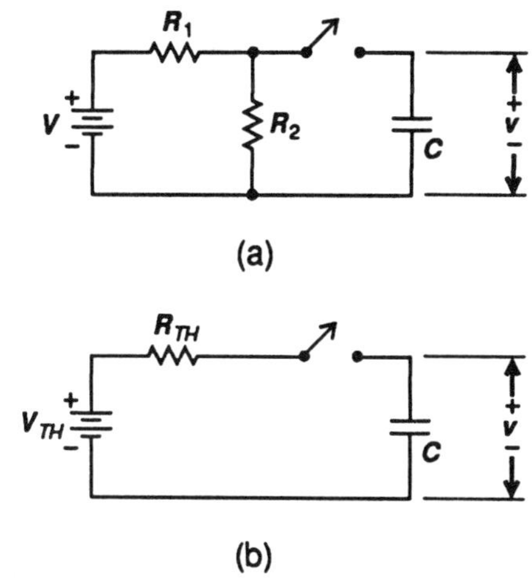

Fig. 12-16.

We know how to find the transients in a series RC circuit. When A circuit is more complicated, we can usually simplify it to a series circuit. Then, we can use the methods discussed earlier. For instance, Fig. 12-16a shows a voltage divider, a switch, and a capacitor. A circuit like this often appears in one form or another in practical electronics. How do you think we can find the transients in a circuit like this? Think about this for a moment before you read on?

The easiest way to find the transients in Fig. 12-16a is by applying the Thevenin theorem to get the equivalent circuit of Fig. 12-16b. Since this equivalent circuit is a series RC circuit, we can calculate the initial voltage, the final voltage, and the time constant. The Thevenin quantities are given by

$$V_{TH} = \frac{R_2}{R_1 + R_2} V \qquad (12\text{-}11)$$

and

$$R_{TH} = R_1 \parallel R_2 \qquad (12\text{-}12)$$

The initial voltage is

$$v_i = 0$$

and the final voltage is

$$v_f = V_{TH}$$

The time constant is

$$\tau = R_{TH}C$$

The following example will demonstrate the use of Thevenin's theorem in finding the transients of a voltage divider.

EXAMPLE 12-5

The capacitor of Fig. 12-17a is initially uncharged. What is the transient voltage across the capacitor when the switch closes?

SOLUTION

The Thevenin voltage is

$$V_{TH} = \frac{3\text{ k}\Omega}{9\text{ k}\Omega} 30\text{ V} = 10\text{ V}$$

and the Thevenin resistance is

$$R_{TH} = 6\text{ k}\Omega \parallel 3\text{ k}\Omega = 2\text{ k}\Omega$$

Figure 12-17b shows the equivalent circuit.

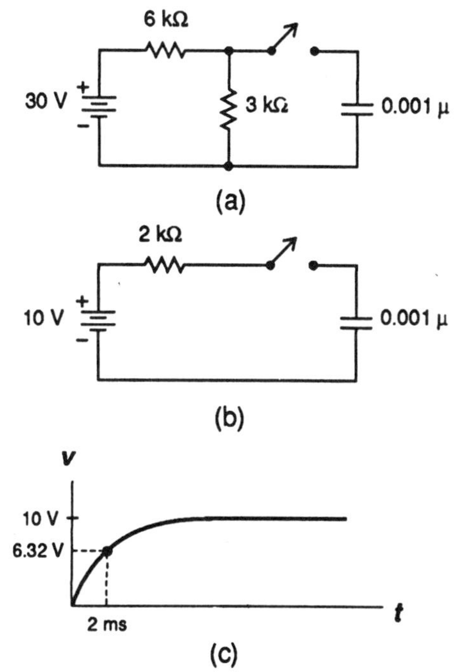

Fig. 12-17.

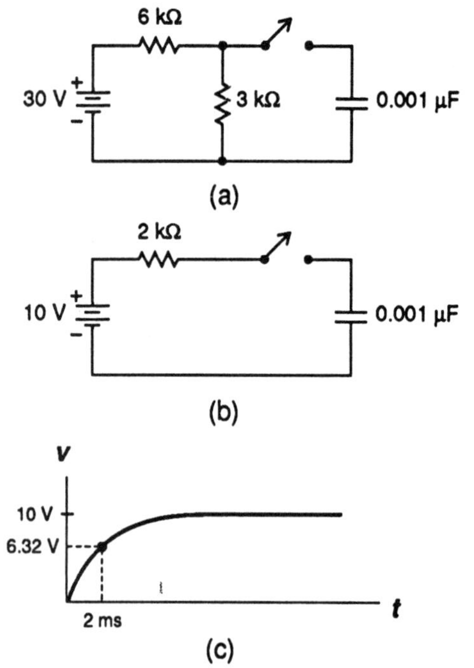

Fig. 12-17.

The initial voltage is

$v_i = 0$

and the final voltage is

$v_f = 10\text{ V}$

and the time constant is

$\tau = (2\text{ k}\Omega)(0.001\ \mu\text{F}) = 2\text{ ms}$

Figure 12-17c shows the transient voltage across the capacitor.

12-6. TRANSIENTS WITH A CURRENT SOURCE

When you study electronics, you will encounter transistors. In some applications, transistors act like current sources. For this reason, it is worthwhile analyzing the transients of a circuit driven by a current source rather than a voltage source.

Figure 12-18a shows a current source and a parallel resistance (a Norton circuit). The current source produces a step in current. When this step occurs, a transient voltage will appear across the capacitor as it charges to its final value.

The easiest way to analyze what happens is to convert the Norton circuit to a Thevenin circuit. Visualize the capacitor temporarily disconnected. Then, you can calculate the open-circuit voltage, which is a voltage step as shown in Fig. 12-18b. The Thevenin voltage at the top of the step equals

$$V = IR \qquad (12\text{-}13)$$

and the Thevenin resistance is

$$R_{TH} = R \qquad (12\text{-}14)$$

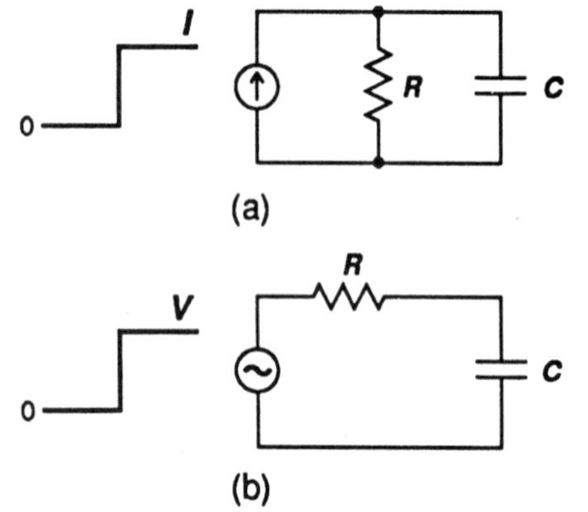

Fig. 12-18.

Since the equivalent circuit of Fig. 12-18b is a series *RC* circuit, we can analyze it using methods discussed earlier. Before the positive voltage step, the initial voltage across the capacitor is zero. After the voltage step, the capacitor charges exponentially toward the final voltage. The following example shows you the idea.

EXAMPLE 12-6

The capacitor of Fig. 12-19a is initially uncharged. What is the transient voltage across the capacitor?

SOLUTION

Visualize the capacitor open. Then, the Thevenin voltage after the step is

$V_{TH} = (2\text{ mA})(5\text{ k}\Omega) = 10\text{ V}$

and the Thevenin resistance is

$R_{TH} = 5\text{ k}\Omega$

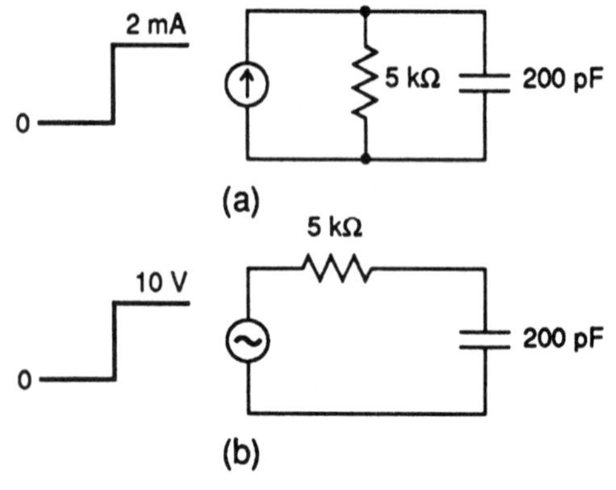

Fig. 12-19.

Figure 12-19b shows the equivalent circuit. It has a time constant of

$\tau = (5\ \text{k}\Omega)(200\ \text{pF}) = 1\ \mu\text{s}$

Therefore, the voltage across the capacitor is a positive exponential between 0 and 10 V. After one time constant (1 μs) the voltage is 6.32 V.

12-7. TRANSISTOR SWITCHING CIRCUIT

When transistors are used in digital computers, they are designed to act like switches. Figure 12-20a shows the equivalent circuit of a circuit where the transistor acts like a switch. When the switch is closed as shown here, the output voltage is zero. On the other hand, when the switch has been open for a long time, the output voltage equals the supply voltage. This type of operation is called *two-state* operation because the output voltage switches between two levels: 0 and $+V$. Figure 12-20b shows the transient voltage after the switch is opened.

Notice the capacitance connected between the output and ground. The dashed lines indicate *stray-wiring* capacitance across the switch. This is unwanted capacitance that exists between the output wire and ground. In other words, a piece of wire acts like one plate of a capacitor and ground acts like the other plate. The resulting capacitance is small but may be important in some application.s Typically, stray-wiring capacitance is in the picofarad region.

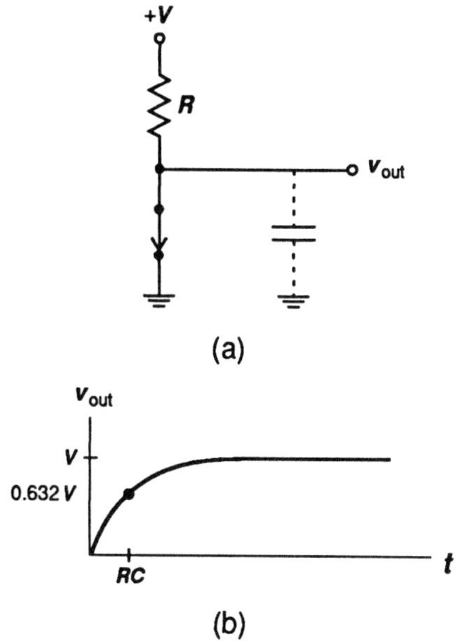

Fig. 12-20.

EXAMPLE 12-7

What is the transient voltage across the capacitor of Fig. 12-21a when the switch opens?

SOLUTION

When the switch opens, the stray-wiring capacitance charges toward 5 V with a time constant of

$\tau = (2.2\ \text{k}\Omega)(10\ \text{pF}) = 22\ \text{ns}$

Figure 12-21b shows the transient voltage across the capacitor. After one time constant, the voltage 3.16 V. After five time constants (110 ns), the output voltage is within 1 percent of 5 V.

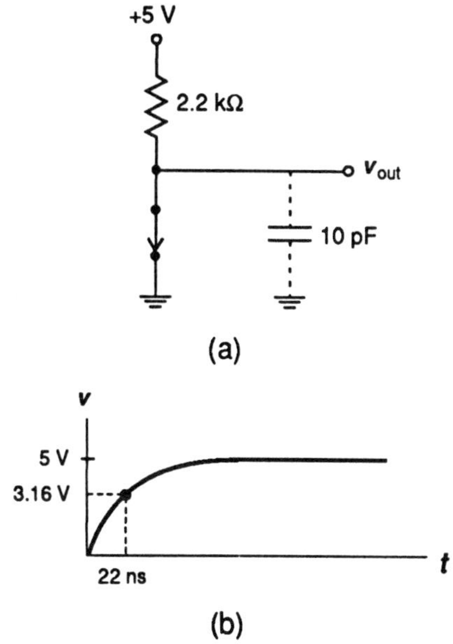

Fig. 12-21.

12-8. RISETIME

Figure 12-22 shows a positive exponential voltage as it would appear on an oscilloscope. The time it takes for the exponential to change from the 10-percent to the 90-percent point is called the *risetime*. The risetime is directly proportional to the time constant. The longer the time constant, the greater the risetime. By an advanced mathematical derivation, we can prove the following useful reation between risetime and time constant:

$$T_R = 2.2\tau \tag{12-15}$$

For instance, Fig. 12-21 has time constant of 22 ns. Therefore, the exponential has a a risetime of 2.2(22 ns), which equals 48.4 ns.

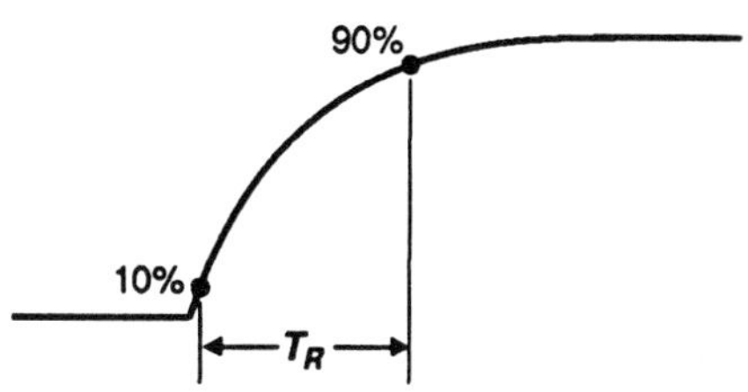

Fig. 12-22.

MULTIPLE-CHOICE TEST

The following questions may have more than one right answer. Select the best answer. This is the one that is always true, or covers more situations, or agrees with the material of this chapter, etc.

1. The current cannot change instantaneously in
   a. a resistor
   b. an inductor
   c. a capacitor
   d. none of the above

2. The waveform of a transient in an *RC* circuit is
   a. sinusoidal
   b. rectangular
   c. exponential
   d. alternating

3. A transient takes more time to reach its final value when inductance
   a. decreases
   b. stays the same
   c. increases
   d. none of the above

4. The transients in a circuit are finished after approximately
   a. one time constant
   b. two time constants
   c. five time constants
   d. none of the above

5. The voltage cannot change instantaneously across
   a. a resistor
   b. an inductor
   c. a capacitor
   d. none of the above

6. The time constant of a series *RC* circuit equals
   a. L/R
   b. R/L
   c. RC
   d. none of the above

7. When a capacitor discharges, its voltage after one time constant is
   a. 0
   b. 36.8 percent of the initial voltage
   c. 63.2 percent of the initial voltage
   d. 90 percent of the initial voltage

8. If the capacitance increases, the time constant
   a. decreases
   b. stays the same
   c. increases
   d. none of the above

9. When a capacitor is across the output of a voltage divider, you can reduce the circuit to a series *RC* circuit by using
   a. Ohm's law
   b. Kirchhoff's laws
   c. total impedance
   d. none of the above

10. When a current source drives a resistor and capacitor in parallel, you can reduce the circuit to a series *RC* circuit by using
   a. Ohm's law
   b. Kirchhoff's laws
   c. total impedance
   d. none of the above

11. One limitation on the speed of computer circuits is the
   a. resistance of wires
   b. stray-wiring capacitance
   c. time it takes to close a switch
   d. Norton's theorem

12. The time constant of a series *RL* circuit equals
   a. L/R
   b. R/L
   c. RC
   d. none of the above

PROBLEMS

Sec. 12-2: *RL* Circuits

12-1. What is the transient current in Fig. 12-23 when the switch closes? The transient voltage across the resistor? the transient voltage across the inductor?

12-2. If the resistance doubles in Fig. 12-23, what is the transient current after the switch closes?

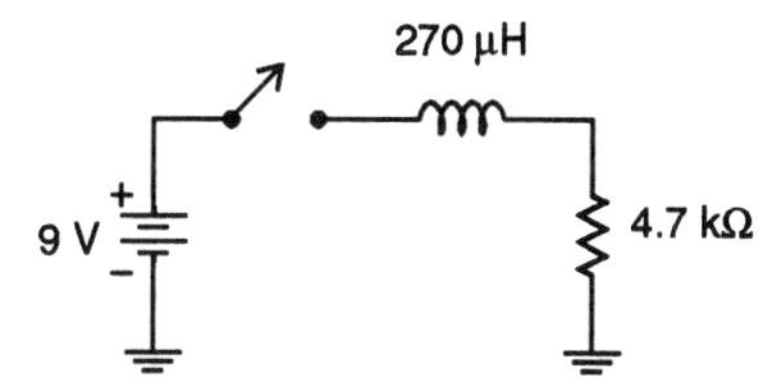

Fig. 12-23.

Sec. 12-3: Charging Transients in *RC* Circuits

12-3. The capacitor of Fig. 12-24 is uncharged. What is the transient voltage across the capacitor when the switch closes? The transient current? the transient voltage across the resistor?

12-4. If the resistance doubles in Fig. 12-24, what is the transient voltage across the capacitor when the switch closes?

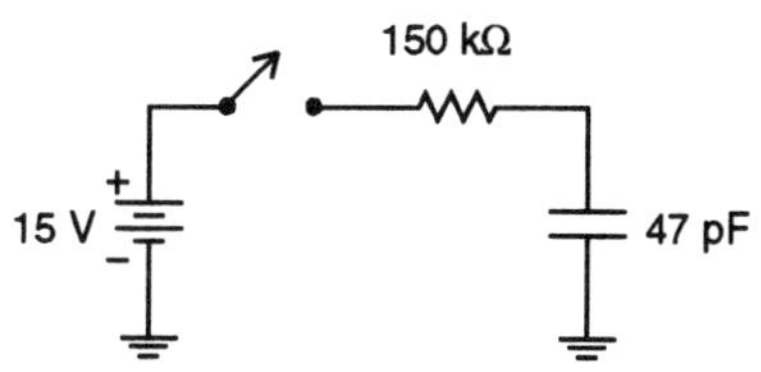

Fig. 12-24.

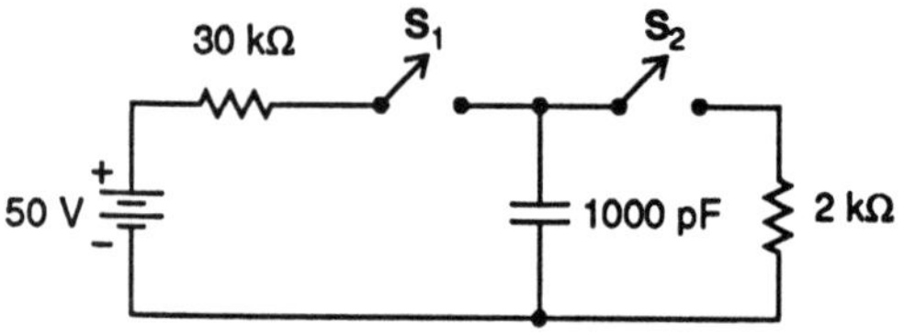

Fig. 12-25.

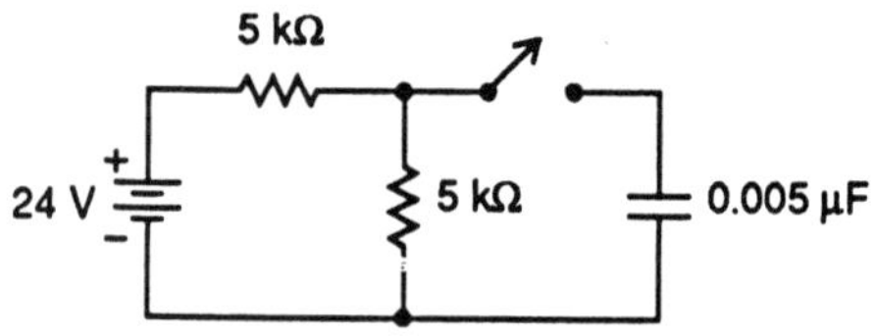

Fig. 12-26.

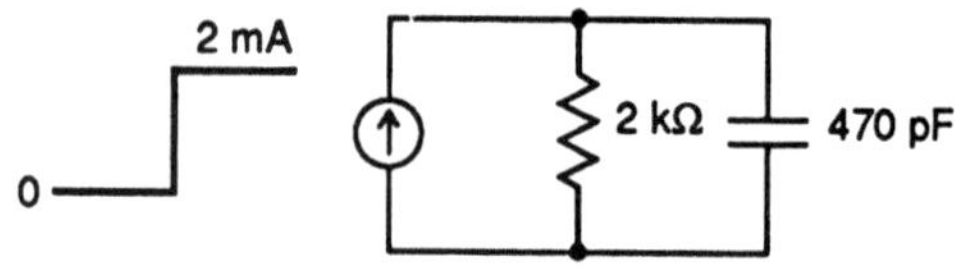

Fig. 12-27.

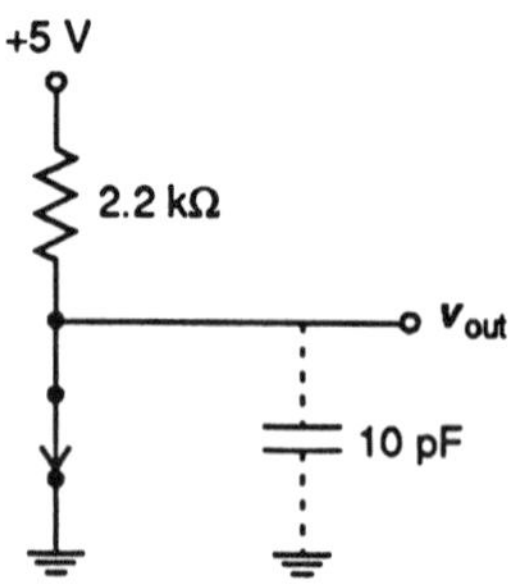

Fig. 12-28.

Sec. 12-4: Discharging Transients in *RC* Circuits

12-5. The capacitor of Fig. 12-25 is uncharged. Switch S_1 is closed. After the capacitor is fully charged, switch S_1 is opened and S_2 is closed. What is the transient voltage across the capacitor?

12-6. Both resistances are doubled in Fig. 12-25. Repeat Prob. 12-5 for this condition.

Sec. 12-5: Transients in a Voltage Divider

12-7. The capacitor of Fig. 12-26 is uncharged. What is the transient voltage across the capacitor when the switch closes?

12-8. Both resistances are doubled in Fig. 12-26. Repeat Prob. 12-7 for this condition.

Sec. 12-6: Transients with a Current Source

12-9. What is the transient voltage across the capacitor of Fig. 12-27?

12-10. The capacitance of Fig. 12-27 is quadrupled. What is the transient voltage across the resistor?

Sec. 12-7: Transistor Switching Circuits

12-11. The stray-wiring capacitance of Fig. 12-28 increases to 50 pF. What is the transient voltage across the output when the switch opens?

12-12. The resistance of Fig. 12-28 triples. What is the transient voltage across the output when the switch opens?

Sec. 12-8: Risetime

12-13. The switch closes in Fig. 12-26. What is the risetime of the transient voltage?

12-14. What is the risetime of the transient voltage in Fig. 12-27?

12-15. The switch closes in Fig. 12-28. What is the risetime of the transient voltage?

ANSWERS TO ODD-NUMBERED PROBLEMS

CHAP. 1

1-1. 0, 5.18 V, 10 V, 14.1 V, 17.3 V, 19.3 V, 20 V
1-3. 127 mA
1-5. 25 kHz
1-7. 40 mV, 41.7 kHz
1-9. 41.4 V
1-11. 18.8 V
1-13. 28.3 mV
1-15. 2.67 μW
1-17. 53.3 mW
1-19. 37.6 μA
1-21. 0.785, 2.36, 3.93, 5.5
1-23. 17.7 V
1-25. 7.39 V
1-27. The graph is a sine wave with a peak of 15 V and a period of 50 μs.
1-29. 17 V, 10.8 V, 12 V, 100 μs, 53.3 mW, –13.8 V

CHAP. 2

2-1. 12.5 V
2-3. 25 V
2-5. 6.25 V
2-7. a. + b. – c. - d. +
2-9. $i = 0.12 \sin 10000\pi t$, $v = V_p \cos 10000\pi t$
2-11. $v = V_p \cos 1000\pi t$, 12 V, 12 V
2-13. 52.5 mH
2-15. 12.5 mH
2-17. 94 V
2-19. 0.0833
2-21. 65 mH
2-23. a. 82 μH b. 42 μH or 122 μH c. 2 μH or 162 μH
2-25. 8.82 mH, 3.02 mH
2-27. 12.9 V, 15.8 V
2-29. 6.49 mA

CHAP. 3

3-1. 4.29 mA
3-3. 2.14 mA

3-5. 1.2 V

3-7. 99.5 mA

3-9. 39.8 mA

3-11. 1.25 V

3-13. 12.4 mA

3-15. 124 mA

3-17. 31.8 mA, 15.9 mA

3-19. 18.8 Ω, 9.4 Ω, 28.3 Ω, 400 mA, 7.53 V, 3.77 V

3-21. 800 mA, 15 , 7.53 V

3-23. 75.4 Ω, 151 Ω, 50.3 Ω, 84.4 mA, 4.24 V, 4.24 V

3-25. 169 mA, 8.48 V, 8.48 V

3-27. 15 V, 10.6 V, 25 mA, 17.7 mA

3-29. The voltage phasor has a length of 42.4 V and points up. The current phasor has a length of 6.06 mA and points to the right.

CHAP. 4

4-1. 2.11 mA, 5.68 V, 14.3 V

4-3. 0.4 mA, 8 V, 4 V

4-5. 3

4-7. 7.06 V

4-9. Approximately 102 V

4-11. 40 V

4-13. 20.1 V

4-15. 1.39 mA, 5.43 V, 14 V

4-17. 3.14 mA, 12.3 V, 15.8 V

4-19. 19.5 V

4-21. 357 μA, 11.8 V, 16.2 V

4-23. 63.4°

4-25. 776 Hz

4-27. 2.97 mA

4-29. 1.73 mA

4-31. 2.23 kΩ

4-33. 2.36 kΩ, 69.7°

4-35. 5.05 kΩ, 42.1°

4-37. 12 kΩ

4-39. 6.67

4-41. 2.65 H

CHAP. 5

5-1. 83.3 kV/m
5-3. 2.4 μF
5-5. 40 nC
5-7. 2.5 V
5-9. 0.0113 m^2
5-11. 10 mA
5-13. - 2.64 mA
5-15. 0.417 μF
5-17. 50 kV/s
5-19. 136 kV/s
5-21. 145 pF
5-23. 14.2 pF
5-25. 1100 pF

CHAP. 6

6-1. 3.33 mA
6-3. 1.67 mA
6-5. 1.2 V
6-7. 0.251 mA
6-9. 0.628 mA
6-11. 49.7 mV
6-13. 33 mA,
6-15. 226 μA
6-17. 22.6 mA, 45.2 mA
6-19. 53.1 kΩ, 106 kΩ, 159 kΩ, 71.1 μA, 3.78 V, 7.54 V
6-21. 142 μA, 7.54 V, 15.1 V
6-23. 169 Ω, 117 Ω, 69.2 Ω, 61.3 mA, 4.24 V, 4.24 V
6-25. 123 mA, 8.48 V, 8.48 V
6-27. 10 V, 7.07 V, 20 mA, 14.1 mA
6-29. The phasor voltage has a length of 4.71 mA and an angle of 0•. The phasor current has a length of 42.4 V and an angle of 90•.
6-31. 12 V, 6 V,
6-33. 1.29 V, 1.29 V, 7.71 V

CHAP. 7

7-1. 6.25 V
7-3. 0.737 V
7-5. 40 V
7-7. 20 V
7-9. 3.84 mA, 15 V, 1.13 V
7-11. 101 μA, 392 mV, 20 V
7-13. 47.2 V
7-15. 472 μA, 15.6 V, 12.5 V
7-17. -63.5°
7-19. 151 Hz
7-21. 1.35 mA
7-23. 2.04 mA
7-25. 2.2 kΩ
7-27. 5.17 kΩ, -40.5°
7-29. 2.47 kΩ, -68.7°
7-31. 12 kΩ
7-33. 6.67
7-35. 265 pF

CHAP. 8

8-1. 1.04 mA, 4.9 V, 7.07
8-3. 3.13 mA, 14.7 V, 21.3 V
8-5. 0.833 mA, 1.67 V, 3.33 V
8-7. 6 mA, 12 V, 24 V
8-9. 1.2 mA, 7.2 V, 10.8 V
8-11. 2.4 A, 14.4 V, 21.6 V
8-13. 4 mA, 60 V, 80 V
8-15. 1 mA, 30 V, 10 V
8-17. 0.5 mA, 1 mA, 0.5 mA
8-19. 0.25 mA, 2 mA, 1.75 mA
8-21. 3.81 mA, 10.3 V, 16 V, 34.3 V
8-23. 4.43 mA, 12 V, 37.2 V, 19.9 V
8-25. 1.8 mA, 1.5 mA, 2.25 mA, 1.95 mA
8-27. 1.8 mA, 0.75 mA, 4.5 mA, 4.16 mA
8-29. 13.9 mA
8-31. 6.43 mA
8-33. 39.3 mW, 80 mVA

CHAP. 9

9-1. $19.2 \angle 38.7^\circ$
9-3. $40.8 \angle -54^\circ$
9-5. $6.88 + j9.83$
9-7. $-29.7 + j29.7$
9-9. $35 + j42$
9-11. $44 - j62$
9-13. $-11 - j26$
9-15. $-11 + j26$
9-17. $240 \angle 65^\circ$
9-19. $300 \angle -5^\circ$
9-21. $2.1 \angle 35^\circ$
9-23. $4 \angle 45^\circ$
9-25. $j25$
9-27. $5.01 \angle 8.2^\circ$
9-29. $800 \angle 40^\circ$
9-31. 80 mVA $\angle 40^\circ$, 61.2 mW, 51.4 mVAR

CHAP. 10

10-1. 0.583 mA $\angle 0^\circ$, 1.17 V $\angle 0^\circ$, 2.33 V $\angle 0^\circ$, 3.5 V $\angle 0^\circ$
10-3. 0.657 mA $\angle -90^\circ$, 1.64 V $\angle 0^\circ$, 4.1 V $\angle 0^\circ$, 1.15 V $\angle 0^\circ$, 2.1 V $\angle 0^\circ$
10-5. 0.255 mA $\angle 90^\circ$, 7.91 V $\angle 0^\circ$, 10.7 V $\angle 0^\circ$, 6.38 V $\angle 0^\circ$
10-7. 42.4 μA $\angle -54^\circ$, 0.148 V $\angle -54^\circ$, 0.2 V $\angle 36^\circ$
10-9. 7.56 V $\angle 41^\circ$
10-11. 74.1 μA $\angle 55^\circ$
10-13. 52 kΩ in parallel with $j10.4$ kΩ
10-15. 14.1 kΩ in parallel with $j13.8$ kΩ
10-17. 962 Ω in series with $j192$ Ω
10-19. 0.735 mA $\angle 54^\circ$, 2.21 V $\angle -36^\circ$, 1.31 V $\angle 81^\circ$, 1.31 V $\angle 81^\circ$
10-21. 0.985 mA $\angle 61^\circ$
10-23. 15.4 Hz, 70.7 mV $\angle -45^\circ$
10-25. 796 kHz, 2.12V $\angle -45^\circ$
10-27. 103 kHz

CHAP. 11

11-1. 535 kHz
11-3. 379 kHz
11-5. 100 kHz, 22.2 kHz

11-7. 62.5

11-9. 124 V $\angle 90°$

11-11. 278 kHz

11-13. 5.56 kHz, 2 kHz

11-15. 15.6 kHz, 424 kHz, 440 kHz

11-17. 64.5 mA

11-19. 158 mA

11-21. 358 μH

11-23. 2.26 MHz, 16.9 V, 21.3 kHz, 79.5 mA

11-25. 705 kHz,11.6 V, 25.3 kHz

CHAP. 12

12-1. Resistor: Positive exponential, v_i of 0, v_f of 9 V,
time constant of 57.4 ns;
Inductor: Negative exponential, v_i of 9, v_f of 0,
time constant of 57.4 ns

12-3. Positive exponential, v_i of 0, v_f of 15 V, time constant of 7.05 μs
Negative exponential, i_i of 0.1 mA, i_f of 0
Negative exponential, v_i of 15 V, v_f of 0

12-5. First transient: Positive exponential, v_i of 0, v_f of 50 V,
time constant of 30 μs;
Second transient: Negative exponential, v_i of 50 V, v_f of 0,
time constant of 2 μs

12-7. Positive exponential, v_i of 0, v_f of 12 V, time constant of 12.5 μs

12-9. Positive exponential, v_i of 0, v_f of 10 V, time constant of 0.94 μs

12-11. Positive exponential, v_i of 0, v_f of 5 V, time constant of 110 ns

12-13. 2.07 μs

12-15. 48.4 ns

INDEX